Mathematical Methods for Oscillations and Waves

Anchored in simple and familiar physics problems, Joel Franklin provides a focused introduction to mathematical methods in a narrative-driven and structured manner. Ordinary and partial differential equation solving, linear algebra, vector calculus, complex variables, and numerical methods are all introduced and bear relevance to a wide range of physical problems. Expanded and novel applications of these methods highlight their utility in less familiar areas, and advertise those areas that will become more important as students continue. This highlights both the utility of each method in progressing with problems of increasing complexity while also allowing students to see how a simplified problem becomes "recomplexified." Advanced topics include nonlinear partial differential equations, and relativistic and quantum mechanical variants of problems like the harmonic oscillator. Physics, mathematics, and engineering students will find 300 problems treated in a sophisticated manner. The insights emerging from Franklin's treatment make it a valuable teaching resource.

Joel Franklin is a professor in the Physics Department of Reed College. His research focuses on mathematical and computational methods with applications to classical mechanics, quantum mechanics, electrodynamics, general relativity, and modifications to general relativity. He is also the author of *Advanced Mechanics and General Relativity* (Cambridge University Press, 2010), *Computational Methods for Physics* (Cambridge University Press, 2013), and *Classical Field Theory* (Cambridge University Press, 2017).

Mathematical Methods for Oscillations and Waves

JOEL FRANKLIN

Reed College

CAMBRIDGE
UNIVERSITY PRESS

CAMBRIDGE
UNIVERSITY PRESS

University Printing House, Cambridge CB2 8BS, United Kingdom

One Liberty Plaza, 20th Floor, New York, NY 10006, USA

477 Williamstown Road, Port Melbourne, VIC 3207, Australia

314–321, 3rd Floor, Plot 3, Splendor Forum, Jasola District Centre, New Delhi – 110025, India

79 Anson Road, #06–04/06, Singapore 079906

Cambridge University Press is part of the University of Cambridge.

It furthers the University's mission by disseminating knowledge in the pursuit of education, learning, and research at the highest international levels of excellence.

www.cambridge.org
Information on this title: www.cambridge.org/9781108488228
DOI: 10.1017/9781108769228

First published 2020

Printed in the United Kingdom by TJ International Ltd, Padstow Cornwall

A catalogue record for this publication is available from the British Library.

ISBN 978-1-108-48822-8 Hardback

For Lancaster, Lewis, Oliver, and Mom

Contents

There are many books on "mathematical methods for physics" [1, 3, 15], including some with that exact title. Most of these are wide-ranging explorations of the physical applications of fairly deep analytic and group-theoretic mathematics. They cover topics that one might encounter anywhere from first-year undergraduate to first-year graduate physics, and remain on our shelves as well-thumbed references and problem libraries. In addition to a plethora of techniques, they cover all sorts of important special cases that can keep the naïve physicist out of trouble in a variety of technical situations.

There is also the Internet, itself a repository for all sorts of human knowledge, including physical, mathematical, and their intersection. The Internet is even more encyclopedic than most mathematical methods books, with more special cases and more specialized examples. Here we can find, in almost equal number, the inspiring, the arcane, and the incorrect. Students of physics, especially early in their studies, need to be sophisticated and wary.

What is missing in both cases (especially the latter) is narrative. A clear description of why we care about these methods, and how they are related to diverse, yet logically connected problems of interest to physicists. Why is it, for example, that the Fourier transform shows up in the analysis of networks of springs and also in the analysis of analog circuits? I suggest the reason is that both involve the characterization of timescales of oscillation and decay, and in some sense, almost all of physics is interested in such timescales, so there is a universality here that is not shared with, say, the Laplace transform. Yet Wikipedia, and other "online resources" fail to make this point – or rather point and counterpoint – effectively, because there is no individual curator deciding what goes in, what stays out, and how much time/space to dedicate to each topic.

This book has such a curator, and I have made choices based on my own research experience, broadened by the teaching I have done at Reed College, and feedback from students I have taught. At a small liberal arts college like Reed, the faculty must teach and advise students in areas beyond the faculty member's expertise and experience. The advantage is a generalist's view, but with the depth that is required to teach extremely curious and talented students confidently. After all, much of what we sell in physics is counterintuitive or even wrong (witness: gravity as one of the fundamental forces of nature). We should expect, delight in, and encourage our students' skepticism. The topics in this book are intended to invite students to ask and answer many of the questions I have been asked by their peers over the past 15 years.

In my department, there is a long tradition of teaching mathematical methods using oscillations and waves as a motivational topic. And there are many appropriate "oscillations and waves" texts and monographs [9, 16]. These are typically short supplemental books that exhaustively treat the topic. Yet few of them attempt to extend their reach fully, to include the mathematical methods that, for example, might be useful to a student of E&M (an exception is [10], which does have a broader mathematical base). I was inspired by Sidney Coleman's remark, "The career of a young theoretical physicist consists of treating the harmonic oscillator in ever-increasing levels of abstraction." I am not sure any physics that I know is particularly far removed from the harmonic oscillator, and embracing that sentiment gives one plenty of room to maneuver. There is no reason that the mathematical methods of oscillations and waves can't serve as a stand-in for "mathematical methods for physics."

I have used chapters of the present volume to teach a one-semester undergraduate course on mathematical physics to second-year physics students. For that audience, I work through the following chapters:

Chapter 1 **Harmonic Oscillator**: A review of the problem of motion for masses attached to springs. That's the physics of the chapter, a familiar problem from high school and introductory college classes, meant to orient and refresh the reader. The mathematical lesson is about series solutions (the method of Frobenius) for ordinary differential equations (ODEs), and the definition of trigonometric special functions in terms of the ODEs that they solve. This is the chapter that reviews complex numbers and the basic properties of ODEs and their solutions (superposition, continuity, separation of variables).

Chapter 2 **Damped Harmonic Oscillator**: Here, we add damping to the harmonic oscillator, and explore the role of the resulting new timescale in the solutions to the equations of motion. Specifically, the ratio of damping to oscillatory timescale can be used to identify very different regimes of motion: under-, critically-, and over-damped. Then driving forces are added, we consider the effect those have on the different flavors of forcing already in place. The main physical example (beyond springs attached to masses in dashpots) is electrical, sinusoidally driven resistor, inductor, capacitor (RLC) circuits provide a nice, experimentally accessible test case. On the mathematical side, the chapter serves as a thinly veiled introduction to Fourier series and the Fourier transform.

Chapter 3 **Coupled Oscillators**: We turn next to the case of additional masses. In one dimension, we can attach masses by springs to achieve collective motions that occur at a single frequency, the normal modes. Building general solutions, using superposition, from this "basis" of solutions is physically relevant and requires a relatively formal treatment of linear algebra, the mathematical topic of the chapter.

Chapter 4 **The Wave Equation**: Taking the continuum limit of the chains of masses from the previous chapter, we arrive at the wave equation, the physical subject of this chapter. The connection to approximate string motion is an additional motivation. Viewed as a manifestation of a conservation law, the wave equation can be extended to other conservative, but nonlinear cases, like traffic flow. Mathematically, we are interested in turning partial differential equations

(PDEs) into ODEs, making contact with some familiar examples. Making PDEs into ODEs occurs in a couple of ways – the method of characteristics, and additive/multiplicative separation of variables are the primary tools.

Chapter 5 **Integration**: With many physical applications already on the table, in this chapter, we return to some of the simplified ones and recomplexify them. These problems require more sophisticated, and incomplete, solutions. Instead of finding the position of the bob for the simple pendulum, we find the period of motion for the "real" pendulum. Instead of the classical harmonic oscillator, with its familiar solution, we study the period of the relativistic harmonic oscillator, and find that in the high-energy limit, a mass attached to a spring behaves very differently from its nonrelativistic counterpart.

The eighth chapter, Numerical Methods, is used as a six-week "lab" component, one section each week. The chapter is relatively self-contained, and consists of numerical methods that complement the analytic solutions found in the rest of the book. There are methods for solving ODE problems (both in initial and boundary value form) approximating integrals, and finding roots. There is also a discussion of the eigenvalue problem in the context of approximate solutions in quantum mechanics and a section on the discrete Fourier transform.

There are two additional chapters that provide content when the book is used in an upper level setting, for third- or fourth-year students. In the sixth chapter, Waves in Three Dimensions, we explore the wave equation and its solutions in three dimensions. The chapter's mathematical focus is on vector calculus, enough to understand and appreciate the harmonic functions that make up the static solutions to the wave equation. Finally, the seventh chapter, Other Wave Equations, extends the discussion of waves beyond the longitudinal oscillations with which we began. Here, we look at the wave equation as it arises in electricity and magnetism (the three-dimensional context is set up in the previous chapter), in Euler's equation and its shallow water approximation, in "realistic" (extensible) strings, and in the quantum mechanical setting, culminating in a quantum mechanical treatment of the book's defining problem, the harmonic oscillator.

There are two appendices to provide review. The first reviews the basic strategy of ODE solving in a step-by-step way – what guesses to try, and when, with references to the motivating solutions in the text. The second appendix is a review of basic vector calculus expressions, like the gradient, divergence, curl, and Laplacian, in cylindrical, spherical, and more general coordinate systems.

My hope is that this book provides a focused introduction to many of the mathematical methods used in theoretical physics, and that the vehicles used to present the material are clear and compelling. I have kept the book as short as possible, yet tried to cover a variety of different tools and techniques. That coverage is necessarily incomplete, and for students going on in physics, a copy of one of the larger [1, 3, 15], and more sophisticated [2, 4, 17] mathematical methods texts will eventually be a welcome necessity, with this book providing some motivating guidance. (I encourage students to have one of these texts on hand as they read, so that when a topic like spherical Bessel functions comes up, they can look at the relevant section for additional information.)

Mary Boas has a wonderful "To the Student" section at the start of [3], an individual call to action that cannot be improved, so I will quote a portion of it:

> To use mathematics effectively in applications, you need not just knowledge, but *skill*. Skill can be obtained only through practice. You can obtain a certain superficial *knowledge* of mathematics by listening to lectures, but you cannot obtain *skill* this way.... The only way to develop the skill necessary to use this material in your later courses is to practice by solving many problems. Always study with pencil and paper at hand. Don't just read through a solved problem – try to do it yourself!

Since I was an undergraduate, I have always followed and benefited from this advice, and so, have included a large number of problems in this text.

Acknowledgments

It is a pleasure to thank the students and my colleagues in the physics department at Reed College. I have benefited from my interactions with them, and in particular, from discussions about teaching our second-year general physics course with Professors Lucas Illing, Johnny Powell, and Darrell Schroeter. A very special thanks to Professor David Latimer, who carefully read and thoughtfully commented on much of this text, his suggestions have added value to the document, and been instructive (and fun) to think about.

My own research background has informed some of the topics and presentation in this book, and that background has been influenced by many talented physicists and physics teachers – thanks to my mentors from undergraduate to postdoctoral, Professors Nicholas Wheeler, Stanley Deser, Sebastian Doniach, and Scott Hughes.

Finally, David Griffiths has, throughout my career been an honest sounding board, a source of clarity and wisdom. He has helped me both think about and present physics far better than I could on my own. I thank him for sharing his insights on this material and my presentation of it.

Harmonic Oscillator

The motivating problem we consider in this chapter is Newton's second law applied to a spring with spring constant k and equilibrium spacing a as shown in Figure 1.1.

The nonrelativistic equation of motion reads

$$m\ddot{x}(t) = -k(x(t) - a) \tag{1.1}$$

and we must specify initial (or boundary) conditions. The point of Newton's second law is the determination of the trajectory of the particle, $x(t)$ in this one-dimensional setting. The initial conditions render the solution unique. As a second-order differential equation (ODE), we expect the general solution to have two constants. Then we need two pieces of information beyond the equation itself to set those constants, and initial or boundary conditions can be used.

1.1 Solution Review

To proceed, we can define $k/m \equiv \omega^2$, so that our equation of motion becomes

$$\ddot{x}(t) = -\omega^2(x(t) - a), \tag{1.2}$$

and finally, we let $y(t) \equiv x(t) - a$ in order to remove reference to a and allow us to identify the solution to this familiar ODE:

$$\ddot{y}(t) = -\omega^2 y(t) \longrightarrow y(t) = A\cos(\omega t) + B\sin(\omega t). \tag{1.3}$$

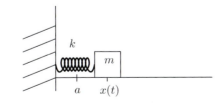

Fig. 1.1   A mass m is attached to a spring with spring constant k and equilibrium spacing a. It moves without friction under the influence of a force $F = -k(x(t) - a)$. We want to find the location of the mass at time t, $x(t)$, by solving Newton's second law.

The constants A and B have no *a priori* physical meaning, they are just the constants we get from a second-order ODE like (1.3).[1] The solution for $x(t)$ is

$$x(t) = y(t) + a = A\cos(\omega t) + B\sin(\omega t) + a. \tag{1.4}$$

Suppose we take the initial value form of the problem. We see the particle at x_0 at time $t = 0$, moving with velocity v_0. This allows us to algebraically solve for A and B:

$$x(0) = A + a = x_0 \longrightarrow A = x_0 - a$$
$$\dot{x}(0) = B\omega = v_0 \longrightarrow B = \frac{v_0}{\omega}. \tag{1.5}$$

When we combine an ODE (like Newton's second law) with constants (the initial position and velocity), we have a well-posed *problem* and a unique solution. Putting it all together, the problem is

$$m\ddot{x}(t) = -k(x(t) - a) \qquad x(0) = x_0 \qquad \dot{x}(0) = v_0 \tag{1.6}$$

with solution

$$x(t) = (x_0 - a)\cos\left(\sqrt{\frac{k}{m}}t\right) + v_0\sqrt{\frac{m}{k}}\sin\left(\sqrt{\frac{k}{m}}t\right) + a. \tag{1.7}$$

There are many physical observations and definitions associated with this solution. Suppose that we start the mass from rest, $v_0 = 0$, with an initial extension x_0, and we set the zero of the x axis at the equilibrium spacing a. Then the solution from (1.7) simplifies to

$$x(t) = x_0 \cos\left(\sqrt{\frac{k}{m}}t\right). \tag{1.8}$$

We call x_0 the "amplitude," the maximum displacement from equilibrium. The "period" of the motion is defined to be the time it takes for the mass to return to its starting point. In this case, we start at $t = 0$, and want to know when the "cosand" (argument of cosine) returns to $2\pi \sim 0$. That is, the period T is defined to be the first time at which

$$x(T) = x_0 \cos\left(\sqrt{\frac{k}{m}}T\right) = x_0 \longrightarrow \sqrt{\frac{k}{m}}T = 2\pi \longrightarrow T = 2\pi\sqrt{\frac{m}{k}}. \tag{1.9}$$

This period is, famously, independent of the initial extension.[2] That makes some sense, physically – the larger the initial extension, the faster the maximum speed of the mass is, so that even though it has to travel a longer distance, it does so at a greater speed. Somehow, magically, the two effects cancel in this special case.

We can also define the "frequency" of the oscillatory motion, that is just the inverse of the period, $f \equiv 1/T$. For the mass on a spring motion,

$$f = \frac{1}{2\pi}\sqrt{\frac{k}{m}}, \tag{1.10}$$

---

[1] These constants are called "constants of integration" and are reminiscent of the constants that appear when you can actually integrate the equation of motion twice. That happens when, for example, a force depends only on time. Then you can literally integrate $\ddot{x}(t) = F(t)/m$ twice to find $x(t)$. There will be two constants of integration that show up in that process.

[2] Don't believe it! See Section 5.5.

and we define the "angular frequency" of the oscillatory motion to be

$$\omega \equiv 2\pi f = \sqrt{\frac{k}{m}}, \qquad (1.11)$$

where ω is the letter commonly used, and we have taken advantage of that in writing (1.2).

Problem 1.1.1   What is the solution to Newton's second law (1.1) with *boundary* values given: $x(0) = x_0$ and $x(t^*) = x_*$ (t^* refers to some specific time at which we are given the position, x_*)?

Problem 1.1.2   What happens to the solution in the previous problem if $\omega t^* = n\pi$ for integer n?

Problem 1.1.3   Solve $m\ddot{x}(t) = F_0$ for constant force F_0 subject to the boundary conditions: $x(0) = x_0$, $x(t^*) = x_*$ with x_0 and x_* given. Solve the same problem for a "mixed" set of conditions: $x(0) = x_0$ and $\dot{x}(t^*) = v_*$ with x_0 and v_* given.

Problem 1.1.4   For the oscillatory function $x(t) = x_0 \cos(\omega t + \phi)$ with constant $\phi \in [0, 2\pi)$ (the "phase"), find the amplitude and period, and sketch one full cycle of this function.

Problem 1.1.5   Suppose Newton's second law read: $\alpha \dddot{x}(t) = F(x(t), t)$ for force F. What are the units of α in this case? Solve the modified Newton's second law if the force is a constant F_0 with initial conditions $x(0) = x_0$ and $\dot{x}(0) = v_0$. What is the problem with this solution? (is it, for example, unique?).

Problem 1.1.6   Solve $m\ddot{x}(t) = F_0 \cos(\omega t)$ (F_0 is a constant with Newtons as it unit, ω is a constant with unit of inverse seconds) for $x(t)$ given $x(0) = x_0$ and $\dot{x}(0) = v_0$.

1.2 Taylor Expansion

To appreciate the role of the harmonic oscillator problem in physics, we need to review the idea of expanding a function $f(x)$ about a particular value x_0 and apply it to minima of a potential energy. We'll start in this section with the former, called "Taylor expansion." The idea is to estimate $f(x_0 + \Delta x)$ for small Δx given the value of the function and its derivatives at x_0. Our first guess is that the function is unchanged at $x_0 + \Delta x$,

$$f(x_0 + \Delta x) \approx f(x_0). \qquad (1.12)$$

That's a fine approximation, but can we improve upon it? Sure: if we knew $f(x_0)$ and

$$f'(x_0) \equiv \frac{df(x)}{dx}\bigg|_{x=x_0}, \qquad (1.13)$$

then we could add in a correction associated with the slope of the line tangent to $f(x_0)$ at x_0:

$$f(x_0 + \Delta x) \approx f(x_0) + f'(x_0)\Delta x. \qquad (1.14)$$

The picture of this approximation, with the initial estimate and the linear refinement is shown in Figure 1.2.

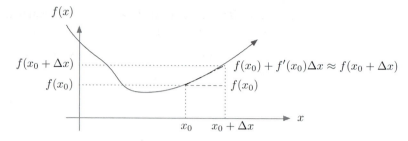

Fig. 1.2 Given a function $f(x)$, if we know the value of the function and its derivative at x_0, then we can estimate the value at a nearby point $x_0 + \Delta x$ for Δx small.

The process continues, we can take the quadratic correction, $f''(x_0)\Delta x^2$, and use it to refine further,

$$f(x_0 + \Delta x) \approx f(x_0) + f'(x_0)\Delta x + \frac{1}{2}f''(x_0)\Delta x^2. \tag{1.15}$$

You can keep going to any desired accuracy, with equality restored when an infinite number of terms are kept,

$$f(x_0 + \Delta x) = \sum_{j=0}^{\infty} \frac{1}{j!}\left(\frac{d^j}{dx^j}f(x)\right)\bigg|_{x=x_0} \Delta x^j, \tag{1.16}$$

although we rarely need much past the first three terms. Note that the first term that you drop gives an estimate of the error you are making in the approximation. For example, if you took only the first two terms in (1.15), you would know that the leading source of error goes like $f''(x_0)\Delta x^2$.

The hardest part of applying Taylor expansion lies in correctly identifying the function $f(x)$, the point of interest, x_0, and the "small" correction Δx. As an example, suppose we wanted to evaluate $\sqrt{102}$, then we have $f(x) = \sqrt{x}$ with $x_0 = 100$ and $\Delta x = 2$. Why pick $x_0 = 100$ instead of 101? Because it is easy to compute $f(100) = \sqrt{100} = 10$, while for $\sqrt{101}$ we have the same basic problem we started out with (i.e. I don't know the value of $\sqrt{101}$ any more than I know $\sqrt{102}$). Using (1.15) gives the estimate

$$f(100 + 2) \approx f(100) + f'(100) \times 2 + \frac{1}{2}f''(100) \times 4$$

$$= \sqrt{100} + \frac{1}{2\sqrt{100}} \times 2 - \frac{1}{2}\frac{1}{4(100)^{3/2}} \times 4 \tag{1.17}$$

$$= 10 + \frac{1}{10} - \frac{1}{2000} = 10.0995$$

while the "actual" value is $\sqrt{102} \approx 10.099505$.

Problem 1.2.1 Evaluate $\sin(\Delta x)$ and $\cos(\Delta x)$ for the values of Δx given in the following table using a calculator. Then use the Taylor expansions of these functions (to second order, from (1.15)) to approximate their value at those same Δx (assuming Δx is small, an assumption that is violated by some of the values so you can see the error in the approximations). Write out your results to four places after the decimal.

Δx	$\sin(\Delta x)$	$\cos(\Delta x)$	$\sin(\Delta x)$ approx.	$\cos(\Delta x)$ approx.
.1				
.2				
.4				
.8				

Problem 1.2.2   Use Taylor expansion to find the approximate value of the function $f(x) = (a + x)^n$ for constants a and n with $x_0 = 0$, i.e. what is $f(\Delta x)$ for Δx small (take only the "leading-order" approximation, in which you just write out the Taylor expansion through the Δx term as in (1.14))? Using your result, give the Taylor expansion approximation near zero (again, you'll write expressions approximating $f(\Delta x)$) for:

$$f(x) = \sqrt{1 + x} \approx$$

$$f(x) = \frac{1}{\sqrt{1 + x}} \approx$$

$$f(x) = \frac{1}{(1 + x)^2} \approx$$

Problem 1.2.3   Estimate the value of $1/121$ using Taylor expansion (to first order in the small parameter) and compare with the value you get from a calculator. Hint: $121 = (10 + 1)^2$.

Problem 1.2.4   For the function $f(\theta) = (1 + \cos \theta)^{-1}$, estimate $f(\pi/2 + \Delta\theta)$ for $\Delta\theta \ll 1$ using (1.15). Try it again by Taylor expanding the $\cos \theta$ function first, then expanding in the inverse polynomial, a two-step process that should yield the same result (up to errors that we have ignored in both cases).

1.3 Conservative Forces

The spring force starts off life as rusty bits of metal providing a roughly linear restoring force. But the model's utility in physics has little to do with the coiled metal itself. Instead, the "harmonic" oscillator behavior is really the dominant response of a particle moving near the equilibrium of a potential energy function. To review, a conservative force F comes from the derivative of a potential energy U via:

$$F(x) = -\frac{dU(x)}{dx}. \tag{1.18}$$

If we have a potential energy $U(x)$ (from whatever physical configuration), then a point of equilibrium is defined to be one for which the force vanishes. For x_e a point of equilibrium,

$$F(x_e) = 0 = -\frac{dU(x)}{dx}\bigg|_{x=x_e} \equiv -U'(x_e). \tag{1.19}$$

Now if we expand the potential energy function $U(x)$ about the point x_e using Taylor expansion:

$$U(x) = U(\underbrace{(x - x_e)}_{\equiv \Delta x} + x_e) = U(x_e) + \Delta x U'(x_e) + \frac{1}{2}\Delta x^2 U''(x_e) + \cdots , \qquad (1.20)$$

then the first term is just a constant, and that will not contribute to the force in the vicinity of x_e (since we take a derivative with respect to x sitting inside Δx to get the force). The second term vanishes by the assumption that x_e is a point of equilibrium, and the first term that informs the dynamics of a particle moving in the vicinity of x_e is the third term $\sim (1/2)U''(x_e)(x - x_e)^2$, leading to a force, near x_e:

$$F(x) = -\frac{dU(x)}{dx} \approx -U''(x_e)(x - x_e) + \cdots \qquad (1.21)$$

The effective force in the vicinity of the equilibrium is just a linear restoring force with "spring constant" $k \sim U''(x_e)$ (assuming $U''(x_e) > 0$ so that the equilibrium represents a local *minimum*) and equilibrium spacing x_e. A picture of a local minimum in the potential energy and the associated force is shown in Figure 1.3. Near x_e, the potential is approximately quadratic, and the force is a linear restoring force of the sort we have been studying. There is also an equilibrium point at the maximum of $U(x)$ in that picture, but the associated force tends to drive motion *away* from this second equilibrium. We call such locations points of "unstable equilibrium," even a small perturbation from the equilibrium location drives masses away.

As an example, suppose we have somehow managed to set up a potential energy of the form $U(x) = U_0 \cos(2\pi x/\ell)$ for a length ℓ and constant $U_0 > 0$. What is the period of motion for a particle that starts out "near" $x_e = \ell/2$? In this case, the equilibrium position

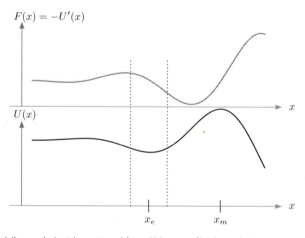

Fig. 1.3 A potential energy $U(x)$ (lower plot) with associated force $F(x) = -U'(x)$ (upper). A minimum in the potential has an approximately linear restoring force associated with it. We can approximate the force, in the vicinity of x_e (bracketed with dotted lines) by $F(x) \approx -U''(x_e)(x - x_e)$. The maximum at x_m is also a point of equilibrium, but this one is "unstable," if a particle starts near x_m it tends to get driven away from x_m (the slope of the force is positive).

is at $x_e = \ell/2$, with $U'(x_e) = 0$ as required. The second derivative sets the effective spring constant, $k = U''(x_e) = (2\pi/\ell)^2 U_0$, and then the period and frequency of the resulting oscillatory motion come from (1.9) and (1.10):

$$T = 2\pi \sqrt{\frac{m}{U''(x_e)}} = \sqrt{\frac{m\ell^2}{U_0}} \qquad f = \sqrt{\frac{U_0}{m\ell^2}}. \tag{1.22}$$

1.3.1 Conservation of Energy

It is worth reviewing the utility of conservation of energy. In particular, while we have a complete solution to our problem, in (1.4), it is not always possible to find such a complete solution. In those cases, we retreat to a "partial" solution, where we can still make quantitative predictions (and hence physical progress), but we might not have the "whole" story.

Let's go back to Newton's second law, this time for an arbitrary potential energy $U(x)$, but with initial values specified. Think of the ODE piece of the "problem":

$$m\ddot{x}(t) = -\frac{dU(x)}{dx}. \tag{1.23}$$

Now, the only ODEs that I can solve in closed form are ones in which both sides are a total derivative, in this case, a total *time*-derivative. Then to integrate, you just remove the $\frac{d}{dt}$ from both sides, and add in a constant – that process returns a function for $\dot{x}(t)$. Then, when possible, you integrate again to get $x(t)$ (picking up another constant). This direct approach will be considered in Section 5.1.

Looking at the left side of (1.23), it is clear that we have a total time derivative: $m\ddot{x}(t) = \frac{d}{dt}(m\dot{x}(t))$, but what about the right-hand side? Is there a function $W(x)$ such that

$$\frac{dW(x(t))}{dt} = -\frac{dU(x)}{dx}? \tag{1.24}$$

The answer is *no*. The reason is clear: If we had a function evaluated at $x(t)$, $W(x(t))$, then the total time derivative of W would look like

$$\frac{dW(x(t))}{dt} = \frac{dW(x)}{dx}\frac{dx(t)}{dt} = \frac{dW(x)}{dx}\dot{x}(t) \tag{1.25}$$

and there is no $\dot{x}(t)$ that appears on the right in (1.23). The fix is easy, just put an $\dot{x}(t)$ on the right-hand side of (1.23), which requires putting one on the left-hand side as well. Then Newton's second law looks like

$$m\dot{x}(t)\ddot{x}(t) = -\frac{dU(x)}{dx}\dot{x}(t). \tag{1.26}$$

The situation on the right is now very good, since we can write the right-hand side as a total time derivative:

$$-\frac{dU(x(t))}{dt} = -\frac{dU(x)}{dx}\dot{x}(t). \tag{1.27}$$

There is potential trouble on the *left*-hand side of Newton's second law, though. Can $\dot{x}(t)\ddot{x}(t)$ be written as a total time derivative? Yes, note that

$$\frac{d}{dt}\left(\dot{x}(t)^2\right) = 2\dot{x}(t)\ddot{x}(t). \tag{1.28}$$

Just multiplying Newton's second law by $\dot{x}(t)$ on both sides has given us the integrable equation

$$\frac{d}{dt}\left(\frac{1}{2}m\dot{x}(t)^2\right) = -\frac{dU(x(t))}{dt} \longrightarrow \frac{1}{2}m\dot{x}(t)^2 = -U(x(t)) + E \tag{1.29}$$

where E is the constant of integration. We could re-write (1.29) as

$$\frac{1}{2}m\dot{x}(t)^2 + U(x(t)) = E. \tag{1.30}$$

This represents an interesting situation – the combination of the time-dependent terms on the left yields a time *in*-dependent term on the right. Thinking about units (or dimensions, if that's what you are into) we have, in the first term of (1.30) a "kinetic" energy (energy because of the units, kinetic because the term is associated with movement through its dependence on $\dot{x}(t)$) and a "potential" energy (again from units, and the dependence on, this time, position). The sum is a constant of the motion of the particle, E, the "total" energy of the system. That is the statement of energy conservation expressed by (1.30). Because E is a constant, we can set its value from the provided initial conditions: $x(0) = x_0$, $\dot{x}(0) = v_0$,

$$\frac{1}{2}mv_0^2 + U(x_0) = E. \tag{1.31}$$

The integration of Newton's second law, in the presence of a "conservative" force, given by (1.30) is notable for its predictive ability – if you tell me where the particle is, its location at time t, $x(t)$, I can tell you how fast it is moving. Using the constant value of E set by the initial conditions for the motion (1.31), we can write (1.30) as

$$\frac{1}{2}m\dot{x}(t)^2 + U(x(t)) = \frac{1}{2}mv_0^2 + U(x_0), \tag{1.32}$$

and then

$$\dot{x}(t) = \pm\left[v_0^2 + \frac{2}{m}\left(U(x_0) - U(x(t))\right)\right]^{1/2} \tag{1.33}$$

gives the speed (taking the positive root), at time t, of the particle at location $x(t)$.

1.3.2 Harmonic Oscillator

The harmonic oscillator potential energy is just the quadratic $U(x) = 1/2k(x - a)^2$ for equilibrium location a. In this case, the quadratic expansion of $U(x)$ about the equilibrium consists of just the one term. We know what happens here: a particle oscillates about the equilibrium value with frequency governed by $\sqrt{k/m}$. If you think of the graph of the potential energy, we have a convex curve, and if you draw a line representing energy E as in Figure 1.4, you can tell the "story of the motion": where the value of E intersects the

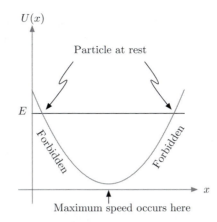

Fig. 1.4   A particle with energy E moving in a quadratic potential well. The particle is at rest where E intersects the potential energy function, and achieves its maximum speed where the difference between E and $U(x)$ is largest.

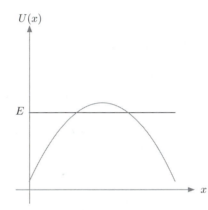

Fig. 1.5   A particle with energy E moving in a quadratic potential with a maximum. The particle cannot exist "underneath" the potential energy curve.

potential energy, the particle must be at rest (all the energy is potential, none is kinetic), the point at which the difference between E and $U(x)$ is largest represents the position at which the maximum speed of the particle occurs. Locations where $U(x) > E$ are impossible to achieve physically, since the kinetic energy would have to be negative.

What if the potential energy had the form $U(x) = -1/2k(x - a)^2$, with a concave graph? This time, a particle tends to move *away* from the equilibrium position, without returning to it. Thinking of motion at a fixed E, as in Figure 1.5, the particle speeds up as it gets further away from the equilibrium location. This is an example of an "unstable" equilibrium, particles that start near a are driven away from it. For the usual harmonic potential, with its + sign, the equilibrium point is "stable," if you start near equilibrium, you remain near it. In a more general potential energy landscape, the sign of the second derivative of the potential energy function, evaluated at a point of equilibrium, determines whether the equilibrium is stable (positive) or unstable (negative).

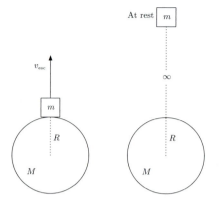

Fig. 1.6 A mass m leaves the surface of the Earth with speed v_{esc} and comes to rest infinitely far away.

1.3.3 Escape Speed

Conservation of energy can be used to make quantitative predictions even when the full position of an object as a function of time is not known. As an example, consider a spherical mass M of radius R. A test particle[3] of mass m at a distance $r \geq R$ from the center of the sphere experiences a force with magnitude

$$F = \frac{GMm}{r^2} \tag{1.34}$$

where G is the gravitational constant. This force comes from a potential energy $U(r) = -GMm/r$. Suppose the test mass starts from the surface of the sphere with a speed v_{esc}. We want to know the minimum value for v_{esc} that allows the test mass to escape the gravitational pull of the spherical body. That special speed is called the "escape speed." Think of the spherical body as the earth, and the test mass is a ball that you throw up into the air. The ball goes up and comes back down. If you throw the ball up in the air a little faster, it takes longer to come down. The escape speed is the (minimum) speed at which you must throw the ball up so that it *never* comes back down.

Formally, we want the test particle to reach $r \to \infty$ where it will be at rest as shown on the right in Figure 1.6.[4] From (1.30), if we take "$x(t) = r \to \infty$" with "$\dot{x}(t) = \dot{r} \to 0$," and use the potential energy associated with gravity, we have $E = 0$. Going back to the initial values, which must of course have the same energy,

$$\frac{1}{2} m v_{\text{esc}}^2 + U(R) = E = 0, \tag{1.35}$$

and then the escape speed can be isolated algebraically

$$v_{\text{sc}} = \sqrt{-\frac{2}{m} U(R)} = \sqrt{\frac{2GM}{R}}. \tag{1.36}$$

---

[3] "Test particle" is a technical term that means "a particle that feels the effect of a force without contributing to it." When we want to probe the gravitational force associated with some external body, we often imagine a test particle's response to that force. The same idea shows up in electricity and magnetism.

[4] That's the "minimum" part of the requirement. You could have the test particle rocketing around at spatial infinity, but that excess speed is overkill.

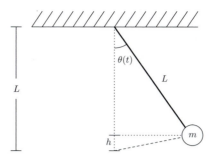

Fig. 1.7   A mass m is attached to a rope of length L. The rope makes an angle of $\theta(t)$ with respect to the vertical axis.

1.3.4 Pendulum

Another case of interest that can be simplified using conservation of energy (1.30) is the pendulum. A pendulum consists of a mass m (the "bob") attached to a rope of length L that is allowed to swing back and forth. We can describe the bob's motion by finding the angle $\theta(t)$ that the rope makes with respect to vertical, as shown in Figure 1.7. The kinetic energy of the pendulum mass is $1/2m(L\dot{\theta})^2$ and its potential energy is $U = mgh$ with $h = L(1 - \cos\theta)$. Conservation of energy can be expressed as

$$E = \frac{1}{2}mL^2\dot{\theta}^2 + mgL(1 - \cos\theta). \tag{1.37}$$

If the pendulum starts from rest at an angle θ_0, then $E = mgL(1 - \cos\theta_0)$, and we can solve for $\dot{\theta}^2$:

$$\dot{\theta}^2 = \frac{2g}{L}(\cos\theta - \cos\theta_0). \tag{1.38}$$

This is interesting, it already tells us that the value of θ must fall between $-\theta_0$ and θ_0 to keep $\dot{\theta}^2 > 0$, so that the largest $|\theta|$ can be, at any given time, is θ_0.

If we take the time-derivative of (1.38), to eliminate the constant and remove the sign ambiguity associated with taking the square root,[5] we get

$$\ddot{\theta} = -\frac{g}{L}\sin\theta. \tag{1.39}$$

Since we know the motion is bounded by θ_0, if the starting angle is small, then θ will be small for all times. In that case, we can make the small angle approximation $\sin\theta \approx \theta$ to write

$$\ddot{\theta} \approx -\frac{g}{L}\theta \tag{1.40}$$

which is of the form of the harmonic oscillator. The solution, with appropriate initial values, is

---

[5] That process of taking a first-order quadratic equation and making it a second-order linear one is almost always a better idea than trying to handle the signs associated with the square roots in the original formulation, see Problem 1.3.11.

Fig. 1.8 For Problem 1.3.5. Two masses are initially separated by a distance d. Find the force and acceleration of each mass, solve Newton's second law for $x_1(t)$ and $x_2(t)$ if $m_2 = -m_1$.

$$\theta(t) = \theta_0 \cos\left(\sqrt{\frac{g}{L}}t\right), \tag{1.41}$$

with period $T = 2\pi\sqrt{L/g}$. This is an exact solution to the approximate problem defined by (1.40), hence an approximate solution to the full problem in (1.39).

Problem 1.3.1 A spring with spring constant k and equilibrium spacing a is hung from the ceiling, and a mass m attached at the bottom. Find the equilibrium position of the mass, and the period of oscillation if you released the mass from rest at some nonequilibrium extension.

Problem 1.3.2 For gravity near the surface of the earth, the magnitude of the force is

$$F = \frac{GMm}{(R+r)^2},$$

where r is the height above the earth's surface. Using Taylor expansion (assuming r is small), find the first nonconstant term in the force expression. Take the radius R and mass M for the earth and evaluate the constant term for a mass m that is 1 kg (it should look familiar). How big is the first-order correction for a 1 kg mass at a distance of 1 m above the surface of the earth?

Problem 1.3.3 One can turn (1.36) around, and ask: "for what radius sphere is a particular v the escape speed?" At what radius is the escape speed the speed of light? Calculate that special radius for the earth and sun. If all of the mass is packed into a sphere of this radius (or less), the resulting object is called a black hole, and the radius is the "event horizon."[6]

Problem 1.3.4 Find the escape speed for a Yukawa potential, $U(r) = -U_0 e^{-\mu r}/(4\pi r)$ for constant μ with dimension of inverse-length and constant U_0 to set the magnitude.

Problem 1.3.5 Two masses, m_1 and m_2 sit a distance d apart (working in one dimension, call it "x"). Find the (Newtonian) gravitational force acting on m_1 and m_2. What is the acceleration of each mass? Suppose we set $m_2 = -m_1$ with $m_1 > 0$. What happens to the forces on each mass? What is the acceleration of each mass? Provide a solution to the equation of motion (i.e. Newton's second law) for the masses assuming they start from rest with $x_1(0) = -d/2$, $x_2(0) = d/2$ in this negative mass case. It is fun to think about the implications for energy and momentum conservation in this problem.

---

[6] This is a nongeneral-relativistic calculation first carried out separately by Michell (1784) and Laplace (1796). The result in general relativity is numerically identical.

Problem 1.3.6   Assuming a force of the form $F = -k/x^3$ for constant k: What are the units of k? What is the potential energy $U(x)$ from which the force comes? What is the escape speed associated with this force (assume you are "blasting off" from a sphere of radius R that exerts this force on your rocket of mass m). Check that your escape speed expression has the units of speed.

Problem 1.3.7   For the potential energy $U(x) = U_0 \sin(x/\ell) \cos(x/\ell)$ with constants U_0 (unit of energy) and ℓ (unit of length), what is the minimum on the domain $x/\ell \in [0, \pi]$? What is the period of oscillation for motion near this minimum (assume mass m for the moving object)?

Problem 1.3.8   The force on a charge Q due to another charge q has magnitude $F = Qq/(4\pi\epsilon_0 r^2)$ where r is the distance between the (point) charges, and ϵ_0 is a proportionality constant. The force is attractive for charges of opposite sign, and repulsive if charges have the same sign. Referring to the configuration, find the net force on $Q > 0$ (at location x) due to the pair of charges $q > 0$ (at 0) and $-q$ (at d) assuming that the individual forces add. For $x \gg d$, find the first nonzero term in the Taylor expansion of the net force on Q.

Problem 1.3.9   Is the period of motion for a "real" pendulum, with motion governed by (1.39), longer or shorter than the small angle "simple" pendulum, with (1.40) as its equation of motion? Hint: plot the right-hand sides of each of these equations of motion and compare magnitudes.

Problem 1.3.10   For the energy curve and particle energy shown in Figure 1.5, identify the regions of the graph where particle motion is classically allowed. In the allowed region, where does the maximum/minimum speed occur (mark the points on the figure)?

Problem 1.3.11   Given a potential energy $U(x)$, conservation of energy in one dimension implies that we can write

$$\dot{x}(t) = \pm \sqrt{\frac{2}{m}} \sqrt{E - U(x(t))},$$

and it would be nice to solve this equation. But we don't know which sign to pick for $\dot{x}(t)$. Show that for either sign, taking the time derivative of this equation yields the same equation for $\ddot{x}(t)$ indicating that we should solve that instead of this first derivative form if we want to get $x(t)$ without worrying about changing signs for velocity.

Problem 1.3.12   In the following plot, we see a potential energy curve $U(x)$.

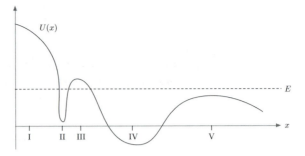

a.   For a particle that has energy E (shown as the dashed line), which of the positions, I–V, are allowed locations for the particle? At which allowed location would the particle be traveling the fastest? How about the location at which the particle is traveling the slowest?

b.   Which locations have oscillatory motion associated with a particle placed in their vicinity (ignore the E line for this and the next part of the problem)?

c.   For the locations that support oscillatory motion, which one has the longest period?

Problem 1.3.13   Gauss's law for electricity and magnetism reads: $\oint \mathbf{E} \cdot d\mathbf{a} = Q_{enc}/\epsilon_0$ where the integration on the left is over a closed surface, and Q_{enc} is the charge enclosed by that surface. There is a similar integral form for Newtonian gravity – the gravitational field[7] $\mathbf{H}$, integrated over a closed surface is related to the mass enclosed by the surface as follows:

$$\oint \mathbf{H} \cdot d\mathbf{a} = -4\pi G M_{enc}$$

where $G \approx 6.67 \times 10^{-11}$ N m²/kg² is the gravitational constant.

a.   Use this relation to find the gravitational field inside a sphere of radius R with uniform mass density ρ_0 (mass-per-unit-volume).

b.   Assuming the earth is a uniform sphere with constant mass density, if we dug a hole from the north pole to the south and dropped a mass m down it (starting from rest), how long, in minutes, would it take to return?

1.4 Series Expansion, Method of Frobenius

We solved (1.3) by "knowing the answer." In a sense, the oscillatory cosine and sine (and their parent, the complex exponential) are defined to be the solution to the differential equation: $\ddot{y}(t) = -\omega^2 y(t)$. But how would we find these given just the ODE itself? One approach, known as the "Method of Frobenius," is to assume a solution of the form

[7] The gravitational field is related to the force on a mass m by: $\mathbf{F} = m\mathbf{H}$, and the field has units of N/kg.

$$y(t) = t^p \sum_{j=0}^{\infty} a_j t^j \tag{1.42}$$

where p and $\{a_j\}_{j=0}^{\infty}$ are constants. This power series solution is motivated by, for example, the Taylor series expansion. For most functions $f(t)$, we know that

$$f(t) = \sum_{j=0}^{\infty} \underbrace{\frac{1}{j!} \frac{d^j f(t)}{dt^j}\bigg|_{t=0}}_{="a_j"} t^j, \tag{1.43}$$

which shares the form of (1.42).

1.4.1 Exponential

To demonstrate the process clearly, let's suppose we were given the first-order ODE: $\dot{y}(t) = y(t)$, with $y(0) = y_0$ also provided. We are asking for a function that is equal to its derivative. There are interesting graphical ways of producing a plot of such a function (where the tangent to the curve $y(t)$, defining the local slope, is itself equal to $y(t)$), but we can use the Method of Frobenius to generate a quantitative solution. Assume $y(t)$ takes the form given in (1.42), then we can calculate the derivative

$$\dot{y}(t) = t^p \sum_{j=0}^{\infty} a_j(j+p) t^{j-1} = t^p \left[a_0 p t^{-1} + \sum_{j=0}^{\infty} a_{j+1}(j+p+1) t^j \right], \tag{1.44}$$

where the sum has been reindexed and written out so that the derivative can be expressed as a sum from $j = 0 \to \infty$, matching the sum limits in $y(t)$ itself.

Now writing the ODE in the form $\dot{y}(t) - y(t) = 0$, and inserting the expansions, we get

$$t^p \left[a_0 p t^{-1} + \sum_{j=0}^{\infty} a_{j+1}(j+p+1) t^j \right] - t^p \sum_{j=0}^{\infty} a_j t^j = 0. \tag{1.45}$$

Because the coefficients $\{a_j\}_{j=0}^{\infty}$ do not depend on t, the only way for the sum in (1.45) to be zero for all values of t is if each power of t vanishes separately. One cannot, for all times t, kill an $a_1 t$ term with an $a_2 t^2$ term, for example, so each power of t must have a zero coefficient in front of it. Writing the equation with the coefficient of t^j isolated, we can explore the implication of requiring that it be zero,

$$t^p \left[a_0 p t^{-1} + \sum_{j=0}^{\infty} (a_{j+1}(j+p+1) - a_j) t^j \right] = 0. \tag{1.46}$$

The coefficient preceding t^j in the sum provide a "recursion relation." By requiring that each term multiplying t^j vanish, we are demanding that

$$a_{j+1} = \frac{a_j}{j+p+1} \tag{1.47}$$

which relates the a_{j+1} coefficient of $y(t)$ to the a_j one. That recursion ensures that the infinite sum in (1.46) vanishes, but we still have the t^{-1} term out front. We could take $a_0 = 0$ to start things off, but then the recursion relation tells us that all the other coefficients are zero (the problem here is that we cannot match the initial value, unless $y_0 = 0$). Instead, we must take $p = 0$, at which point the recursion becomes

$$a_{j+1} = \frac{a_j}{j+1}. \tag{1.48}$$

We can "solve" the recursion by writing a_j in terms of the starting value, a_0, and this can be done by inspecting terms. The first few look like,

$$a_1 = a_0$$
$$a_2 = \frac{a_1}{2} = \frac{a_0}{2} \tag{1.49}$$
$$a_3 = \frac{a_2}{3} = \frac{a_0}{6}$$

from which it is pretty clear that $a_j = a_0/(j!)$. The sum, with these coefficients, is

$$y(t) = a_0 \sum_{j=0}^{\infty} \frac{t^j}{j!}. \tag{1.50}$$

Finally, $y(0) = y_0$, so we pick $a_0 = y_0$ to match the provided initial value. The sum in (1.50) comes up so often it has its own name,[8] the "exponential" of x is defined as:

$$e^x \equiv \sum_{j=0}^{\infty} \frac{x^j}{j!}, \tag{1.51}$$

and the familiar properties of exponentials follow from this definition. As an example of one of these, we'll sketch the proof that $e^{x+y} = e^x e^y$. We can expand the right-hand side from the product of the sums:

$$\begin{aligned} e^x e^y &= \left(\sum_{j=0}^{\infty} \frac{x^j}{j!} \right) \left(\sum_{k=0}^{\infty} \frac{y^k}{k!} \right) \\ &= \left(1 + x + \frac{1}{2}x^2 + \cdots \right) \left(1 + y + \frac{1}{2}y^2 + \cdots \right) \\ &= 1 + (x+y) + \frac{1}{2} \left(x^2 + 2xy + y^2 \right) + \cdots \\ &= \sum_{m=0}^{\infty} \frac{(x+y)^m}{m!} \end{aligned} \tag{1.52}$$

which is e^{x+y}.

---

[8] Certain functions show up so much that we identify them by a common name rather than the more precise set of coefficients in the infinite sum.

1.4.2 Harmonic Oscillator

The starting point in (1.42), used in the equation of motion for the harmonic oscillator, $\ddot{y}(t) = -\omega^2 y(t)$, will allow us to solve for the coefficients $\{a_j\}_{j=0}^\infty$ (and p). This time, we need to write the second derivative in terms of the unknown coefficients. Working from (1.44), we can differentiate to get the second derivative of $y(t)$,

$$\ddot{y}(t) = t^p \sum_{j=0}^\infty a_j (j+p)(j+p-1) t^{j-2}$$

$$= t^p \left[a_0 p(p-1) t^{-2} + a_1 (p+1)pt^{-1} + \sum_{j=0}^\infty a_{j+2}(j+p+2)(j+p+1) t^j \right]$$

(1.53)

where we have again re-indexed and extracted terms so as to start all sums at $j = 0$ with t^j in the summand.

The ODE of interest here can be written as $\ddot{y}(t) + \omega^2 y(t) = 0$, and then inserting the sums for $\ddot{y}(t)$ and $y(t)$, and collecting in powers of t, we have

$$a_0 p(p-1) t^{-2} + a_1 (p+1)pt^{-1} + \sum_{j=0}^\infty \left(a_{j+2}(j+p+2)(j+p+1) + a_j \omega^2 \right) t^j = 0.$$

(1.54)

Looking at the t^{-2} term, we can get this to be zero if: $p = 0$, $p = 1$ or $a_0 = 0$. Let's take $p = 0$, then (1.54) becomes

$$\sum_{j=0}^\infty \left(a_{j+2}(j+2)(j+1) + a_j \omega^2 \right) t^j = 0,$$

(1.55)

so that in order for each coefficient of t^j to vanish separately, we must have

$$a_{j+2} = -\frac{a_j \omega^2}{(j+2)(j+1)}.$$

(1.56)

This is a recursion relation that links the coefficient a_{j+2} to the coefficient a_j. Given a_0, the first few terms are

$$a_2 = -\frac{a_0 \omega^2}{2}$$

$$a_4 = -\frac{a_2 \omega^2}{12} = \frac{a_0 \omega^4}{24}$$

(1.57)

$$a_6 = -\frac{a_4 \omega^2}{30} = -\frac{a_0 \omega^6}{720}.$$

We can now see the pattern:

$$a_{2k} = (-1)^k \frac{\omega^{2k}}{(2k)!} a_0 \text{ for } k = 0, 1, \ldots.$$

(1.58)

For the odd coefficients, take a_1 as given:

$$a_3 = -\frac{a_1\omega^2}{6}$$

$$a_5 = -\frac{a_3\omega^2}{20} = \frac{a_1\omega^4}{120} \tag{1.59}$$

$$a_7 = -\frac{a_5\omega^2}{42} = -\frac{a_1\omega^6}{5040}$$

from which

$$a_{2k+1} = (-1)^k \frac{\omega^{2k}}{(2k+1)!}a_1. \tag{1.60}$$

The full solution, obtained by putting the expressions for the coefficients $\{a_j\}_{j=0}^\infty$ into (1.42), is

$$y(t) = a_0 \sum_{k=0}^\infty (-1)^k \frac{\omega^{2k}}{(2k)!}t^{2k} + \frac{a_1}{\omega}\sum_{k=0}^\infty (-1)^k \frac{\omega^{2k+1}}{(2k+1)!}t^{2k+1}$$

$$= a_0 \cos(\omega t) + \frac{a_1}{\omega}\sin(\omega t), \tag{1.61}$$

where the sums themselves define the cosine and sine functions. We have a two-parameter family of solutions here, with a_0 and a_1 available to set initial or boundary conditions. This is to be expected given the starting second-order ODE. If we had taken $p = 1$ to eliminate the first term in (1.54), we would have to set $a_1 = 0$ to get rid of the second term, and we would recover the even powers of t in the sum (the cosine term). Similarly, if we took $p = -1$ to kill the second term, we'd be forced to take $a_0 = 0$ and would recover the odd powers of t in the sum, defining sine.

From the current point of view, what we have is a pair of infinite sums that represent the independent solutions to the ODE $\ddot{y}(t) + \omega^2 y(t) = 0$. These sums are given special names (because they show up a lot)

$$\cos\theta \equiv \sum_{k=0}^\infty (-1)^k \frac{\theta^{2k}}{(2k)!} \qquad \sin\theta \equiv \sum_{k=0}^\infty (-1)^k \frac{\theta^{2k+1}}{(2k+1)!}. \tag{1.62}$$

All of the properties of cosine and sine are contained in these expressions. For example, we can take the derivative of the terms in the sum[9] to evaluate the derivatives of cosine and sine:

$$\frac{d}{d\theta}\cos\theta = \sum_{k=1}^\infty (-1)^k \frac{2k\theta^{2k-1}}{(2k)!}$$

$$= \sum_{\ell=0}^\infty (-1)^{\ell+1} \frac{\theta^{2\ell+1}}{(2\ell+1)!} \tag{1.63}$$

$$= -\sum_{k=0}^\infty (-1)^k \frac{\theta^{2k+1}}{(2k+1)!} = -\sin\theta.$$

---

[9] Throughout this book, we will take the physicist's view that "all is well" – sums converge, and we can interchange summation and differentiation, etc.

Problem 1.4.1 By differentiating each term in

$$f(t) = f_0 \sum_{j=0}^{\infty} \frac{t^j}{j!},$$

show that $\frac{df(t)}{dt} = f(t)$ explicitly.

Problem 1.4.2 Use the series approach to find the solution to the ODE:

$$\frac{df(t)}{dt} = \alpha f(t)$$

(for constant α) with $f(0) = f_0$ given.

Problem 1.4.3 Take the derivative of $\sin \theta$ from the defining sum in (1.62).

Problem 1.4.4 Using the Frobenius series solution method, solve:

$$t^2 \ddot{f}(t) + t\dot{f}(t) + t^2 f(t) = 0$$

starting from

$$f(t) = t^p \sum_{j=0}^{\infty} a_j t^j.$$

You will end up with two terms (from the t^0 and t^1 powers) that can be used to set p and a_1, take $a_1 = 0$. Find the "even" coefficients using the recursion relation you get from demanding that all terms in the infinite sum (which goes from 2 to ∞) vanish individually. Use the initial value $f(0) = f_0$ to set a_0 and write the infinite sum in terms of an index $k = 0, 1, 2, \ldots$ (because you are finding the "even" coefficients, only t^{2k} will appear in your sum). Be on the lookout for functions of factorials of k (like, for example, $(k!)^2$). Treat $f(t)$ and t as dimensionless variables (i.e. don't worry about units, if you normally worry about units!). The function you get is called a "Bessel function" (in particular, the "zeroth" Bessel function).

Problem 1.4.5 Solve the ODE

$$\frac{d}{dx}\left(x^2 \frac{df(x)}{dx}\right) + (x^2 - 2) f(x) = 0$$

using the Frobenius method: Start from the usual sum, set $p = 0$, and solve the recursion relation to find the odd (in powers of x) solution. This function is an example of a "spherical Bessel function."

1.5 Complex Numbers

There is a connection between exponentials and the cosine and sine functions. To establish that connection, we need to review some definitions and properties of complex numbers. A complex number can be represented by a pair of real numbers: $z = a + ib$ for real a and b, and where the imaginary $i \equiv \sqrt{-1}$. The "real" part of z is a, the "imaginary" part is b. We can think of z as a location in the two-dimensional plane, where a represents the

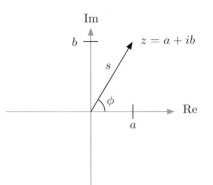

Fig. 1.9 The complex plane is spanned by a horizontal real ("Re") axis, and a vertical imaginary ("Im") axis. We locate points z in the plane by specifying their real and imaginary pieces. We can do this in Cartesian coordinates, where a and b represent distances along the horizontal and vertical axes, or in polar coordinates where s is the distance from the origin to the point z and ϕ is the angle that z makes with the real axis.

horizontal distance to the origin, b the vertical. Then the angle that a line drawn from the origin to the point (a, b) makes, with respect to the horizontal axis, is $\phi = \tan^{-1}(b/a)$. The distance from the origin to the point (a, b) is $s = \sqrt{a^2 + b^2}$. In terms of s and ϕ, we can write the "Cartesian" components of the complex number, a and b, in terms of the "polar" components, s and ϕ: $a = s \cos \phi$, and $b = s \sin \phi$. We have

$$z = a + ib = s(\cos \phi + i \sin \phi), \tag{1.64}$$

the two-dimensional picture to keep in mind is shown in Figure 1.9.

Using the summation form from (1.51), we can write the exponential of a complex argument as

$$e^{ix} = \sum_{j=0}^{\infty} i^j \frac{x^j}{j!} = \sum_{j=0}^{\infty} (-1)^j \frac{x^{2j}}{(2j)!} + i \sum_{j=0}^{\infty} (-1)^j \frac{x^{2j+1}}{(2j+1)!} = \cos(x) + i \sin(x), \tag{1.65}$$

where we have identified the sums that define cosine and sine from (1.62). This result, that $e^{ix} = \cos(x) + i \sin(x)$, is known as "Euler's formula." We can use it to neatly write the polar form of the complex number z from (1.64)

$$z = s(\cos \phi + i \sin \phi) = s e^{i\phi}. \tag{1.66}$$

Addition for complex numbers is defined in terms of addition for the real and imaginary parts. For $z_1 = a + ib$ and $z_2 = c + id$,

$$z_1 + z_2 = (a + c) + i(b + d). \tag{1.67}$$

Geometrically, this is like vector addition in two dimensions, with its usual "head-to-tail" visualization as shown in Figure 1.10.

Multiplication proceeds by treating z_1 and z_2 as polynomials in i, so that

$$z_1 z_2 = ac + i(ad + bc) + i^2 bd = ac - bd + i(ad + bc), \tag{1.68}$$

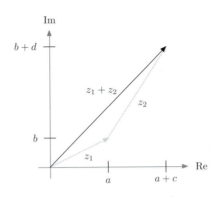

Fig. 1.10   We add vectors by components. The light gray vectors are $z_1 = a + ib$ and $z_2 = c + id$, then the sum is $z_1 + z_2 = (a + c) + i\,(b + d)$.

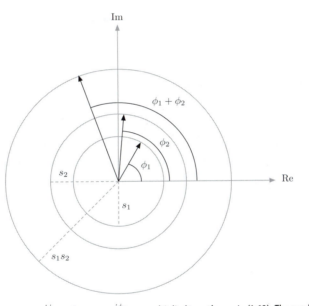

Fig. 1.11   Two complex numbers, $z_1 = s_1 e^{i\phi_1}$ and $z_2 = s_2 e^{i\phi_2}$ are multiplied together as in (1.69). The resulting product has magnitude $s_1 s_2$ and makes an angle $\phi_1 + \phi_2$ with the real axis.

and the real part of the product is $ac - bd$, with imaginary part $ad + bc$. The polar form makes it easy to see the effect of multiplication. Suppose $z_1 = s_1 e^{i\phi_1}$ and $z_2 = s_2 e^{i\phi_2}$, then using the multiplicative properties of the exponential, we have

$$z_1 z_2 = s_1 s_2 e^{i\,(\phi_1 + \phi_2)} \tag{1.69}$$

so that the product has magnitude $s_1 s_2$ (a stretching) and makes an angle of $\phi_1 + \phi_2$ with respect to the horizontal axis (a rotation). The stretch and rotation are shown in Figure 1.11.

Finally, there is a new operation that is not inherited directly from the real numbers, "conjugation," which is defined, for $z = a + ib = se^{i\phi}$, as

$$z^* \equiv a - ib = se^{-i\phi}. \tag{1.70}$$

The two-dimensional plane operation here is reflection about the horizontal axis. As a practical matter, when we take the "complex conjugate" of expressions involving complex numbers and/or variables, we just flip the sign of i wherever it appears (see Problem 1.5.5 for partial justification of this procedure). We can use the conjugate to solve for cosine and sine by algebraically inverting Euler's formula:

$$\cos\phi = \frac{1}{2}\left(e^{i\phi} + e^{-i\phi}\right) \qquad \sin\phi = \frac{1}{2i}\left(e^{i\phi} - e^{-i\phi}\right). \tag{1.71}$$

Let's return to our model problem, $\ddot{y}(t) = -\omega^2 y(t)$. The two independent solutions, cosine and sine, are linear combinations of exponentials. So for a general solution like

$$y(t) = A\cos(\omega t) + B\sin(\omega t), \tag{1.72}$$

if we use (1.71), we could write

$$y(t) = \bar{A}e^{i(\omega t)} + \bar{B}e^{-i(\omega t)} \tag{1.73}$$

for new (complex) constants $\bar{A}$ and $\bar{B}$.

This alternate form can be quite useful, and it is important to get comfortable moving back and forth from the trigonometric form of the solution to the exponential form. One advantage of the exponential solution is that it makes the original problem of *solving* for $y(t)$ easier by turning $\ddot{y}(t) + \omega^2 y(t) = 0$ into an algebraic equation. To see this, suppose we guess $y(t) = \alpha e^{\beta t}$ for constants α and β, motivated by the fact that the derivatives of exponentials are proportional to themselves. Inserting this into the ODE gives

$$\alpha\left(\beta^2 + \omega^2\right)e^{\beta t} = 0 \tag{1.74}$$

from which we learn that $\beta = \pm i\omega$. There are two value of β, hence two solutions here, so the most general case is a linear combination of the two, and we are led immediately to

$$y(t) = \bar{A}e^{i(\omega t)} + \bar{B}e^{-i(\omega t)}. \tag{1.75}$$

Problem 1.5.1   Find the relationship between $\bar{A}$, $\bar{B}$ from (1.73) and the original A and B from (1.72).

Problem 1.5.2   Evaluate the products i^2, i^3, i^4, and i^5.

Problem 1.5.3   For a complex number, like $p = u + iv$ the "real" part of p is u and the "imaginary" part of p is v (both u and v are themselves real numbers). For $z = a + ib$ (with a and b both real) and $p = u + iv$, what are the real and imaginary parts of z/p?

Problem 1.5.4   For $z = a + ib$, $p = u + iv$ (with a, b, u, and v all real), show that:

   a. $(z + p)^* = z^* + p^*$

   b. $(zp)^* = z^* p^*$

   c. $\left(\frac{z}{p}\right)^* = \frac{z^*}{p^*}$

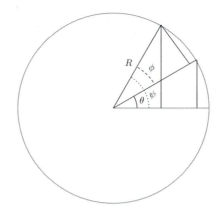

Fig. 1.12   For Problem 1.5.8.

Problem 1.5.5   Use part b. of the previous problem to show that for $z = a + ib$ we have $\left(z^j\right)^* = \left(z^*\right)^j$. Now consider a function $f(z)$ that we can expand as

$$f(z) = \sum_{j=0}^{\infty} a_j z^j,$$

for real constants $\{a_j\}_{j=0}^{\infty}$. Show that $f(z)^* = f(z^*)$ so that if you have a function of $a + ib$, to take the conjugate, you just evaluate f at $a - ib$.

Problem 1.5.6   From Euler's formula: $e^{i\theta} = \cos(\theta) + i\sin(\theta)$, write $\cos(\theta)$ and $\sin(\theta)$ in terms of exponentials (and any constants you need).

Problem 1.5.7   Convert $\cos(\theta + \phi)$ to complex exponentials and use that representation to prove that

$$\cos(\theta + \phi) = \cos\phi\cos\theta - \sin\phi\sin\theta.$$

Problem 1.5.8   In Figure 1.12, we have a circle of radius R with three angles shown: θ, ϕ, and $\psi \equiv \theta + \phi$. Using Figure 1.12, identify lengths and angles to show

$$\sin\psi = \sin\phi\cos\theta + \cos\phi\sin\theta.$$

You are provding a pictorial version, for sine, of what you did in Problem 1.5.7 for cosine, providing a complementary point of view.

1.6 Properties of Exponentials and Logarithms

We have developed the exponential as the function that has derivative equal to itself at every point, so that the exponential $y(t) = e^t$ satisfies $\dot{y}(t) = y(t)$ for all t by construction. Suppose we have an exponential that involves a *function* of t, $f(t)$, then we can use the chain rule to find the derivative of $e^{f(t)}$. Let $y(p) = e^p$, then

$$\frac{d}{dt}e^{f(t)} = \frac{dy(f(t))}{dt} = \frac{dy(f)}{df}\frac{df(t)}{dt} = y(f(t))\dot{f}(t) = e^{f(t)}\dot{f}(t). \tag{1.76}$$

Any time you have an ODE of the form

$$\dot{y}(t) = g(t)y(t) \tag{1.77}$$

where $g(t) = \dot{f}(t)$ for some function $f(t)$, the solution is

$$y(t) = \bar{y}e^{f(t)} \tag{1.78}$$

where $\bar{y}$ is a constant that can be used to set the initial value (if $f(0) = 0$, then $\bar{y} = y_0$, the provided value at $t = 0$).

As an example, suppose we have

$$\dot{y}(t) = t^p y(t) \text{ with } y(0) = y_0 \text{ given,} \tag{1.79}$$

and p some arbitrary constant. Here, $t^p = g(t)$, so that $f(t) = t^{p+1}/(p+1)$, and then

$$y(t) = y_0 e^{\frac{t^{p+1}}{p+1}}. \tag{1.80}$$

The "logarithm" function un-does the exponential. If you have $e^a = b$ and you want to know what a is, you take the "log" of both sides,[10]

$$\log(e^a) \equiv a = \log(b). \tag{1.81}$$

The log plays the same inverse role as arcsine and arccosine. It inherits properties directly from properties of the exponential. For example, suppose we have $e^a = b$ and $e^c = d$, so that $a = \log(b)$ and $c = \log(d)$. We know that

$$e^a e^c = bd \rightarrow e^{a+c} = bd \tag{1.82}$$

since $e^a e^c = e^{a+c}$. Then taking the log of both sides,

$$a + c = \log(bd) = \log(b) + \log(d), \tag{1.83}$$

and the sum of the log terms becomes the log of the product.

We can also find the derivative (and anti-derivative) of the logarithm function, again from its definition. We have $\log(e^t) = t$, and let $f(t) = e^t$, then $\log(f(t)) = t$, and taking the t-derivative of both sides gives

$$\frac{d\log(f)}{df}\frac{df(t)}{dt} = 1 \tag{1.84}$$

using the chain rule. The derivative of $f(t)$ is itself, by the definition of exponential, so that

$$\frac{d\log(f)}{df}f = 1 \longrightarrow \frac{d\log(f)}{df} = \frac{1}{f}. \tag{1.85}$$

---

[10] For my purposes, the function "$\log(x)$" refers to the "natural logarithm" (base e). This is sometimes written $\ln(x)$ to remind us of the appropriate base, and to differentiate it from the base 10 log which, in some contexts, is called log. In physics, we rarely have uses for bases other than e, so I'll use $\log(x)$ to refer exclusively to the base e form.

You can call f anything you like, we have

$$\frac{d \log(t)}{dt} = \frac{1}{t}$$ (1.86)

and if you have a generic function $p(t)$, then the chain rule gives

$$\frac{d \log(p(t))}{dt} = \frac{\dot{p}(t)}{p(t)}.$$ (1.87)

Going the other direction, we can integrate both sides of (1.86) to get (omitting the constant of integration)

$$\log(t) = \int \frac{1}{t}\, dt.$$ (1.88)

We can do this to the defining relation for the exponential, too: $\dot{y}(t) = y(t)$ means $y(t) = \int y(t)\, dt$ so that

$$e^t = \int e^t\, dt,$$ (1.89)

and the exponential is its own integral. This result can be extended for integrals of the form $e^{\alpha t}$ for constant α,

$$\int e^{\alpha t}\, dt = \frac{1}{\alpha} \int e^q\, dq = \frac{e^{\alpha t}}{\alpha},$$ (1.90)

where we set $q \equiv \alpha t$ in a change of variables. For definite limits of integration, $t_0 \to t_f$, we have

$$\int_{t_0}^{t_f} e^{\alpha t}\, dt = \frac{1}{\alpha}\left(e^{\alpha t_f} - e^{\alpha t_0}\right).$$ (1.91)

Problem 1.6.1   Show that $\log(x^p) = p \log(x)$ for constant p (not necessarily an integer).

Problem 1.6.2   Using the chain rule, take the t-derivative of $h(t) = e^{g(t)}$ and write your expression for $\frac{dh(t)}{dt}$ in terms of $h(t)$ itself and the derivative of $g(t)$. Using this result (or any other technique you like short of looking it up) find the solution to the ODE:

$$\frac{df(t)}{dt} = i\omega t^2 f(t),$$

with $f(0) = f_0$ given. What is the real part of your solution?

Problem 1.6.3   The "hyperbolic" cosine and sine are defined by the sums:

$$\cosh(x) \equiv \sum_{j=0}^{\infty} \frac{x^{2j}}{(2j)!} \qquad \sinh(x) \equiv \sum_{j=0}^{\infty} \frac{x^{2j+1}}{(2j+1)!}.$$

Write the infinite sum for the exponential in terms of these two sums (i.e. write e^x in terms of $\cosh(x)$ and $\sinh(x)$), arriving at "Euler's formula" for hyperbolic cosine and sine. "Invert" the relation to find $\cosh(x)$ and $\sinh(x)$ in terms of exponentials (as in Problem 1.5.6) and sketch $\cosh(x)$ and $\sinh(x)$ (include positive and negative values for x).

Problem 1.6.4   Evaluate sine and sinh with complex arguments: What are $\sin(i\theta)$ and $\sinh(i\eta)$ for real θ and η?

Problem 1.6.5   When we solve Newton's second law for the harmonic oscillator:

$$m\frac{d^2x(t)}{dt^2} = -kx(t),$$

we get cosine and sine solutions. Suppose we "complexify time" (people do this) by letting $t = is$ for a new "temporal" parameter s. Write Newton's second law in terms of s and solve that equation for $x(s)$ using $x(0) = x_0$ and $\frac{dx(s)}{ds}\big|_{s=0} = 0$. What do your solutions look like?

1.7 Solving First-Order ODEs

We will be taking the harmonic oscillator that started us off, with its near universal applicability to motion near the minima of any potential energy function, and adding additional forcing, capturing new and interesting physics. In the next chapter, we will think about the effects of friction and driving (adding an external force to the spring system), and will be focused on solving more and more general second-order ODEs. As a warmup and advertisement of some of the techniques we will encounter, we'll close the chapter by discussing the solutions to

$$\frac{df(x)}{dx} = G(x,f(x)) \quad f(0) = f_0 \tag{1.92}$$

for some provided "driving" function $G(x,f(x))$ and initial value f_0.

1.7.1 Continuity

As a first observation, we can show that solutions to (1.92) are continuous provided G is finite. Continuity means that for all points x in the domain of the solution, we have

$$\lim_{\epsilon \to 0} (f(x + \epsilon) - f(x - \epsilon)) = 0. \tag{1.93}$$

If the function $G(x,f(x))$ is itself continuous, we can be assured that $f(x)$ is continuous since integrating a continuous function returns a continuous function. So we'll focus on the case where the right-hand side of (1.92) is discontinuous at some point. Then integrating both sides of the ODE across the discontinuity serves to "smooth" it out, leading to continuous $f(x)$.

To be concrete, suppose our function $G(x,f(x))$ takes the following discontinuous but simple form,[11] for constant G_0,

$$G(x,f(x)) = \begin{cases} 0 & x < a \\ G_0 & x \geq a \end{cases} \tag{1.94}$$

---

[11] We could have two separate constants on the "left" and "right" of the discontinuity, but it is simplest to pick zero for one of them.

with a discontinuity at $x = a$. Now integrate both sides of (1.92) from $a - \epsilon \rightarrow a + \epsilon$:

$$\int_{a-\epsilon}^{a+\epsilon} \frac{df(x)}{dx} dx = \int_{a-\epsilon}^{a+\epsilon} G(x, f(x)) dx$$

$$f(a + \epsilon) - f(a - \epsilon) = (a + \epsilon) G_0 - aG_0 = \epsilon G_0,$$

(1.95)

and taking the limit as $\epsilon \rightarrow 0$ gives us continuity of $f(x)$, reproducing (1.93) at $x = a$. Continuity of the solution is a property of the ODE in (1.92), unless $G(x, f(x))$ becomes infinite at the discontinuity, that case requires a more nuanced limit on the right of (1.95).

1.7.2 Separation of Variables

For a generic $G(x, f(x))$, it is not necessarily possible to solve (1.92), at least analytically.[12] If the function can be "separated" into a function of x and a separate function of $f(x)$, then we can make progress. We'll consider a separation of the form $G(x, f(x)) = g(x)h(f(x))$, a product of a function $g(x)$ and $h(y)$, the latter of which is evaluated at $y = f(x)$. Now for the mnemonic, we can take the fraction $\frac{df}{dx}$ seriously (à la Leibniz), and multiply both sides of $\frac{df}{dx} = g(x)h(f)$ by dx while dividing by $h(f)$ (less problematic)

$$\frac{df}{h(f)} = g(x)dx.$$

(1.96)

Now we just integrate both sides between the relevant limits, starting from $x = 0$ on the right, and $f = f_0$ on the left

$$\int_{f_0}^{f} \frac{d\bar{f}}{h(\bar{f})} = \int_{0}^{x} g(\bar{x}) d\bar{x}.$$

(1.97)

All that's left is to perform the integrals (which may or may not be difficult), and invert to find $f(x)$ (which will almost certainly be difficult).

As an example of the process, take $g(x) = 1$, $h(y) = y$, so that we are solving

$$\frac{df(x)}{dx} = f(x),$$

(1.98)

and we multiply by dx and divide by $f(x)$ on both sides,

$$\frac{df}{f} = dx,$$

(1.99)

then integrating both sides yields

$$\int_{f_0}^{f} \frac{d\bar{f}}{\bar{f}} = \int_{0}^{x} d\bar{x} \longrightarrow \log\left(\frac{f}{f_0}\right) = x.$$

(1.100)

Now for the inversion, we exponentiate both sides and isolate f:

$$f(x) = f_0 e^{x}.$$

(1.101)

---

[12] Meaning, here, without the use of a computer.

Let's try it again for an ODE whose solution we haven't been focused on, take $g(x) = x$ and $h(y) = 1/y$, we want to solve

$$\frac{df(x)}{dx} = \frac{x}{f(x)} \qquad f(0) = f_0. \tag{1.102}$$

Separating the x and f variables as before, we have the integrals

$$\int_{f_0}^{f} \bar{f} d\bar{f} = \int_0^x \bar{x} \, d\bar{x} \longrightarrow \frac{1}{2}(f^2 - f_0^2) = \frac{1}{2}x^2 \tag{1.103}$$

giving the pair of solutions

$$f(x) = \pm\sqrt{x^2 + f_0^2}. \tag{1.104}$$

The "multiplication" by dx that is the basis of the technique can be justified. For a function $f(x)$, we know that the differential df is related to dx by $df = \frac{df}{dx}dx$, telling us how f changes due to a change in x. So really, it is a property of the differential df that is being exploited here. Since $\frac{df}{dx} = G$ from the start, we always had

$$df = \frac{df(x)}{dx} dx = G(x, f(x)) dx, \tag{1.105}$$

and in the special separable case $G(x, f(x)) = g(x)h(f(x))$, we get

$$df = g(x)h(f)dx \longrightarrow \frac{df}{h(f)} = g(x)dx \tag{1.106}$$

as before.

What sort of ODE would *not* be solved by this approach (at least, up to integration and inversion)? We need an example in which the multiplicative separation fails, which is easy to imagine. Take $G(x, f(x)) = \sin(xf(x))$, for example. The separation technique will not allow us to make progress here. Fortunately, this right-hand side has no physical significance, and in practice, many first-order differential equations can be solved using separation of variables. In the broader context of partial differential equations, there is a related class of solutions obtained by "separation of variables," and we shall see these later on in Section 4.3.2.

1.7.3 Superposition

As a special case of separation of variables, consider the most general *linear* form of the model problem:

$$\frac{df(x)}{dx} = g(x)f(x) \qquad f(0) = f_0 \tag{1.107}$$

where $g(x)$ is any function of x and f_0 is given. The problem is linear in the sense that only $f(x)$ and its derivative show up, there are no terms like $f(x)^2$, for example. When an ODE is linear, we can add together solutions and the result is also a solution. To see this, suppose we have $f_1(x)$ and $f_2(x)$ both satisfying the ODE in (1.107), then let $h(x) = Af_1(x) + Bf_2(x)$ for constants A and B, and we have

$$\frac{dh(x)}{dx} = A\frac{df_1(x)}{dx} + B\frac{df_2(x)}{dx} = Ag(x)f_1(x) + Bg(x)f_2(x) = g(x)h(x), \qquad (1.108)$$

so that $h(x)$ is also a solution.

Once the initial value is introduced, we have fewer choices, and in many cases, satisfying the initial value constraint on the problem renders the solution unique. Given that this problem is separable, we can also write down an explicit integral form[13] of the solution,

$$\int_{f_0}^{f} \frac{d\bar{f}}{\bar{f}} = \int_0^x g(\bar{x})\,d\bar{x} \longrightarrow f(x) = f_0 e^{\int_0^x g(\bar{x})\,d\bar{x}}, \qquad (1.109)$$

as we saw in Section 1.6.

1.7.4 Homogeneous and Sourced Solutions

Finally, there are linear problems of the form

$$\frac{df(x)}{dx} = \alpha f(x) + g(x) \qquad f(0) = f_0 \qquad (1.110)$$

for some function of x only, the "source," $g(x)$ (here, α is just a constant). In order to solve these, the general approach is to separate the solution $f(x)$ into a piece that solves the source-free ($g(x) = 0$) "homogeneous" problem, $h(x)$, and a piece associated with the source, call it $\bar{f}(x)$. Then we can add the two solutions to get a solution to the full problem, $f(x) = h(x) + \bar{f}(x)$. The homogeneous piece is normally associated with the constant that will allow us to set the initial value.

The homogeneous solution, $h(x)$, solves

$$\frac{dh(x)}{dx} = \alpha h(x) \longrightarrow h(x) = h_0 e^{\alpha x} \qquad (1.111)$$

where h_0 is the promised constant. To get the source piece, $\bar{f}(x)$, we can use the "variation of parameters" approach (see [2], for example). Let $\bar{f}(x) = u(x)h(x)$ where $u(x)$ is an unknown function, and $h(x)$ is the homogeneous solution. Running this through the ODE gives

$$\frac{d\bar{f}(x)}{dx} = \alpha\bar{f}(x) + g(x) \longrightarrow \frac{du(x)}{dx}h(x) + u(x)\frac{dh(x)}{dx} = \alpha u(x)h(x) + g(x), \qquad (1.112)$$

and we know the derivative of $h(x)$ is $\alpha h(x)$ from (1.111). We can use that to cancel a term on the right. The ODE for $u(x)$ is now

$$\frac{du(x)}{dx} = \frac{g(x)}{h(x)} = \frac{1}{h_0}e^{-\alpha x}g(x) \qquad (1.113)$$

which can be solved by integration,

$$u(x) = \frac{1}{h_0}\int_0^x e^{-\alpha\bar{x}}g(\bar{x})d\bar{x}, \qquad (1.114)$$

---

[13] Of course, the integral may or may not exist, depending on $g(x)$ and the relevant domain.

and

$$\bar{f}(x) = u(x)h(x) = e^{ax} \int_0^x e^{-a\bar{x}} g(\bar{x}) \, d\bar{x} \tag{1.115}$$

with no constants for setting initial values. The full solution is the sum

$$f(x) = h(x) + \bar{f}(x) = h_0 e^{ax} + e^{ax} \int_0^x e^{-a\bar{x}} g(\bar{x}) \, d\bar{x} \tag{1.116}$$

and we can even set $h_0 = f_0$ to satisfy the initial value.

Problem 1.7.1 Find $f(x)$ solving the following first-order ODEs:

$$\frac{df(x)}{dx} = x^2 f(x) \quad f(0) = 1$$

$$\frac{df(x)}{dx} = \cos(x)f(x) \quad f(0) = -2$$

$$\frac{df(x)}{dx} = xf(x)^2 \quad f(0) = 1$$

$$\frac{df(x)}{dx} = \frac{\sin(x)}{f(x)} \quad f(0) = 0.$$

Problem 1.7.2 Find the homogeneous (where the "source" function on the right is zero) and sourced solution to the ODE

$$\frac{df(x)}{dx} + f(x) = g_0 \sin(x).$$

Notice that it is the homogeneous solution that comes with the constant of integration. Use that to set $f(0) = 5$.

Problem 1.7.3 Suppose you had the second-order differential equation for a driven harmonic oscillator

$$m\ddot{x}(t) = -m\omega^2 x(t) + F(t)$$

for a given driving force $F(t)$. This is a second-order version of (1.111) with "source" $F(t)$. Find the two homogeneous solutions, $h_1(t)$ and $h_2(t)$ and try the variation of parameters procedure starting with $\bar{x}(t) = u(t)h_1(t) + v(t)h_2(t)$ to write an integral solution like (1.116). The functions $u(t)$ and $v(t)$ can be related in a variety of ways (there aren't two independent functions here – you only get one solution, not two), use the constraint $\dot{u}(t)h_1(t) + \dot{v}(t)h_2(t) = 0$ to relate the two (that will simplify the expressions you get in the variation of parameters).

2   Damped Harmonic Oscillator

Let's complicate matters – suppose we take our mass-on-a-spring, and introduce some friction. We'll start with the familiar kinetic friction which opposes motion with constant magnitude. Calling that magnitude α, we can include the friction force in Newton's second law,

$$m\ddot{x}(t) = -kx(t) - \alpha\,\text{sign}(\dot{x}), \tag{2.1}$$

and we have set the equilibrium position of the spring to be at $a = 0$, as shown in Figure 2.1.

We'll start the mass off from rest with initial extension $p_0 > 0$. The mass moves to the left, so that the sign of $\dot{x}$ is -1, and we start by solving

$$\ddot{x}_1(t) = -\frac{k}{m}x_1(t) + \frac{\alpha}{m} = -\frac{k}{m}\left(x_1(t) - \frac{\alpha}{k}\right). \tag{2.2}$$

Again letting $\omega^2 \equiv k/m$, and defining the length $z \equiv \alpha/k$, we have solution

$$x_1(t) = A\cos(\omega t) + B\sin(\omega t) + z, \tag{2.3}$$

and the initial conditions, $x_1(0) = p_0$, $\dot{x}_1(0) = 0$, give $A = (p_0 - z)$ and $B = 0$. The solution is

$$x_1(t) = (p_0 - z)\cos(\omega t) + z. \tag{2.4}$$

This solution only holds until the mass comes to rest on the other side of its equilibrium location. That happens when $t_1 = \pi/\omega$, at which point the mass is at $x_1(\pi/\omega) = -p_0 + 2z$. Now we have a mass that starts from rest at $p_1 \equiv -p_0 + 2z$ and travels to the right, so that the sign of $\dot{x}(t)$ is $+1$. We must solve

$$\ddot{x}_2(t) = -\omega^2(x_2(t) + z) \text{ with } x_2(\pi/\omega) = -p_0 + 2z \text{ and } \dot{x}_2(\pi/\omega) = 0. \tag{2.5}$$

Since this is the same problem as in (2.2), with $z \to -z$, and $p_0 \to p_1$, we have solution

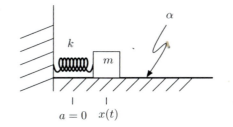

Fig. 2.1   A mass moves under the influence of both a spring and kinetic friction.

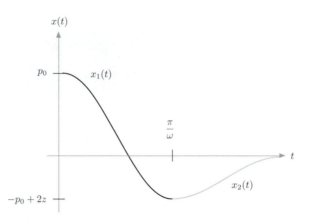

Fig. 2.2   The first full "cycle" of the frictionally damped harmonic oscillator.

$$x_2(t) = (p_1 + z)\cos(\omega(t - \pi/\omega)) - z$$
$$= (-p_0 + 3z)\cos(\omega(t - \pi/\omega)) - z. \tag{2.6}$$

The process continues, each time the mass stops to turn around, we pick up a new value for "p" and the sign of z changes. Notice that at any point, the position and its derivative are continuous, that's a requirement of Newton's second law (see Section 1.7.1). A solution like this, which is pieced together from individual solutions, is known as a "piecewise solution." For this constant friction case, it is the only reasonable way to express the solution to the problem. The first two iterations of the solution are shown in Figure 2.2.

Problem 2.0.1   Finish the job – what is the nth solution to (2.1), $x_n(t)$ assuming that at the points where the mass is at rest, $|x_n(t)| > z$ (that way the force reliably switches direction each time the mass comes to rest).

Problem 2.0.2   Show, from Newton's second law (in one dimension, for simplicity), that if a force is discontinuous but finite, the velocity and position of a mass subject to that force are continuous.

Problem 2.0.3   For the piecewise force:

$$F(t) = \begin{cases} 0 & t < 0 \\ F_0 & t \geq 0 \end{cases},$$

solve Newton's second law for a particle of mass m for both $t < 0$ and $t > 0$ given $x(0) = 0$, $\dot{x}(0) = 0$. Use the continuity and derivative continuity you established in the last problem, applied at the discontinuity in the force, to set any constants in your solution.

2.1 Damping

Imagine the resistance you feel moving a hand through water – the faster you try to move your hand, the harder you must push. The response of the water is to generate a force that

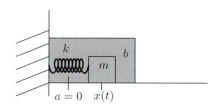

Fig. 2.3 A mass is attached to a spring and immersed in a fluid to generate a force that opposes the motion of the mass with magnitude proportional to the mass's speed and proportionality constant mb.

always opposes the motion of your hand, as in the previous example, but the magnitude of the force the water generates is itself dependent on how fast your hand is moving. In general, a force that is meant to model "drag" has the following general form (for velocity vector $\mathbf{v}$ with magnitude v, and constants $\{\alpha_j\}_{j=0}^{\infty}$)

$$\mathbf{F}_v = -\left(\alpha_0 + \alpha_1 v + \alpha_2 v^2 + \cdots\right)\frac{\mathbf{v}}{v}. \tag{2.7}$$

The minus sign out front, and velocity unit vector at the end enforce the notion that the force always opposes the current motion of the object on which it acts. Inside the parentheses, we see a (Taylor) series expansion of the magnitude of $\mathbf{F}$ in powers of the speed v, and we can tune the coefficients to account for a wide array of "drag" behaviors. For kinetic friction, only the constant α_0 is nonzero. Suppose we take the first correction to that, let's take α_1 to be nonzero with all other terms vanishing. Using this form in Newton's second law, we have (setting $\alpha_1 = 2mb$ for new constant b and letting $k/m \equiv \omega^2$ as in the last chapter)

$$m\ddot{x}(t) = -kx(t) - 2mb\dot{x}(t) \longrightarrow \ddot{x}(t) + 2b\dot{x}(t) + \omega^2 x(t) = 0. \tag{2.8}$$

This new drag force can be generated for the spring setup by attaching a "dashpot," basically some oil or other viscous medium that slows down the motion of the mass in a manner that is proportional to its speed. The physical setup is sketched in Figure 2.3.

How should we solve (2.8), together with initial conditions (to make the problem well posed and uniquely solvable): $x(0) = x_0$ and $\dot{x}(0) = v_0$? We could "guess" (or just write down a solution and check it), or use the series technique from the previous chapter. Thinking of the limiting cases, if $b = 0$ (no damping), we have $x(t) \sim e^{\pm i\omega t}$, a complex exponential. On the other hand, if $\omega = 0$ (no oscillation), we get $x(t) \sim e^{-2bt}$, a decaying exponential. These limiting cases suggest that for the full problem, we try the general form $x(t) = pe^{qt}$ for constant p and q, then each term in (2.8) will have a factor of pe^{qt} and we can cancel those, and in addition, we know we can capture the limiting cases.[1] Inserting our ansatz gives

$$q^2 + 2bq + \omega^2 = 0 \longrightarrow q = -b \pm \sqrt{b^2 - \omega^2}. \tag{2.9}$$

There are two independent values for q here, and so we know the general solution will be a linear combination of these two

---

[1] Interestingly, the case $b = 0$ and $\omega = 0$ has $x(t) = At + B$ for constants A and B, and while $B = Be^0$, the At term is not so easily expressed as an exponential. That will prove to be an issue in the case of "critical damping" as we shall see in Section 2.1.2.

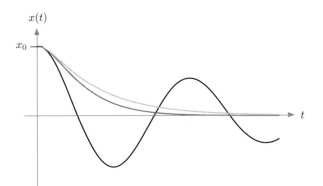

Fig. 2.4   A mass is released from rest with the same ω, but varying b to demonstrate the underdamped (black), critically damped (gray), and overdamped (light gray) behaviors.

$$x(t) = Ae^{\left(-b+\sqrt{b^2-\omega^2}\right)t} + Be^{\left(-b-\sqrt{b^2-\omega^2}\right)t} = e^{-bt}\left(Ae^{\sqrt{b^2-\omega^2}t} + Be^{-\sqrt{b^2-\omega^2}t}\right).$$

$$(2.10)$$

Fine, but what about the physics of these solutions? It's pretty clear that we have some sort of decaying exponential, since e^{-bt} is sitting out front, and for $b > 0$, this will go to zero as $t \to \infty$. But the relative values of b and ω are also important in setting the physical behavior. For example, if $b/\omega < 1$, then the square root in (2.10) introduces a factor of i in the exponential, and we have oscillatory solutions (sines and cosines), an "underdamped" motion. If $b/\omega > 1$, the square root is real, so we pick up growing and decaying exponentials, the motion is "overdamped." Finally, if $b/\omega = 1$, we just get decay and the motion is said to be "critically damped."

We can look at the various cases together to see what sort of behavior to expect in general. In Figure 2.4, we see the underdamped (black), critically damped (gray), and overdamped (light gray) motion that occurs for a mass that begins from rest at some x_0. To make these plots, I used the same ω and varied b to explore the three regimes.

2.1.1 Underdamped

Each of these cases is of interest, so we'll consider them one at a time. For $b/\omega < 1$ the physical system is "underdamped." In this case, we write $\sqrt{b^2 - \omega^2} = i\sqrt{\omega^2 - b^2}$ and can introduce sines and cosines explicitly from Euler's formula. The solution, written in terms of x_0 and v_0, is

$$x(t) = e^{-bt}\left(x_0 \cos\left(\sqrt{\omega^2 - b^2}t\right) + \frac{bx_0 + v_0}{\sqrt{\omega^2 - b^2}} \sin\left(\sqrt{\omega^2 - b^2}t\right)\right). \qquad (2.11)$$

This clearly reduces to our familiar expression in the $b = 0$ limit. The oscillation occurs with modified (as compared with $b = 0$) period and frequency

$$T = \frac{2\pi}{\sqrt{\omega^2 - b^2}} \qquad f = \frac{1}{T} = \frac{\sqrt{\omega^2 - b^2}}{2\pi}. \qquad (2.12)$$

The definition of period here is different when compared with the purely oscillatory $b = 0$ case. There, we defined the period to be the amount of time it took, starting from rest, to return to the initial location. Here, the mass never returns to the initial location, so we revert to a weaker definition, "the period is the time it takes for the mass, starting from rest, to come to rest twice," matching our definition in the undamped case (the mass stops at maximal extension on the other side of the equilibrium, then returns).

2.1.2 Critically Damped

If $b/\omega = 1$, the two constants are equal, the motion is "critically damped." Referring to the general (2.10), we get

$$x(t) = e^{-bt}(A + B). \tag{2.13}$$

This is a problem, since the combination $A + B$ is just a single constant. We have lost a constant of integration, and cannot set both $x(0) = x_0$ and $\dot{x}(0) = v_0$ .... The issue is that just as the quadratic equation loses a root when its discriminant vanishes (the two distinct roots becoming a single *double* root), we have lost our ability to describe the two independent solutions to the ODE. Let's go back to the ODE, with $b = \omega$,

$$\ddot{x}(t) + 2\omega\dot{x}(t) + \omega^2 x(t) = 0. \tag{2.14}$$

We'll try an ansatz of the form $x(t) = e^{-\omega t}y(t)$ since we know that the decaying exponential is common to all of our solutions. The goal is to "peel off" the portion of the ODE that is setting the decaying exponential (which we already know about) leaving us with a simplified ODE governing the auxiliary function $y(t)$. This works well, running the ansatz through the ODE gives

$$e^{-\omega t}\ddot{y}(t) = 0 \longrightarrow y(t) = A + Bt \tag{2.15}$$

and now we can set A and B using the initial conditions for $x(t)$

$$x(t) = e^{-\omega t}(x_0 + (v_0 + \omega x_0)\, t)\,. \tag{2.16}$$

You may be worried about the linear growth in t found inside the parenthesis here, but you can show, in Problem 2.1.1, that $x(t \to \infty) \to 0$.

2.1.3 Overdamped

Finally, for $b/\omega > 1$, we are in the "overdamped" regime, where $\sqrt{b^2 - \omega^2}$ is a real, positive number. The exponentials are no longer oscillatory, but rather growing and decaying. We can express the solution in terms of the hyperbolic cosh and sinh functions. Those satisfy a sort of "real" version of Euler's formula and can be inverted providing a relation similar to (1.71)

$$\cosh \eta = \frac{1}{2}\left(e^{\eta} + e^{-\eta}\right) \qquad \sinh \eta = \frac{1}{2}\left(e^{\eta} - e^{-\eta}\right). \tag{2.17}$$

Then we have

$$x(t) = e^{-bt}\left(x_0 \cosh\left(\sqrt{b^2 - \omega^2}t\right) + \frac{bx_0 + v_0}{\sqrt{b^2 - \omega^2}}\sinh\left(\sqrt{b^2 - w^2}t\right)\right) \tag{2.18}$$

which can be compared to (2.11).

In the critically damped and overdamped cases, there is no oscillation, so our notion of periodicity is missing. For exponential functions of the form $f(t) = f_0 e^{\pm \alpha t}$, a characteristic time can be defined as the time it takes for the function to grow (decay) by a factor of e from its initial value: $t = 1/\alpha$ in this case. For the critically damped solution, the characteristic time is

$$\tau_{cd} = \frac{1}{\omega} = \frac{1}{b} \tag{2.19}$$

while for the overdamped case, taking the negative root (since it leads to the larger value)

$$\tau_{od} = \frac{1}{b - \sqrt{b^2 - w^2}} > \tau_{cd}. \tag{2.20}$$

The critically damped solution approaches its equilibrium value (zero here) faster than the overdamped solution, as is evident in Figure 2.4.

When we have both damping and oscillation, as in the underdamped case, there are two natural timescales that are defined, the period $T = 2\pi/\sqrt{\omega^2 - b^2}$ from the oscillation, and $\tau = 1/b$ from the decay. The ratio of those two defines a dimensionless constant called the "Q"-factor of the system (or the "quality" factor). Formally,

$$Q \equiv 2\pi\frac{\tau}{T} = \frac{1}{b}\sqrt{\omega^2 - b^2} = \sqrt{\left(\frac{\omega}{b}\right)^2 - 1}, \tag{2.21}$$

where the 2π in the definition is just for aesthetics (there are other conventions that omit the 2 or the π or both). The value of Q is large when there are many oscillations within the decay envelope ($T \ll \tau$), and it is small when the decay envelope is small compared to the oscillation, $\tau \ll T$. A sketch of two decaying, oscillatory solutions, together with the exponential decay envelope is shown in Figure 2.5.

Problem 2.1.1   Show that

$$\lim_{t \to \infty} e^{-t}t \to 0.$$

A convincing indication is enough; we're not looking for a rigorous mathematical proof here.

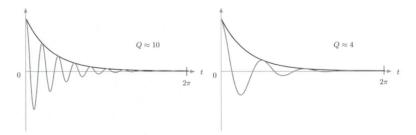

Fig. 2.5   Two oscillatory solutions with their respective decay envelopes plotted over the same time interval. The left solution has $Q \approx 10$ while the right has $Q \approx 4$.

Problem 2.1.2   Solve

$$\ddot{x}(t) = -\omega^2 x(t) - 2b\dot{x}(t)$$

with initial values $x(0) = 0$, $\dot{x}(0) = v_0$ and $b^2 < \omega^2$ (underdamped). Evaluate the energy of the system:

$$E(t) = \frac{1}{2}m\dot{x}(t)^2 + \frac{1}{2}m\omega^2 x(t)^2.$$

What happens as $t \to \infty$? What happens to $E(t)$ if $b = 0$?

Problem 2.1.3   A potential energy function $U(x)$ has a minimum at $x = 0$. A particle of mass m moves under the influence of the potential while inside a viscous medium with coefficient b (our usual oil-filled cylinder). If the mass starts near zero, find the value of b that leads to critical damping – write your expression in terms of the value of $U(x)$ and its derivatives evaluated at zero.

Problem 2.1.4   For the potential energy curve below, imagine starting a bunch of particles off with energy E, but with different initial locations along the x axis. Assuming a realistic physical situation (no perpetual motion machines), indicate where, on the x axis, the particles end up.

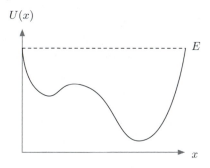

Problem 2.1.5   Given the equation of motion governing a mass m,

$$m\ddot{x}(t) = -kx(t) - 2mb\dot{x}(t) + F_0$$

for constant F_0 (and positive constants b, k with $\sqrt{k/m} > b$), find the angular frequency of oscillation and equilibrium location of the mass.

Problem 2.1.6   The following ODE shows up in a study of the electromagnetic "self force" (see, for example, [8, 12]):

$$\ddot{x}(t) = -2b\dot{x}(t) - \sigma\dddot{x}(t).$$

What are the units of σ? Solve this ODE for $x(t)$ – you should have three constants of integration that you could set using initial values (don't worry about setting those constants).

Problem 2.1.7   Suppose you have a charge $2q$ at $-a$, and a charge q at a. These are fixed charges, they are pinned down. In between these charges, there is a charge q with mass m that is free to move. Write the equation of motion for $x(t)$, the location of the moving charge. If there is some damping mechanism in place, where will the central charge end up? If you moved the charge a little from this position, what will the period of small oscillation be (assume no damping for this piece)?

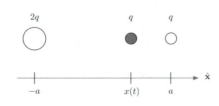

Fig. 2.6   The charges at $-a$ and a are fixed, the charge at $x(t)$ is free to move.

2.2 Driven Harmonic Oscillator

Now suppose we take our mass attached to a spring, and drive it with a force $F(t)$ (I attach a motor, or just exert a time-varying force with my hand). Letting $f(t) \equiv F(t)/m$, Newton's second law reads

$$\ddot{x}(t) = -\omega^2 x(t) + f(t). \tag{2.22}$$

There will be two pieces to the solution here, the homogeneous and sourced solutions from Section 1.7.4. Remember the point of this decomposition, as a second-order ODE (2.22) requires two constants to set initial (or boundary) conditions. The homogeneous piece, $h(t)$, obtained by setting $f(t) = 0$ in (2.22), will give us those constants, while the source piece, $\bar{x}(t)$ allows us to focus on the forcing term in isolation. The homogeneous $h(t)$ solves

$$\ddot{h}(t) = -\omega^2 h(t). \tag{2.23}$$

The sourced piece of (2.22), $\bar{x}(t)$, solves

$$\ddot{\bar{x}}(t) = -\omega^2 \bar{x}(t) + f(t). \tag{2.24}$$

Then the solution to (2.22) is $x(t) = h(t) + \bar{x}(t)$. We know $h(t) = A\cos(\omega t) + B\sin(\omega t)$ is the homogeneous solution, and expect that the constants A and B will be involved in setting the initial values for $x(t)$. To find $\bar{x}(t)$, we could go back to Problem 1.7.3, but we will take a more physically inspired approach.

For the source piece, suppose we take $f(t) = f_0 e^{i\sigma t}$ for constants σ and f_0. Why have we introduced a complex driving force here? How do we produce a complex driving force? We could have used $f_0 \cos(\sigma t)$ or $f_0 \sin(\sigma t)$, but the exponential form is easier to work with. This means that $\bar{x}(t)$ itself will be complex, so how do we extract the physical (i.e. real) solution from the $\bar{x}(t)$ we will get upon solving? Take $\bar{x}(t) = p(t) + iq(t)$, for real functions $p(t)$ and $q(t)$. Then we can write the now complex (2.24) as the real pair:

$$\begin{aligned}
\ddot{p}(t) &= -\omega^2 p(t) + f_0 \cos(\sigma t) \\
\ddot{q}(t) &= -\omega^2 q(t) + f_0 \sin(\sigma t).
\end{aligned} \tag{2.25}$$

The sum of the first equation with i times the second gives us back (2.24) with complex $f(t)$. Conversely, if we solve (2.24) with the complex force in place, we can take either the real or imaginary part of $\bar{x}(t)$ (depending on the phase of our driving force) to recover a real signal. The other solution, we just throw out.

Issues of complex forcing aside, we might also be concerned that the force we're using is too highly specialized. It turns out that it suffices to consider this very simple source, since we can build *any* more complicated source out of these (that notion is the subject of the next section). We have a linear ODE with a (complex) driving "force" – our first guess at the solution must be $\bar{x}(t) = X_0 e^{i\alpha t}$ for constants α and X_0 to be determined. Inserting the ansatz into (2.24), we have

$$-\alpha^2 X_0 e^{i\alpha t} = -\omega^2 X_0 e^{i\alpha t} + f_0 e^{i\sigma t} \tag{2.26}$$

and the only way for this to hold, for all values of t, is if $\alpha = \sigma$. That choice allows us to clear out the exponentials in (2.26). Then we can solve for X_0,

$$X_0 = \frac{f_0}{\omega^2 - \sigma^2} \tag{2.27}$$

and we have pinned down $\bar{x}(t)$. Putting $\bar{x}(t)$ together with $h(t)$, our $x(t)$ is

$$x(t) = A\cos(\omega t) + B\sin(\omega t) + \frac{f_0}{\omega^2 - \sigma^2} e^{i\sigma t}. \tag{2.28}$$

To get a physically relevant solution, we could take either the real or imaginary piece of this $x(t)$, depending on the desired phase. Taking the real part,

$$x_R(t) = A\cos(\omega t) + B\sin(\omega t) + \frac{f_0}{\omega^2 - \sigma^2}\cos(\sigma t). \tag{2.29}$$

The values of A and B can be set once we have been given initial or boundary values. Take $x_R(0) = x_0$ and $\dot{x}_R(0) = v_0$, we start the mass off from rest with some initial extension. Then we have to solve the pair of algebraic equations

$$x_0 = A + \frac{f_0}{\omega^2 - \sigma^2}$$
$$v_0 = \omega B, \tag{2.30}$$

where we can isolate A and B and put them back in to (2.28) to obtain a complete solution

$$x_R(t) = \frac{f_0 + x_0(\sigma^2 - \omega^2)}{\sigma^2 - \omega^2}\cos(\omega t) + \frac{v_0}{\omega}\sin(\omega t) + \frac{f_0}{\omega^2 - \sigma^2}\cos(\sigma t). \tag{2.31}$$

Notice, in passing, the denominator in the terms in (2.31). What would happen if $\sigma = \omega$ (see Problem 2.2.1)?

Problem 2.2.1 Solve

$$\ddot{x}(t) = -\omega^2 x(t) - 2b\dot{x}(t) + f_0 e^{i\sigma t},$$

but set the homogeneous portion of the solution that solves $\ddot{h}(t) = -\omega^2 h(t) - 2b\dot{h}(t)$ to zero (i.e. don't worry about the familiar under/over/critically damped exponential solutions that we have been studying, set all constants of integration to zero). Take the real part of $x(t)$ to obtain an actual position. What is the real part of $x(t)$ if you take $\sigma = \omega$?

2.3 Fourier Series

First, the result: For a function[2] $p(t)$ that is periodic with period T, so that $p(t + T) = p(t)$, there exist complex coefficients $\{a_j\}_{j=0}^{\infty}$ such that

$$p(t) = \sum_{j=-\infty}^{\infty} a_j e^{i2\pi jt/T}. \tag{2.32}$$

This is, in a sense, the same idea as the one behind Taylor series expansion, but we're working in a different "basis" (here, functions in which to expand, exponentials instead of powers of t). Still, to the extent that the exponentials themselves represent infinite sums, the shift from powers of t to special (exponential) collections of powers of t is not such a dramatic one.[3] We could expand each of the exponentials in the sum (2.32) using

$$e^{\alpha t} = \sum_{k=0}^{\infty} \frac{(\alpha t)^k}{k!} \tag{2.33}$$

to get a sum that only involved powers of t,

$$p(t) = \sum_{j=-\infty}^{\infty} c_j t^j \tag{2.34}$$

where the coefficients $\{c_j\}_{j=-\infty}^{\infty}$ include the a_j and elements of the exponential decomposition.

Why would we pick one type of decomposition, exponentials, or polynomials, over another? The Fourier series, which uses exponentials, is well adapted to periodic functions, while a power series is not. Notice that every term in the sum from (2.32) is periodic with period T since

$$e^{i2\pi j(t+T)/T} = e^{i2\pi jt/T} \underbrace{e^{i2\pi j}}_{=1} = e^{i2\pi jt/T}. \tag{2.35}$$

As a property of exponentials, we also note the integral identity, for integers j and k (from (1.91)):

$$\int_0^T e^{i2\pi jt/T} e^{-i2\pi kt/T}\, dt = \int_0^T e^{i2\pi(j-k)t/T}\, dt = \frac{T}{i2\pi(j-k)}\left(e^{-i2\pi(j-k)} - 1\right) \tag{2.36}$$

which is zero provided $j \neq k$ (because $e^{-i2\pi(j-k)} = 1$ for j and k integers). If $j = k$, the original integration is a little off, we would have:

$$\left.\int_0^T e^{i2\pi jt/T} e^{-i2\pi kt/T}\, dt\right|_{j=k} = \int_0^T dt = T. \tag{2.37}$$

---

[2] Nice, well-behaved, smooth, continuous, infinitely differentiable, etc.
[3] Indeed, there are any number of basis "functions" that one can use to decompose a general function: polynomials, exponentials, etc.

We can encapsulate these results using the "Kronecker delta" symbol, defined to be

$$\delta_{jk} \equiv \left\{ \begin{array}{ll} 1 & j = k \\ 0 & j \neq k \end{array} \right. . \tag{2.38}$$

In terms of this symbol, the integrals for $j = k$ and $j \neq k$ can be handily combined:

$$\frac{1}{T} \int_0^T e^{i2\pi jt/T} e^{-i2\pi kt/T} \, dt = \delta_{jk}. \tag{2.39}$$

Going back to the decomposition in (2.32), we want to know how to extract the coefficients $\{a_j\}_{j=-\infty}^{\infty}$ from the sum. Multiply both sides of (2.32) by $e^{-i2\pi kt/T}/T$ and integrate over one full period:[4]

$$\frac{1}{T} \int_0^T p(t) e^{-i2\pi kt/T} \, dt = \sum_{j=-\infty}^{\infty} a_j \frac{1}{T} \int_0^T e^{i2\pi jt/T} e^{-i2\pi kt/T} \, dt$$

$$= \sum_{j=-\infty}^{\infty} a_j \delta_{jk} \tag{2.40}$$

$$= a_k.$$

The moral of the story: Given $p(t)$, the coefficients in the decomposition from (2.32) are obtained by integrating,

$$a_j = \frac{1}{T} \int_0^T p(t) e^{-i2\pi jt/T} \, dt. \tag{2.41}$$

2.3.1 Example: Square Wave

All of the functions in this section are periodic with period T, so we only need to specify values for one cycle, $t \in [0, T)$. One of the simplest periodic functions is the square wave,

$$p(t) = \left\{ \begin{array}{ll} p_0 & 0 \leq t < T/2 \\ -p_0 & T/2 \leq t < T \end{array} \right. . \tag{2.42}$$

We can find the coefficients in the Fourier series decomposition using (2.41),

$$a_j = \frac{1}{T} \left(\int_0^{\frac{T}{2}} p_0 e^{-i2\pi jt/T} \, dt + \int_{\frac{T}{2}}^T (-p_0) e^{-i2\pi jt/T} \, dt \right)$$

$$= -\frac{p_0}{i2\pi j} \left(e^{-i2\pi jt/T} \Big|_{t=0}^{\frac{T}{2}} - e^{-i2\pi jt/T} \Big|_{t=\frac{T}{2}}^T \right) \tag{2.43}$$

$$= -\frac{p_0}{i\pi j} \left(e^{-i\pi j} - 1 \right).$$

For j even, we have $a_j = 0$ because in that case, $e^{-i\pi j} = 1$. For j odd, we have $a_j = -2p_0 i/(\pi j)$. The infinite sum representing the decomposition of $p(t)$ into exponentials is then

---

[4] Assuming, as always, that the integration and summation operations are interchangeable.

$$p(t) = -\frac{2p_0 i}{\pi} \sum_{j=-\infty}^{\infty} \frac{e^{i2\pi(2j+1)t/T}}{2j+1}. \tag{2.44}$$

If we use a summation index k that is only odd, we can simplify the decomposition:

$$
\begin{aligned}
p(t) &= -\frac{2p_0 i}{\pi} \sum_{\substack{k=-\infty \\ \text{odd}}}^{\infty} \frac{e^{i2\pi kt/T}}{k} \\
&= -\frac{2p_0 i}{\pi} \sum_{\substack{k=-\infty \\ \text{odd}}}^{\infty} \frac{\cos(2\pi kt/T) + i\sin(2\pi kt/T)}{k}.
\end{aligned}
\tag{2.45}
$$

In the sum, we will have pairs of positive and negative integer terms that look like $a_k \cos(2\pi kt/T) + a_{-k}\cos(2\pi(-k)t/T) = 0$ since $a_{-k} = -a_k$ here. All of the cosine terms will vanish, leaving only the sine terms, and these show up in pairs as well: $a_k i \sin(2\pi kt/T) + a_{-k}i\sin(2\pi(-k)t/T) = 2a_k i\sin(2\pi kt/T)$. The sum in (2.44) can be written entirely in terms of the sine terms:

$$p(t) = \frac{4p_0}{\pi} \sum_{\substack{k=1 \\ \text{odd}}}^{\infty} \frac{\sin(2\pi kt/T)}{k} = \frac{4p_0}{\pi} \sum_{j=0}^{\infty} \frac{\sin(2\pi(2j+1)t/T)}{2j+1} \tag{2.46}$$

where in the second sum, we revert to $j = 0, 1, 2, \ldots$.

Even vs. Odd Functions

The cosine terms dropped out of the exponential sum in (2.46) because the square wave we were trying to match was "odd," meaning that $p(-t) = -p(t)$. Technically, the odd-ness of the square wave here is manifest in the $[0, T)$ range, so that what we really have is $p(t + T/2) = -p(t)$ for $t \in [0, T/2]$ (although that makes the function odd about $t = 0$, as well, $p(-t) = -p(t)$). Examples of odd and even (with $p(-t) = p(t)$) functions are shown in Figure 2.7. When we multiply an odd function by an even function, the resulting product is an odd function, so for $p(t)$ even, $p(t)\cos(2\pi jt/T)$ is odd. When we integrate that over the "symmetric" limits of the a_j integral in (2.41) (with the exponential written in terms of trigonometric functions), we get zero, so that odd functions have Fourier series decompositions involving only sine.

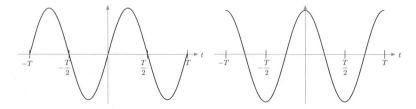

On the left, we have an odd function, with $p(t + T/2) = -p(t)$ (but then, by periodicity, $p(-t) = -p(t)$). The function on the right is even, $p(t + T/2) = p(t)$ (or, about zero, $p(-t) = p(t)$).

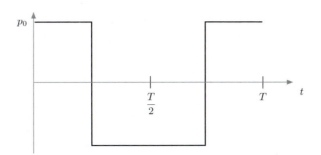

Fig. 2.8 An even square wave, this function has $p(t + T/2) = p(t)$.

Similarly, we expect an *even* function $p(t)$ to have vanishing sine terms since $p(t) \sin(2\pi jt/T)$ is now odd. Let's see how that works out explicitly. To make an even square wave, take

$$p(t) = \begin{cases} p_0 & 0 \leq T/4 \text{ and } 3T/4 < t < T \\ -p_0 & T/4 < t \leq 3T/4, \end{cases} \tag{2.47}$$

shown in Figure 2.8, and using (2.41), we get (for $j \neq 0$, that case will give $a_0 = 0$, but must be handled separately)

$$a_k = \frac{1}{T} \left(\int_0^{T/4} p_0 e^{-i2\pi kt/T} \, dt - \int_{T/4}^{3T/4} p_0 e^{-i2\pi kt/T} \, dt + \int_{3T/4}^T p_0 e^{-i2\pi kt/T} \, dt \right)$$

$$= \frac{ip_0}{k\pi} \left(e^{-ik\pi/2} - e^{-i3k\pi/2} \right), \tag{2.48}$$

and the complex exponentials are $e^{-ik\pi/2} = (-i)^k$ and $e^{-i3k\pi/2} = (-i)^{3k}$ which means that for k even, the coefficient vanishes, while for k odd, we get

$$a_k = (-1)^{\frac{k-1}{2}} \frac{2p_0}{k\pi} \text{ for } k = \pm 1, \pm 3, \pm 5, \ldots. \tag{2.49}$$

Now what is interesting is that here, $a_k = a_{-k}$, so that this time, when we write the entire sum out, we get $a_k \cos(2\pi kt/T) + a_{-k} \cos(2\pi(-k)t/T) = 2a_k \cos(2\pi kt/T)$ while $a_k \sin(2\pi kt/T) + a_{-k} \sin(2\pi(-k)t/T) = 0$. Then

$$p(t) = \frac{4p_0}{\pi} \sum_{k=1,3,\ldots}^{\infty} (-1)^{\frac{k-1}{2}} \frac{\cos(2\pi kt/T)}{k}, \tag{2.50}$$

or in terms of an index $j = 0 \to \infty$,

$$p(t) = \frac{4p_0}{\pi} \sum_{j=0}^{\infty} (-1)^j \frac{\cos(2\pi(2j+1)t/T)}{2j+1}. \tag{2.51}$$

The moral of the story is that even functions can be written entirely in terms of cosine (with positive j-integers only), while odd functions can be written in terms of sine (again, for positive integers).

Finally, what if we wanted to decompose the function $p(t) = p_0$ for $t \in [0, T)$? This is clearly an even function, and we expect to decompose it into a "cosine" series. The problem is that $a_j = 0$ unless $j = 0$, in which case $a_0 = p_0$. This suggests that the cosine series should include $j = 0$ to complete the story. So, for even $p(t)$, we have

$$p(t) = \sum_{j=0}^{\infty} a_j \cos\left(\frac{2\pi j t}{T}\right) \tag{2.52}$$

and for odd $p(t)$,

$$p(t) = \sum_{j=1}^{\infty} a_j \sin\left(\frac{2\pi j t}{T}\right). \tag{2.53}$$

For a function $p(t)$ that is neither purely even nor purely odd, you should go back to the full exponential decomposition (2.32).

Alternative

Many people (see, for example, [3, 15]) will define the Fourier series as

$$p(t) = \sum_{j=-\infty}^{\infty} f_j e^{i\pi j t/S} \tag{2.54}$$

with coefficients $\{f_j\}_{j=-\infty}^{\infty}$, where $t = 0 \to S$ is the temporal domain of interest. This is easily related to the form we are using by noting that $S = T/2$ recovers our decomposition. The advantage of using the half-domain is that the even or odd-ness of the signal is unspecified, and this allows us to use either of the sine/cosine series for decomposition. The sine series is

$$p(t) = \sum_{j=1}^{\infty} s_j \sin\left(\frac{j\pi t}{S}\right) \tag{2.55}$$

and the cosine series is

$$p(t) = \sum_{j=0}^{\infty} c_j \cos\left(\frac{j\pi t}{S}\right). \tag{2.56}$$

If the extension of the signal from $S = T/2 \to T$ is even, you should use the cosine form, if it is odd, use the sine form. If you don't know or care, use either one, and the behavior in the unspecified domain will be fixed by your choice (referring to Figure 2.8, if you took only up to $S = T/2$, you wouldn't know what type of function you have, even or odd, for $t \in [T/2, T)$). I prefer to specify the signal over the entire interval, and use basis functions that have the same periodicity as the signal. In addition to this natural adaptation, it is easier to develop the Fourier transform with the 2π in place, as we shall see in Section 2.6.

2.3.2 Gibb's Phenomenon

We started by assuming that

$$p(t) = \sum_{j=-\infty}^{\infty} a_j e^{i2\pi j t/T} \tag{2.57}$$

for a function $p(t)$ with $p(t + T) = p(t)$. Then we identified the coefficients:

$$a_j = \frac{1}{T} \int_0^T p(\bar{t}) e^{-i2\pi j\bar{t}/T} \, d\bar{t}. \tag{2.58}$$

But what can we say about the self-consistency of the procedure? What happens if we try to evaluate the sum in (2.57) with the coefficients from (2.58) in place? Let that function be called $q(t)$:

$$q(t) = \sum_{j=-\infty}^{\infty} \left(\frac{1}{T} \int_0^T p(\bar{t}) e^{-i2\pi j\bar{t}/T} \, d\bar{t} \right) e^{i2\pi jt/T}$$

$$= \frac{1}{T} \int_0^T p(\bar{t}) \left[\sum_{j=-\infty}^{\infty} e^{i2\pi j(t-\bar{t})/T} \right] d\bar{t}. \tag{2.59}$$

This *should* be equal to $p(t)$ (almost everywhere, at least).

The sum of the exponentials comes from Problem 2.3.15:

$$\sum_{j=-N}^{N} e^{i2\pi j(t-\bar{t})/T} = \frac{\sin\left(\left(N + \frac{1}{2}\right) 2\pi (t - \bar{t})/T \right)}{\sin(2\pi (t - \bar{t})/(2T))}. \tag{2.60}$$

If we identify the sum in (2.59) as the $N \to \infty$ limit of the sum in (2.60), then we can write

$$q(t) = \lim_{N\to\infty} \left(\frac{1}{T} \int_0^T p(\bar{t}) \frac{\sin\left(\left(N + \frac{1}{2}\right) 2\pi (t - \bar{t})/T \right)}{\sin(\pi (t - \bar{t})/T)} \, d\bar{t} \right). \tag{2.61}$$

The limit must be evaluated carefully, for specific cases, to ensure the convergence of $q(t)$ to $p(t)$ for all values of t. Limiting integrals of this basic form will be considered again in Section 2.6.1, but there must be something special about them since somehow we must recover $q(t) = p(t)$ (almost everywhere). There are subtleties, and as an example of what can go wrong, let's go back to our square wave example, with $p(t)$ given by (2.42). You will see, in Problem 2.3.3, that truncating the sum in (2.44) leads to overshoot at the boundaries, a sort of "ringing" that can be reduced by including more terms in the sum. What we will establish here is that while the ringing can be reduced, the overshoot remains at the discontinuity, even in the limit as $N \to \infty$. That remaining overshoot is known as the "Gibb's phenomenon." There are many ways to establish its existence, and we will take a truncated numerical experiment approach. For a more formal proof of the Gibb's phenomenon that starts with (2.61), see [1], for example.

Go back to the square wave signal from (2.42), with decomposition (using the sine series) in (2.46) and take $p_0 = 1$. Define the truncated form

$$p_N(t) = \frac{4}{\pi} \sum_{j=0}^{N} \frac{\sin(2\pi (2j + 1)t/T)}{2j + 1}, \tag{2.62}$$

and we'll probe values near $t = 0$, where the discontinuity in the full $p(t)$ occurs. Take $t = \epsilon T$ where $\epsilon \ll 1$, then we have

$$p_N(t = \epsilon T) = \frac{4}{\pi} \sum_{j=0}^{N} \frac{\sin(2\pi (2j + 1)\epsilon)}{2j + 1}. \tag{2.63}$$

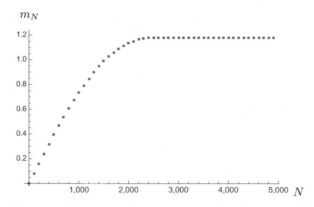

Fig. 2.9 The maximum value of the truncated Fourier series for a square wave evaluated near zero as defined in (2.64). As the number of terms in the truncated sum, N, is increased, the maximum converges to ≈ 1.179.

We are looking for the maximum value of $p_N(t)$ for t near zero, so we'll probe the maximum value over a small grid in ϵ. To be concrete, let the maximum value of the truncated signal, evaluated near zero, be defined by

$$m_N \equiv \max_{k \in [1,100]} \left(p_N(k\epsilon T) \right), \qquad (2.64)$$

and then we can probe m_N for a variety of N. A plot for $N = 1 \rightarrow 5{,}000$ (in steps of 50) is shown in Figure 2.9. It is clear that the maximum value in this region near zero is converging to some number greater than one as N gets large. That represents an overshoot since the maximum value of the original signal is 1, and we would have expected $m_N \rightarrow 1$ as $N \rightarrow \infty$. The actual convergent value for this "data" is ≈ 1.179, agreeing with the analytic result (from [1] again) of $1.1789797\ldots$.

Problem 2.3.1 Show that

$$\frac{2}{T} \int_0^T \sin\left(\frac{j\pi t}{T}\right) \sin\left(\frac{k\pi t}{T}\right) dt = \delta_{jk}$$

by expressing sine in terms of exponentials.

Problem 2.3.2 For a function $p(t) = p_0 \sin(\omega t + \phi)$, for what value of $\phi \in [0, 2\pi)$ is $p(t) = p_0 \cos(\omega t)$?

Problem 2.3.3 Truncate the sum in (2.46) at integer values $n = 1$, 10, and 100. Plot the resulting approximations to $p(t)$ for $p_0 = 1$, $T = 2$.

Problem 2.3.4 We have $p(t) = p_0 t/T$ for $t \in [0, T]$ – assuming $p(t)$ is periodic with period T, sketch $p(t)$ for $t = -T \rightarrow 3T$ (this is just to get an idea of what the function looks like over multiple periods). Find the coefficients $\{a_j\}_{j=-\infty}^{\infty}$ in the Fourier series expansion of this $p(t)$ using (2.41).[5] Be careful, the $j = 0$ case must be handled separately. If you have access to Mathematica, download the "Fourier.nb"

---

[5] The "integration by parts" formula (see Section 5.4.2) may prove useful: for $u(t)$ and $v(t)$ functions of t,

$$\int_a^b \frac{du(t)}{dt} v(t)\, dt = u(t)v(t)\big|_{t=a}^{b} - \int_a^b u(t) \frac{dv(t)}{dt}\, dt.$$

notebook from the book website, and put in your values for a_j in the "Homework Problem Template" section – taking $p_0 = 1/2$ and $T = 2$, plot the truncated form of your series for $N = 25$ (all of this is set up in the notebook, except for your coefficients). Does the truncated plot look like your sketch?

Problem 2.3.5   Find the coefficients $\{a_j\}_{j=-\infty}^{\infty}$ in the decomposition of the "triangle wave" function, with constant p_0,

$$p(t) = \begin{cases} p_0 t & 0 \le t < T/2 \\ -p_0(t - T) & T/2 \le t < T \end{cases}.$$

Problem 2.3.6   Find the coefficients $\{a_j\}_{j=-\infty}^{\infty}$ in the decomposition of the "sawtooth" function

$$p(t) = p_0 - \alpha t \text{ for } 0 \le t < T,$$

with constants p_0 and α.

Problem 2.3.7   For

$$p(t) = \begin{cases} p_0 & 0 \le t < T/4 \\ -p_0 & T/4 \le t < 3T/4 \\ p_0 & 3T/4 \le t < T \end{cases}$$

find the cosine series decomposition:

$$p(t) = \sum_{k=0}^{\infty} c_k \cos\left(\frac{2\pi k t}{T}\right) \text{ with } c_k = \frac{2}{T} \int_0^T p(t) \cos\left(\frac{2\pi k t}{T}\right) dt.$$

(watch out for $k = 0$).

Problem 2.3.8   Show that any signal $p(t)$ can be written as a piece that is even in t and a piece that is odd in t, i.e. show that $p(t) = e(t) + o(t)$ with $e(-t) = e(t)$ and $o(-t) = -o(t)$ (Hint: construct $e(t)$ and $o(t)$ explicitly using $p(t)$ and $p(-t)$).

Problem 2.3.9   Show that for even functions $e_1(t)$, $e_2(t)$, and odd $o_1(t)$, $o_2(t)$, the product $e_1(t)o_1(t)$ is an odd function while $o_1(t)o_2(t)$ and $e_1(t)e_2(t)$ are even.

Problem 2.3.10   Given an even function $e(t)$ and an odd function $o(t)$, show that:

$$\int_{-T}^{T} e(t)o(t) \, dt = 0$$

where T is a constant.

Problem 2.3.11   For $p(t) = p_0 \sin(2\pi k t/T) \cos(2\pi \ell t/T)$, a signal with period T and constant integers k and ℓ, is $p(t)$ even or odd? Decompose $p(t)$ into the appropriate (sine if odd, cosine if even) Fourier series (i.e. find the coefficients in the relevant infinite sum).

Problem 2.3.12   Given the periodic signal (with period T):

$$p(t) = \begin{cases} p_0 & 0 \le t \le T/4 \\ 0 & T/4 < t < T \end{cases}$$

where p_0 is a positive constant, and $p(t + T) = p(t)$. Is $p(t)$ even, odd, or neither? Find the Fourier series decomposition of $p(t)$ (use the cosine series if the function is even, the sine series if odd, and the exponential if neither).

Problem 2.3.13   Decompose the square wave signal

$$p(t) = \begin{cases} p_0 & 0 < t < S/2 \\ 0 & S/2 < t < S \end{cases}$$

into both the sine (2.55) and cosine (2.56) series (using the orthogonality of sine and cosine to find the coefficients). In each case, plot the reconstruction of the signal from $t = 0 \to 2S$ to see how the even/odd behavior is enforced (use $S = 2$ with $p_0 = 3/2$ and truncate the sums at $N = 25$).

Problem 2.3.14   For the finite sum,

$$T_N = \sum_{j=1}^{N} r^j,$$

find an expression for the sum that involves just r and N (and any numerical constants like 1) i.e. "solve" for the sum (Hint: write rT_N in terms of T_N and powers of r, then solve for T_N as in [3]). Do the same for

$$S_N = \sum_{j=-N}^{-1} r^j.$$

Put the pieces together to write an expression for

$$\sum_{j=-N}^{N} r^j.$$

Problem 2.3.15   Using your result from the previous problem, show that

$$\sum_{k=-N}^{N} e^{ik\theta} = \frac{\cos(N\theta) - \cos((N+1)\theta)}{1 - \cos\theta}.$$

The right-hand side can be simplified,

$$\frac{\cos(N\theta) - \cos((N+1)\theta)}{1 - \cos\theta} = \frac{\sin\left(\left(N + \frac{1}{2}\right)\theta\right)}{\sin\left(\frac{\theta}{2}\right)}.$$

Problem 2.3.16   Produce the plot shown in Figure 2.9 as follows: take the truncated form of the Fourier series expansion of the square wave, and, using Mathematica (or other), evaluate (2.64) with $\epsilon = .000001$ for values of N ranging from $1 \to 5,000$ (in steps of 50, say). Plot the values you get as a function of N, you should see convergence to the appropriate number.

2.4 Fourier Series and ODEs

The Fourier series can be used to turn ODEs into algebraic equations, possibly infinite in number, but easy to solve individually. This is the formal justification for the "guesses" we have made previously, leading to (1.74) and (2.9).

2.4.1 First-Order Example

We'll start by applying the Fourier series approach to a first-order ODE. To retain maximal overlap with our driven harmonic oscillator, take

$$\dot{x}(t) + i\omega x(t) = f(t) \qquad x(0) = x_0 \tag{2.65}$$

which has homogeneous solution $h(t) = h_0 e^{-i\omega t}$. We'll focus on finding the "sourced" solution, $\bar{x}(t)$, and add in $h(t)$ at the end. Assuming $f(t)$ is periodic with period T, we can write it as a Fourier series:

$$f(t) = \sum_{k=-\infty}^{\infty} c_k e^{i2\pi kt/T} \qquad c_j = \frac{1}{T} \int_0^T f(t) e^{-i2\pi jt/T} \, dt \tag{2.66}$$

and assume that $\bar{x}(t)$ can also be expanded

$$\bar{x}(t) = \sum_{k=-\infty}^{\infty} a_k e^{i2\pi kt/T}. \tag{2.67}$$

We want to use (2.65) to find the unknown coefficients $\{a_k\}_{k=-\infty}^{\infty}$. Inserting the expansions, from (2.66) and (2.67), into (2.65), we have

$$\sum_{k=-\infty}^{\infty} \left[a_k \left(\frac{i2\pi k}{T} + i\omega \right) - c_k \right] e^{i2\pi kt/T} = 0. \tag{2.68}$$

Multiply this equation by $e^{-i2\pi jt/T}/T$ and integrate from $t = 0 \rightarrow T$. The resulting Kronecker delta can be used to isolate the jth coefficient equation

$$a_j \left(\frac{i2\pi j}{T} + i\omega \right) - c_j = 0 \longrightarrow a_j = \frac{c_j}{i(\omega + 2\pi j/T)}. \tag{2.69}$$

Then the solution for $\bar{x}(t)$ is

$$\bar{x}(t) = \sum_{j=-\infty}^{\infty} \frac{c_j}{i(\omega + 2\pi j/T)} e^{i2\pi jt/T}. \tag{2.70}$$

This $\bar{x}(t)$ contains the piece that is responsible for $f(t)$ in the derivative $\dot{x}(t)$, but we can add in the homogeneous piece, giving us a constant of integration that we can use to set the initial value, so the full solution, $x(t) = h(t) + \bar{x}(t)$, is

$$x(t) = h_0 e^{-i\omega t} + \sum_{j=-\infty}^{\infty} \frac{c_j}{i(\omega + 2\pi j/T)} e^{i2\pi jt/T}$$

$$c_j \equiv \frac{1}{T} \int_0^T f(t) e^{-i2\pi jt/T} \, dt, \tag{2.71}$$

and we can use h_0 to set $x(0) = x_0$.

2.4.2 Driven Oscillator

Let's go back to our second-order driven oscillator problem,

$$\ddot{x}(t) + \omega^2 x(t) - f(t) = 0, \tag{2.72}$$

and repeat the procedure. This time, we know the homogeneous solution is $h(t) = A\cos(\omega t) + B\sin(\omega t)$, and we can add this back in at the end.

We will expand $f(t)$ in its Fourier series as in (2.66) with coefficients $\{c_k\}_{k=-\infty}^{\infty}$ and then use (2.72) to find the coefficients $\{a_k\}_{k=-\infty}^{\infty}$ in the expansion of $\bar{x}(t)$, the sourced solution. Inserting

$$f(t) = \sum_{k=-\infty}^{\infty} c_k e^{i2\pi kt/T}$$

$$\bar{x}(t) = \sum_{k=-\infty}^{\infty} a_k e^{i2\pi kt/T} \tag{2.73}$$

into (2.72), and collecting terms, we get:

$$\sum_{k=-\infty}^{\infty} \left[\left(-\left(\frac{2\pi k}{T} \right)^2 + \omega^2 \right) a_k - c_k \right] e^{i2\pi kt/T} = 0. \tag{2.74}$$

Now multiply both sides by $e^{-i2\pi jt/T}/T$ and integrate from $0 \to T$, only the jth term in the sum contributes, so we have

$$\left[\left(-\left(\frac{2\pi j}{T} \right)^2 + \omega^2 \right) a_j - c_j \right] = 0 \tag{2.75}$$

and we can solve for a_j (note the similarity between this coefficient and the one appearing in (2.28) with $\sigma \equiv 2\pi j/T$),

$$a_j = \frac{c_j}{\omega^2 - (2\pi j/T)^2}. \tag{2.76}$$

The coefficients c_j come from (2.66) in the usual way. We can write out $x(t)$ explicitly, adding back in the homogeneous solution

$$x(t) = A\cos(\omega t) + B\sin(\omega t) + \sum_{j=-\infty}^{\infty} \frac{\frac{1}{T}\int_0^T f(\bar{t})e^{-i2\pi j\bar{t}/T}\, d\bar{t}}{\omega^2 - (2\pi j/T)^2} e^{i2\pi jt/T}, \tag{2.77}$$

and this solution is just a sum of the individual (2.28) solutions with coefficients tuned appropriately. It is the second-order ODE version of the solution in (2.71).

There is a physical problem with the solution here: What if the natural spring frequency ω was equal to one of the driving frequencies? If $\omega = 2\pi j/T$ for some j, then the solution blows up due to the denominator in the sum in (2.77). This extreme "resonance" between the driving frequency and the spring frequency is not what we observe. We could argue that it is impossible (in the Platonic sense) to achieve the exact equality of driving and spring frequency. Still, an arbitrary approach to infinite amplitude is not what we expect from a driven oscillator. The issue is that no physical system can be driven without some sort of damping loss, and the damping puts a finite cutoff on the resonance between driving and natural frequency.

Problem 2.4.1   Find $x(t)$ solving (2.72) for a driving force that is a square wave

$$f(t) = \begin{cases} f_0 & 0 \le t < T/2 \\ -f_0 & T/2 \le t < T \end{cases}$$

using (2.77) with $A = B = 0$ to focus on the driven piece of the solution.

2.5 Damped Driven Harmonic Oscillator

We'll now add back in the damping term to consider the full damped, driven harmonic oscillator problem

$$\ddot{x}(t) + 2b\dot{x}(t) + \omega^2 x(t) = f(t). \tag{2.78}$$

We already know the pair of solutions to the homogeneous ($f(t) = 0$) problem: $\ddot{h}(t) + 2b\dot{h}(t) + \omega^2 h(t) = 0$, those are just

$$h_\pm(t) = e^{-bt} e^{\pm\sqrt{b^2 - \omega^2}\, t}, \tag{2.79}$$

and we could use the same Fourier series approach as for the driven harmonic oscillator to find a solution analogous to (2.77).

Instead we'll focus on the common situation in which a single driving frequency is used. This happens in, for example, driven electrical (capacitor-inductor-resistor, CLR) circuits all the time (a function generator can be enticed to produce a sinusoidal voltage of well-defined frequency). In this specialized case

$$\ddot{x}(t) + 2b\dot{x}(t) + \omega^2 x(t) = f_0 e^{i\sigma t}, \tag{2.80}$$

where we again have a complex driving force, and will eventually take either the real or imaginary piece of the complex solution $x(t)$.

To solve (2.80), we can use the same approach as in Section 2.2. For the full solution $x(t) = A h_+(t) + B h_-(t) + \bar{x}(t)$, where $\bar{x}(t)$ satisfies (2.78), we'll take it to be of the form $\bar{x}(t) = X_0 e^{i\alpha t}$. With this guess in place, the ODE reads

$$-\alpha^2 X_0 e^{i\alpha t} + 2ib\alpha X_0 e^{i\alpha t} + \omega^2 X_0 e^{i\alpha t} = f_0 e^{i\sigma t}. \tag{2.81}$$

Once again, we must set $\alpha = \sigma$ to render the equation true for all times t. Then we can solve for X_0:

$$X_0 = \frac{f_0}{\omega^2 + 2ib\sigma - \sigma^2} \tag{2.82}$$

and the full solution is

$$x(t) = Ae^{-bt}e^{\sqrt{b^2-\omega^2}t} + Be^{-bt}e^{-\sqrt{b^2-\omega^2}t} + \frac{f_0 e^{i\sigma t}}{\omega^2 + 2ib\sigma - \sigma^2}. \quad (2.83)$$

Regardless of the relative values of b and ω, it is clear that the terms associated with the homogeneous solutions will eventually go away, so that the new piece of the solution is the one that survives as $t \to \infty$. For that reason, we can focus on it (and specifically, its real part in the end, since that will be relevant to an actual physical driving force). The solution obtained by setting $A = B = 0$ (alternatively, letting $t \to \infty$) is called the "steady state" solution, and dominates after the "transients" (the decaying bits) have died down.

Let's identify the real and imaginary pieces of the steady state solution,

$$x(t) = \frac{f_0 e^{i\sigma t}}{\omega^2 + 2ib\sigma - \sigma^2}, \quad (2.84)$$

by writing the numerator in terms of identifiable real values. Multiply the top and bottom of the fraction by $\omega^2 - 2ib\sigma - \sigma^2$ (the complex conjugate of the denominator):

$$x(t) = \left(\frac{\omega^2 - \sigma^2 - 2ib\sigma}{(\omega^2 - \sigma^2)^2 + 4b^2\sigma^2} \right) \left(f_0 e^{i\sigma t} \right). \quad (2.85)$$

This looks like the product of a pair of complex numbers, and it is useful to write the first term in polar notation to go along with the second. Let

$$\phi \equiv \tan^{-1}\left(\frac{2b\sigma}{\omega^2 - \sigma^2} \right), \quad (2.86)$$

so that in polar form, we have

$$\left(\frac{\omega^2 - \sigma^2 - 2ib\sigma}{(\omega^2 - \sigma^2)^2 + 4b^2\sigma^2} \right) = \frac{1}{\sqrt{(\omega^2 - \sigma^2)^2 + 4b^2\sigma^2}} e^{-i\phi}. \quad (2.87)$$

Using this polar form in (2.85) gives a fully polar representation of the solution:

$$x(t) = \frac{f_0}{\sqrt{(\omega^2 - \sigma^2)^2 + 4b^2\sigma^2}} e^{i(\sigma t - \phi)}. \quad (2.88)$$

There are a few things to note here. First, the solution and the driving force are not in phase, they differ by ϕ, a constant set by ω and b (the physical parameters governing the oscillation and damping) in addition to σ. If you are given a driving frequency σ, and want to know the frequency ω that will maximize the amplitude of $x(t)$, then we want to minimize the denominator with respect to ω:

$$\frac{d}{d\omega}\left((\omega^2 - \sigma^2)^2 + 4b^2\sigma^2 \right) = 4\omega(\omega^2 - \sigma^2) = 0 \longrightarrow \omega = \sigma, \quad (2.89)$$

and the "resonant" frequency is $\omega = \sigma$. Turning it around, if we are given ω, the natural frequency of the spring, what driving frequency σ maximizes the amplitude?

$$\frac{d}{d\sigma}\left((\omega^2 - \sigma^2)^2 + 4b^2\sigma^2 \right) = 8b^2\sigma - 4\sigma(\omega^2 - \sigma^2) = 0 \longrightarrow \sigma = \sqrt{\omega^2 - 2b^2}. \quad (2.90)$$

The solution (2.83) is for a single driving frequency, but if we knew the Fourier series decomposition of a more general driving function, $f(t) = F(t)/m$, with

$$f(t) = \sum_{j=-\infty}^{\infty} f_j e^{i2\pi jt/T} \tag{2.91}$$

then we can just add up the solutions for the individual driving frequencies "$\sigma \sim 2j\pi/T$":

$$x(t) = Ae^{-bt}e^{\sqrt{b^2-\omega^2}t} + Be^{-bt}e^{-\sqrt{b^2-\omega^2}t} + \sum_{j=-\infty}^{\infty} \frac{f_j e^{i2\pi jt/T}}{\omega^2 + 4ibj\pi/T - (2j\pi/T)^2}. \tag{2.92}$$

Electrical Example

Suppose we have a signal generator outputting $V(t) = V_0 e^{i\sigma t}$ (the real part is all that will matter in the end), and current runs through a capacitor (capacitance C), an inductor (inductance L) and a resistor (with resistance R) as shown in Figure 2.10. Using Kirchoff's voltage law with $V_R = IR$, $V_C = Q/C$ and $V_L = -\dot{I}L$ as the voltage drops across each device, we have

$$V_0 e^{i\sigma t} - \frac{Q}{C} - \dot{I}L - IR = 0. \tag{2.93}$$

Positive current is bringing charge to the positive capacitor plate, so here $I(t) = \dot{Q}(t)$, and we can write everything in terms of the charge on the positive capacitor plate and its derivatives:

$$\ddot{Q}(t) + \frac{1}{LC}Q(t) + 2\frac{R}{2L}\dot{Q}(t) = \frac{V_0}{L}e^{i\sigma t}. \tag{2.94}$$

We can solve this ODE by comparing it with (2.78), where we identify "ω"$\sim 1/\sqrt{LC}$, "b"$\sim R/(2L)$, "f_0"$\sim V_0/L$, and, of course, "$x(t)$"$\sim Q(t)$. Using these in the solution (2.88) gives (taking, finally, the real part)

$$Q(t) = \frac{\frac{V_0}{L}}{\sqrt{\left(\left(\frac{1}{LC}\right) - \sigma^2\right)^2 + \frac{R^2}{L^2}\sigma^2}} \cos(\sigma t - \phi)$$

$$\phi = \tan^{-1}\left(\frac{\frac{R}{L}\sigma}{\frac{1}{LC} - \sigma^2}\right). \tag{2.95}$$

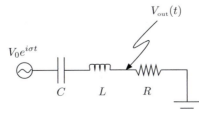

$V_{\text{out}}(t)$

$V_0 e^{i\sigma t}$

C L R

Fig. 2.10   An oscillatory driving voltage powers a circuit with a capacitor (C), an inductor (L), and a resistor (R). The output voltage is taken across the resistor.

The current flowing through the circuit is

$$I(t) = \frac{dQ(t)}{dt} = -\frac{\frac{V_0}{L}\sigma}{\sqrt{\left(\left(\frac{1}{LC}\right) - \sigma^2\right)^2 + \frac{R^2}{L^2}\sigma^2}} \sin(\sigma t - \phi).$$

(2.96)

Finally, the power in the circuit is

$$P = I^2 R = \frac{\left(\frac{V_0}{L}\sigma\right)^2 R}{\left(\left(\frac{1}{LC}\right) - \sigma^2\right)^2 + \frac{R^2}{L^2}\sigma^2} \sin^2(\sigma t - \phi).$$

(2.97)

To maximize the power, we take the σ derivative of the term in front of $\sin^2$ and set it equal to zero to get $\sigma = 1/\sqrt{LC}$, the maximum power is obtained by tuning the driving frequency to the natural frequency of the circuit, $\omega \equiv 1/\sqrt{LC}$. The maximum power delivered at this resonant frequency is V_0^2/R.

There are a variety of ways to characterize these "resonant circuits," and the particular language depends on the application (see [9] for further examples). On the electrical side, if we take the voltage drop across the resistor (the final circuit element before ground) to be the "output" voltage, then

$$V_{\text{out}}(t) = I(t)R = -\frac{\frac{\sigma R}{L}V_0}{\sqrt{\left(\left(\frac{1}{LC}\right) - \sigma^2\right)^2 + \frac{R^2}{L^2}\sigma^2}} \sin(\sigma t - \phi).$$

(2.98)

The amplitude here is

$$|V_{\text{out}}(t)| = \frac{\frac{\sigma R}{L}V_0}{\sqrt{\left(\left(\frac{1}{LC}\right) - \sigma^2\right)^2 + \frac{R^2}{L^2}\sigma^2}} = \frac{V_0}{\sqrt{1 + \frac{1}{R^2}\left(\frac{1}{\sigma C} - L\sigma\right)^2}}.$$

(2.99)

The amplitude for the input voltage (the driving signal) is V_0, and the "gain" of the circuit is defined to be the ratio of these amplitudes:

$$g(\sigma) \equiv \frac{|V_{\text{out}}(t)|}{V_0} = \frac{1}{\sqrt{1 + \frac{1}{R^2}\left(\frac{1}{\sigma C} - L\sigma\right)^2}}.$$

(2.100)

You can plot $g(\sigma)$, the gain of a circuit as a function of the input driving frequency, such a plot is called a "resonance curve." The curve is peaked about $\sigma = 1/\sqrt{LC}$, the natural frequency of the circuit. How sharply peaked is the curve? One way to characterize the width of the resonance curve is to pick the values of σ for which the gain has dropped from its maximum at 1 (for $\sigma = 1/\sqrt{LC}$) to $1/\sqrt{2} \approx .7$. There are two values of σ, one on either side of the peak (σ_- to the left, σ_+ to the right), at which the gain is roughly 70 percent of the maximum. Given those two frequencies, define $\Delta\sigma \equiv \sigma_+ - \sigma_-$, the width at $1/\sqrt{2}$ of max. In Figure 2.11 we see an example gain curve, $g(\sigma)$, with $\sigma_\pm$ marked.

If we write the gain in terms of the frequency $\omega = 1/\sqrt{LC}$, and use the Q-factor from (2.21),

$$Q = \sqrt{\left(\frac{\omega}{b}\right)^2 - 1} = \sqrt{\left(\frac{2}{\omega RC}\right)^2 - 1}$$

(2.101)

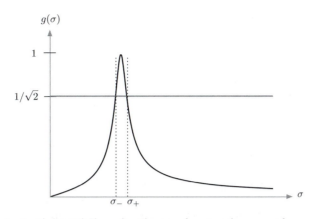

Fig. 2.11 A plot of $g(\sigma)$ (for a circuit with $Q \approx 12$). The peak, with a gain of one, is at the resonant frequency of the circuit, and $\sigma_\pm$ represent the frequencies at which the gain is $1/\sqrt{2}$.

to eliminate C, the capacitance, then the gain can be written as

$$g = \frac{2}{\sqrt{2(1 - Q^2) + \left(\frac{\sigma^2}{\omega^2} + \frac{\omega^2}{\sigma^2}\right)(1 + Q^2)}}. \tag{2.102}$$

We want the pair of σ values for which $g = 1/\sqrt{2}$. The positive roots end up being:

$$\sigma_\pm = \omega\sqrt{\frac{(3 + Q^2) \pm 2\sqrt{(2 + Q^2)}}{1 + Q^2}} \tag{2.103}$$

so that

$$\frac{\Delta\sigma}{\omega} = \sqrt{\frac{(3 + Q^2) + 2\sqrt{(2 + Q^2)}}{1 + Q^2}} - \sqrt{\frac{(3 + Q^2) - 2\sqrt{(2 + Q^2)}}{1 + Q^2}}. \tag{2.104}$$

Remember that Q is defined entirely in terms of the ratio of decay to oscillation time-scale. It has nothing to do with driving or anything else, yet it governs the circuit's response to driving frequencies. As Q gets large (so that oscillations dominate decay), we have

$$\frac{\Delta\sigma}{\omega} \sim \frac{2}{Q} \tag{2.105}$$

and large Q means small width, a sharply peaked resonance curve (or, in terms of its definition (2.21), many cycles of oscillation within the decay envelope of the undriven motion).

Problem 2.5.1   What is the amplitude associated with the optimizations in (2.89) and (2.90)?

Problem 2.5.2   A capacitor is charged up so that it has charge Q_0 on its positive plate ($-Q_0$ on its negative plate). At time $t = 0$, we close the switch in the following circuit and current flows. At time t, what is the charge $Q(t)$ on the positive capacitor plate? What

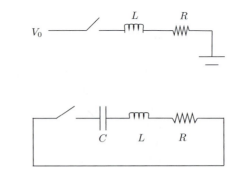

Fig. 2.12
Circuit for Problem 2.5.3.

Fig. 2.13
Circuit for Problem 2.5.4.

is the timescale that governs the decay of charge? Does it have the appropriate unit? What is the current through the resistor at time $t = 0$?

Problem 2.5.3   We have an inductor and a resistor in series. At time $t = 0$, we close the switch, connecting the inductor and resistor to a constant potential V_0. Find the current, $I(t)$, flowing through the resistor as a function of time, taking $I(0) = 0$.

Problem 2.5.4   We have a capacitor, inductor, and resistor in series. We charge up the positive plate of the capacitor to Q_0 (the negative plate has $-Q_0$) and then close the switch at $t = 0$ so that current can flow. Kirchoff's law, applied around the loop, gives the following ODE:

$$\frac{Q(t)}{C} + L\ddot{Q}(t) + \dot{Q}(t)R = 0$$

with $Q(0) = Q_0$ and $\dot{Q}(0) = 0$. Solve this ODE for $Q(t)$, and impose the initial condition. What relations between the values R, L, and C give an "underdamped" solution? What is the exponential timescale τ for this solution? What is the oscillatory period T for the underdamped solution? Using these expressions, write the Q-factor: $Q \equiv 2\pi\tau/T$ in terms of the R, L, and C values.

Problem 2.5.5   We run a "complex" voltage $V(t) = V_0 e^{i\sigma t}$ through a circuit element, represented by the ? box in the following diagram. Find the current, $I(t)$, flowing "through" the circuit element if it is a resistor (with resistance R), a capacitor (with capacitance C), and an inductor (with inductance L). In each case, write the potential drop across these elements as $I(t)Z$ and determine Z, the "complex impedance" (generalizing the notion of resistance) for the resistor, capacitor, and inductor.

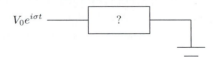

2.6 Fourier Transform

There is an extension of the Fourier series decomposition appropriate for functions with no definite periodicity. Our various harmonic oscillator problems have had a notion of periodicity (except the critically damped and overdamped cases), and so the Fourier series made a certain amount of sense. But how should we think about decomposing a function that has no natural period?

The fundamental shift from fixed period (for the Fourier series) and "infinite period" (for the Fourier transform) is in going from the discrete to the continuous. For the Fourier series, we decompose a function $p(t)$ with period T so that $p(t + T) = p(t)$ using periodic "basis" functions $e^{i2\pi jt/T}$ for integer j:

$$p(t) = \sum_{j=-\infty}^{\infty} a_j e^{i2\pi jt/T} \tag{2.106}$$

with

$$a_k = \frac{1}{T} \int_0^T p(t) e^{-i2\pi kt/T} \, dt. \tag{2.107}$$

From (2.106), we can see that the coefficients $\{a_k\}_{k=-\infty}^{\infty}$ tell us, for a given k, how much[6] $e^{i2\pi kt/T}$ is in $p(t)$. In the Fourier transform, the coefficients themselves become continuous, so that the "k" is no longer just an integer.

The Fourier transform of a function $p(t)$, periodic or not, is defined by

$$\tilde{p}(k) = \int_{-\infty}^{\infty} p(t) e^{i2\pi kt} \, dt \tag{2.108}$$

which is analogous to (2.107). The function $\tilde{p}(k)$ is the coefficient that tells us "how much" $e^{-i2\pi kt}$ is in $p(t)$, for a given k (now continuous). The only problem is that $\tilde{p}(k)$ is complex, but we'll address that later.

To recreate the function $p(t)$, we sum (in the continuous sense, so that $\sum \rightarrow \int$) up the coefficients $\tilde{p}(k)$ together with their contributing[7] $e^{-i2\pi kt}$

$$p(t) = \int_{-\infty}^{\infty} \tilde{p}(k) e^{-i2\pi kt} \, dk, \tag{2.109}$$

a continuous version of (2.106).

The Fourier pair of t and k correspond to time and frequency. I used the letter k to remind us of the discrete index k in the Fourier series decomposition coefficients, a_k. But for Fourier transforms, the frequency variable is generally called f, and k is reserved for the

---

[6] This interpretation is difficult, since a_k could be complex – as we shall see in a moment, a better measure is $|a_k|^2$.

[7] There is a conventional sign change between the Fourier series and Fourier transform. For the series, the functions we use to create $p(t)$, are $e^{i2\pi kt/T}$, while for the transform, the functions from which we build $p(t)$ are $e^{-i2\pi kt}$. It couldn't matter less, since we sum over both positive and negative values, but it is a *notational* pitfall that spoils the otherwise pristine discrete-to-continuous story.

Fourier transform of spatial variables (like x). So in what follows, I'll revert to the standard notation: the Fourier transform of $p(t)$ will be called $\tilde{p}(f)$ with frequency variable f. The transform and its inverse are, then

$$\tilde{p}(f) = \int_{-\infty}^{\infty} p(t) e^{i2\pi ft}\, dt$$

$$p(t) = \int_{-\infty}^{\infty} \tilde{p}(f) e^{-i2\pi ft}\, df. \tag{2.110}$$

2.6.1 A Surprising "Orthogonality"

If you are given $p(t)$, and you compute its Fourier transform $\tilde{p}(f)$ using the top equation in (2.110), how do we know that taking that $\tilde{p}(f)$ and inserting it into the bottom equation of (2.110), we really recover $p(t)$? Let's try it out: take the output of $\tilde{p}(f)$ from (2.110) and use it in the equation for $p(t)$:

$$p(t) \overset{?}{=} \int_{-\infty}^{\infty} \left(\int_{-\infty}^{\infty} p(s) e^{i2\pi fs}\, ds \right) e^{-i2\pi ft}\, df$$

$$= \int_{-\infty}^{\infty} p(s) \left(\int_{-\infty}^{\infty} e^{i2\pi f(s-t)}\, df \right)\, ds, \tag{2.111}$$

where we have once again performed mathematical surgery (interchanging the integral orders) with the physicists' impunity (although there are important times when that interchange would be invalid).

Now the question: What is the integral in parenthesis in the second line of (2.111)? It looks innocuous enough, few functions are easier to integrate than exponentials, but the evaluation of the oscillatory integral at $\pm\infty$ is troubling. What is $\cos(\infty)$? Let's take a more formal view and consider

$$g(s,t) \equiv \int_{-\infty}^{\infty} e^{i2\pi f(s-t)}\, df = \lim_{R\to\infty} \int_{-R}^{R} e^{i2\pi f(s-t)}\, df$$

$$= \lim_{R\to\infty} \frac{\sin(2\pi R(s-t))}{\pi(s-t)}. \tag{2.112}$$

For any values of s and t, the numerator is bounded by one (in absolute value) while the denominator gets very large for $s \to t$ (the numerator has the R in it, which is running off to infinity, so even if $s \to t$, the argument of sine is not necessarily zero). The function $h(s,t) \equiv \sin(2\pi R(s-t))/(\pi(s-t))$ with $g(s,t) = \lim_{R\to\infty} h(s,t)$ for various values of R is shown in Figure 2.14. Notice that the peak grows with R, while the off-peak values oscillate faster and faster. If we integrate $g(s,t)$ in t, letting $z \equiv 2\pi R(s-t)$, we get the "sine integral" (defined to be $\int \sin(z)/z\, dz$, see Problem 8.3.5 for a numerical approach to the sine integral),

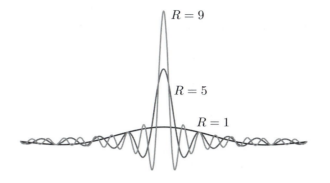

Fig. 2.14 The function $h(0, t)$ plotted for $R = 1$ (black), $R = 5$ (gray), and $R = 9$ (light gray).

$$\int_{-\infty}^{\infty} g(s, t)\, dt = \lim_{R \to \infty} \int_{-\infty}^{\infty} \frac{\sin(2\pi R(s - t))}{\pi(s - t)}\, dt$$

$$= \lim_{R \to \infty} \frac{1}{\pi} \int_{-\infty}^{\infty} \frac{\sin(z)}{z}\, dz \qquad (2.113)$$

$$= \lim_{R \to \infty} 1 = 1$$

where the integral of $h(s, t)$ is R-independent.

We want a function $g(s, t)$ that is infinite when $s = t$, zero elsewhere (looking at what we need to get equality in (2.111)) and integrates to 1. That "function" is known as the "Dirac delta function," and is denoted $\delta(s - t)$. It has the property that for "any" function $p(t)$:

$$\int_{-\infty}^{\infty} p(s)\delta(s - t)\, ds = p(t), \qquad (2.114)$$

just what we wanted to establish the consistency of the Fourier Transform in (2.111) which now reads

$$p(t) \stackrel{?}{=} \int_{-\infty}^{\infty} p(s) \left(\int_{-\infty}^{\infty} e^{i2\pi f(s-t)}\, df \right) ds = \int_{-\infty}^{\infty} p(s)\delta(s - t)\, ds = p(t). \qquad (2.115)$$

2.6.2 Dirac Delta Function

Motivated by the discussion in the last section, we posit the existence of a function $\delta(x)$ defined by the following properties:

$$\delta(x) = \begin{cases} 0 & x \neq 0 \\ \infty & x = 0 \end{cases} \quad \text{with (for positive constants } a \text{ and } b) \int_{-a}^{b} \delta(x)\, dx = 1. \quad (2.116)$$

That is, a symmetric function that is zero except at $x = 0$ where it is infinite, but such that any integral of $\delta(x)$ that fully encloses $x = 0$ is unity. Then we clearly have, for a well-behaved function $f(x)$,

$$\int_{-a}^{b} f(x)\delta(x)\, dx = \int_{-a}^{b} f(0)\delta(x)\, dx = f(0)\int_{-a}^{b} \delta(x)\, dx = f(0), \tag{2.117}$$

and by change of variables, we can shift the zero of the argument of the δ wherever we like:

$$\int_{-a}^{b} f(s)\delta(s-t)\, ds = \begin{cases} f(t) & -a < t < b \\ 0 & \text{else} \end{cases} \tag{2.118}$$

How could we possibly make such a function (beyond the maddening definition of $g(s,t)$ in (2.112))? There are a variety of ways to build a function[8] that behaves according to (2.116). As an example, take a Gaussian

$$d(x,\sigma) \equiv \frac{1}{\sigma\sqrt{2\pi}} e^{-\frac{x^2}{2\sigma^2}} \tag{2.119}$$

normalized so that its integral over all x is one. The Gaussian becomes more sharply peaked as $\sigma \to 0$, while retaining its integral normalization. So one way to achieve $\delta(x)$ is to take the limit of the Gaussian as $\sigma \to 0$,

$$\delta(x) = \lim_{\sigma \to 0} d(x,\sigma) = \lim_{\sigma \to 0} \frac{1}{\sigma\sqrt{2\pi}} e^{-\frac{x^2}{2\sigma^2}}. \tag{2.120}$$

In the end, any tunable sharply peaked function with unit integral will do.

Connection to Kronecker Delta

The integral form of the delta function, sketched in Section 2.6.1 is

$$\delta(x-y) = \int_{-\infty}^{\infty} e^{i2\pi(x-y)t}\, dt. \tag{2.121}$$

This form was necessary just to get a consistent Fourier transform/inverse transform pair. The integral was motivated by looking at the limiting behavior of a truncated (and hence defined) version. We can also argue for the plausibility of the equality in (2.121) by appealing to the Kronecker delta, which appears in a well-defined orthogonality integral for exponentials. Referring to (2.39), we have

$$\delta_{jk} = \frac{1}{T}\int_0^T e^{i2\pi(j-k)t/T}\, dt = \frac{1}{T}\int_{-T/2}^{T/2} e^{i2\pi(j-k)t/T}\, dt. \tag{2.122}$$

which is similar to (2.121) if we identify $x \sim j/T$ and $y \sim k/T$. Then we could even write[9]

$$\delta(x-y) = \lim_{T \to \infty} \left(T\delta_{\lfloor xT \rfloor \lfloor yT \rfloor}\right), \tag{2.123}$$

from which it is clear that, at least up to normalization, the Dirac delta function is the continuum limit of the Kronecker delta.

---

[8] Or, more appropriately, a "distribution."

[9] Here, we are using the "floor" operator that takes a number and rounds it down to the integer below it.

2.6.3 Properties of the Fourier Transform

Given a function $p(t)$ with Fourier transform $\tilde{p}(f)$ from (2.110), we are tempted to interpret $\tilde{p}(f)$ as the "amount" (or relative amount) of $e^{-i2\pi ft}$ in $p(t)$ from (2.109). That's fine, except that $\tilde{p}(f)$ is complex. What does it mean to have "$5i$" of something? Instead, it is really the magnitude (squared) $|\tilde{p}(f)|^2$ that we think of as giving the fractional amount of the contribution of $e^{-i2\pi ft}$ to $p(t)$. Fractional, because we don't know what the integral

$$P \equiv \int_{-\infty}^{\infty} |\tilde{p}(f)|^2 \, df \tag{2.124}$$

is. If it were 1, then $|\tilde{p}(f)|^2 df$ tells us the amount of $p(t)$ that depends on the df-vicinity of f. If P in (2.124) is not one, we can just divide,

$$\frac{|\tilde{p}(f)|^2}{P} df \tag{2.125}$$

is the amount of $p(t)$ that has frequencies in the vicinity of f.

The "power spectrum" associated with $p(t)$ is just the function $P(f) \equiv |\tilde{p}(f)|^2$, a density in frequency space. The total "power" in an interval f_a to f_b is defined to be

$$P(f_a, f_b) \equiv \int_{f_a}^{f_b} P(f) \, df = \int_{f_a}^{f_b} |\tilde{p}(f)|^2 \, df \tag{2.126}$$

and again, our interpretation is that $P(f_a, f_b)$ captures how much of $p(t)$ depends on the frequencies in the range $f_a \to f_b$. The "total power" is obtained by sampling across the entire frequency range, just P from (2.124) again,

$$P = P(-\infty, \infty) = \int_{-\infty}^{\infty} |\tilde{p}(f)|^2 \, df. \tag{2.127}$$

If we write the total power in terms of the original function $p(t)$, using (2.109), we find

$$P = \int_{-\infty}^{\infty} |p(t)|^2 \, dt, \tag{2.128}$$

so that we can calculate the total power without ever computing the Fourier transform. This result, that

$$\int_{-\infty}^{\infty} |p(t)|^2 \, dt = \int_{-\infty}^{\infty} |\tilde{p}(f)|^2 \, df \tag{2.129}$$

is an example of "Parseval's relation" (see [1] for the general case).

If the function $p(t)$ is real, then we can show that there is a relation between the positive and negative frequencies in the Fourier transform (see Problem 2.6.3) that allows us to interpret the negative frequencies. Those could otherwise be confusing: if "concert A" is 440 Hz, what note is represented by -440 Hz? Another confounding frequency, the "zero," $f = 0$, value of the Fourier transform,

$$\tilde{p}(0) = \int_{-\infty}^{\infty} p(t) \, dt, \tag{2.130}$$

is just the total area under $p(t)$.

Problem 2.6.1   For $p(t) = p_0 \sin(2\pi j t/T)$ for integer j, find the Fourier series coefficients, $\{a_k\}_{k=-\infty}^{\infty}$ (stick with the full exponential form). Write your expression for a_k in terms of Kronecker deltas and compare with the Fourier transform of this $p(t)$.

Problem 2.6.2   Find the Fourier transform of $p(t) = p_0 \cos(2\pi q t)$ for (real) constants p_0 and q. What is the relationship between $\tilde{p}(f)$ and $\tilde{p}(-f)$ in this case?

Problem 2.6.3   For a real signal $p(t)$ (with $p(t)^* = p(t)$) what is the relationship between the positive and negative frequency values in the Fourier transform? (i.e. what is the relation between $\tilde{p}(f)$ and $\tilde{p}(-f)$).

Problem 2.6.4   Find the Fourier transform of

$$p(t) = \begin{cases} 0 & t < 0 \\ p_0 e^{-\alpha t} & t \geq 0 \end{cases}$$

for constant real α. Does the relationship between $\tilde{p}(f)$ and $\tilde{p}(-f)$ that you found for the previous problem hold here? Sketch the power spectrum $|\tilde{p}(f)|^2$.

Problem 2.6.5   The Fourier transform, and in particular the power spectrum, tells us "how much" of a particular frequency f is in our signal $p(t)$. The zero-frequency case is interesting, show that for odd functions $p(t)$, $\tilde{p}(f = 0) = 0$.

Problem 2.6.6   Evaluate the integral

$$\int_{-\infty}^{\infty} \delta(kt) p(t)\, dt$$

for constant k (Hint: use change of variables). Make sure your expression holds for both positive and negative values of k.

Problem 2.6.7   For a function $f(t)$ with roots at the set of values $\{t_j\}_{j=1}^{n} : f(t_j) = 0$ but with $f'(t_j) \neq 0$, show that

$$\int_{-\infty}^{\infty} p(t)\delta(f(t))\, dt = \sum_{j=1}^{n} \frac{p(t_j)}{|f'(t_j)|}. \tag{2.131}$$

One way to proceed is to isolate the integrals near the roots of $f(t)$, then you can Taylor expand $f(t_j + t)$ for each integral and use your result from the previous problem. Because of its behavior under the integral sign, we sometimes write (2.131) more generally as

$$\delta(f(t)) = \sum_{j=1}^{n} \frac{\delta(t - t_j)}{|f'(t_j)|}. \tag{2.132}$$

Problem 2.6.8   The "Heaviside step function" is defined as

$$\theta(t) = \begin{cases} 0 & t < 0 \\ 1 & t > 0 \end{cases}. \tag{2.133}$$

Show that $\frac{d\theta(t)}{dt} = \delta(t)$ by establishing that the derivative of the step function behaves like the delta function under an integral. Usually, we leave the value of $\theta(0)$ undefined, but in order for the derivative to be identified with the Dirac delta function, $\theta(0)$ must take on a specific value, what is that value?

Problem 2.6.9   Show that

$$\int_{-\infty}^{\infty} p(t)\frac{d}{dt}\delta(t)\, dt = -p'(0).$$

Problem 2.6.10 What is the Fourier transform of $p(t) = p_0\delta(t - t_0)$ for constants p_0 and t_0? What is the power spectrum, $|\tilde{p}(f)|^2$?

Problem 2.6.11 Show that for a real signal $p(t)$, if $p(-t) = p(t)$, then $\tilde{p}(f)$ is real and also symmetric, $\tilde{p}(-f) = \tilde{p}(f)$.

Problem 2.6.12 For the function

$$p(t) = \begin{cases} p_0 e^{\alpha t} & t < 0 \\ p_0 e^{-\alpha t} & t \geq 0 \end{cases},$$

with $\alpha > 0$ a real constant, sketch $p(t)$ and find its Fourier transform, $\tilde{p}(f)$. Make a sketch of $\tilde{p}(f)$. For what values (large or small) of α is $p(t)$ sharply peaked about the origin? For what values of α is $\tilde{p}(f)$ sharply peaked?

2.7 Fourier Transform and ODEs

Just as the Fourier series can be used to simplify the problem of solving ODEs, we can use the Fourier transform to do so when the target functions are not periodic (think of the overdamped motion in this chapter). Take an ODE like the damped harmonic oscillator:

$$\ddot{x}(t) + \omega^2 x(t) + 2b\dot{x}(t) = 0. \tag{2.134}$$

Assume that $x(t)$ has a Fourier transform, $\tilde{x}(f)$, then we can write $x(t)$, $\dot{x}(t)$, and $\ddot{x}(t)$ in terms of $\tilde{x}(f)$:

$$x(t) = \int_{-\infty}^{\infty} \tilde{x}(\bar{f})e^{-i2\pi\bar{f}t}\, d\bar{f}$$

$$\dot{x}(t) = \int_{-\infty}^{\infty} (-i2\pi\bar{f})\,\tilde{x}(\bar{f})e^{-i2\pi\bar{f}t}\, d\bar{f} \tag{2.135}$$

$$\ddot{x}(t) = -\int_{-\infty}^{\infty} (2\pi\bar{f})^2\,\tilde{x}(\bar{f})e^{-i2\pi\bar{f}t}\, d\bar{f},$$

and we can insert these into the equation of motion (2.134) to get

$$\int_{-\infty}^{\infty} \left[-(2\pi\bar{f})^2 + \omega^2 - 4i\pi\bar{f}b \right] \tilde{x}(\bar{f})e^{-i2\pi\bar{f}t}\, d\bar{f} = 0. \tag{2.136}$$

Multiply this equation by $e^{i2\pi ft}$ and integrate from $t = -\infty \to \infty$ to pick up a $\delta(f - \bar{f})$ under the original $\bar{f}$ integral.[10] Using that delta to turn all $\bar{f}$s into fs with the $\bar{f}$ integration, we have an algebraic equation

$$\left(\omega^2 - 4i\pi bf - 4\pi^2 f^2 \right) \tilde{x}(f) = 0 \tag{2.137}$$

which is solved by

$$f_\pm = \frac{-ib \pm \sqrt{\omega^2 - b^2}}{2\pi}, \tag{2.138}$$

---

[10] The usual caveats apply.

and $\tilde{x}(f)$ should be zero everywhere except at these two "frequencies," where it should be ... sharply peaked. The natural form for $\tilde{x}(f)$ is

$$\tilde{x}(f) = A\delta(f - f_+) + B\delta(f - f_-) \tag{2.139}$$

for constants A and B. In this case, we can perform the inverse Fourier transform

$$x(t) = \int_{-\infty}^{\infty} \tilde{x}(f)e^{-i2\pi ft}\, df = Ae^{-i2\pi f_+ t} + Be^{-i2\pi f_- t}$$
$$= e^{-bt}\left(Ae^{-i\sqrt{\omega^2 - b^2}\,t} + Be^{i\sqrt{\omega^2 - b^2}\,t}\right), \tag{2.140}$$

just as we got in (2.10). Of course, the procedure we have just outlined is the *motivation* for making exponential guesses as we did in Section 2.1.

If you have a driving term, the starting point is

$$\ddot{x}(t) + \omega^2 x(t) + 2b\dot{x}(t) = \underbrace{F(t)/m,}_{\equiv a(t)} \tag{2.141}$$

and we assume that $a(t)$, in addition to $x(t)$, has a Fourier transform,

$$a(t) = \int_{-\infty}^{\infty} \tilde{a}(\bar{f})e^{-i2\pi \bar{f}t}\, d\bar{f}. \tag{2.142}$$

Then going through the same process as before, we get

$$\left(\omega^2 - 4i\pi bf - 4\pi^2 f^2\right)\tilde{x}(f) = \tilde{a}(f), \tag{2.143}$$

and this time, we don't just pick up two coefficients (unless $\tilde{a}(f)$ has a special form, like a delta function at a particular frequency, for example). We can still solve for $\tilde{x}(f)$ algebraically,

$$\tilde{x}(f) = \frac{\tilde{a}(f)}{\omega^2 - 4i\pi bf - 4\pi^2 f^2} \tag{2.144}$$

but now we need to Fourier transform back to obtain $x(t)$, and the difficulty of that operation is determined by the form of $\tilde{a}(f)$.

Problem 2.7.1 For a charged particle of mass m attached to a spring with spring constant $m\omega^2$, Newton's second law reads (see [8, 12] for the development of this equation)

$$m\ddot{x}(t) = -m\omega^2 x(t) + m\tau \dddot{x}(t)$$

for constant τ. Find the algebraic equation governing the Fourier transform of $x(t)$ and solve it for $\tilde{x}(f)$.

Problem 2.7.2 The "cross-correlation" of two real functions $p(t)$ and $q(t)$ is defined to be

$$(p \star q)(\tau) \equiv \int_{-\infty}^{\infty} p(t)q(t + \tau)\, dt.$$

This is a measure of the "overlap," as a function of time τ, of the "signals" $p(t)$ and $q(t)$. Show that for $c(\tau) \equiv (p \star q)(\tau)$, the Fourier transform of $c(\tau)$ has $\tilde{c}(f) = \tilde{p}(f)^* \tilde{q}(f)$, so that the cross-correlation is just a product on the Fourier transform side.

3 Coupled Oscillators

Suppose we now include additional masses in our mass-on-a-spring problem. If we have two masses, m_1 and m_2 at respective locations $x_1(t)$ and $x_2(t)$ (in one dimension) attached by a spring with spring constant k and equilibrium spacing a as in Figure 3.1, the equations of motion are

$$m_1\ddot{x}_1(t) = k(x_2(t) - x_1(t) - a)$$
$$m_2\ddot{x}_2(t) = -k(x_2(t) - x_1(t) - a). \tag{3.1}$$

It is useful, in this situation, to change variables, since $m_1\ddot{x}_1 + m_2\ddot{x}_2 = 0$ (by Newton's third law), the coordinate $z \propto m_1 x_1 + m_2 x_2$ has $\ddot{z} = 0$. To give $z(t)$ the appropriate dimension of length, we must divide by a mass, and to be democratic, we'll treat m_1 and m_2 symmetrically, let

$$z(t) \equiv \frac{m_1 x_1(t) + m_2 x_2(t)}{m_1 + m_2}. \tag{3.2}$$

If we define the difference coordinate $d(t) \equiv x_2(t) - x_1(t) - a$, then the equations of motion can be combined to give

$$\ddot{z}(t) = 0$$
$$\ddot{d}(t) = -k\left(\frac{1}{m_1} + \frac{1}{m_2}\right) d(t). \tag{3.3}$$

The angular frequency ω can be read off from the equation for $d(t)$:

$$\omega^2 \equiv k\left(\frac{1}{m_1} + \frac{1}{m_2}\right), \tag{3.4}$$

and we know the general solution to the problem is

$$z(t) = At + B \qquad d(t) = C\cos(\omega t) + D\sin(\omega t) \tag{3.5}$$

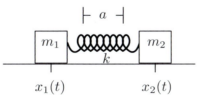

Fig. 3.1 Two masses connected by a spring.

with four constants of integration, $\{A, B, C, D\}$, just right for a pair of second-order ODEs. We could set the constants given the initial position and velocity for each mass. The "center of mass" motion, $z(t)$, and relative motion $d(t)$ have been decoupled, each has its own ODE in (3.3) that makes no reference to the other. Our goal is to perform that same decoupling for a system of n masses connected by $n - 1$ springs, but in order to do this, we need to review some linear algebra.

3.1 Vectors

As we add masses and springs, we add equations to (3.1), and the language of linear algebra allows us to think about those equations of motion as a whole. A "vector" is a collection of n numbers that could be real or complex. For a vector with n real entries, we write $\mathbf{v} \in \mathbb{R}^n$, "the vector vee is in arr-enn," and n is called the "dimension."[1] We can represent the collection as a vertical list. Given the entries $\{v_i\}_{i=1}^n$, we write

$$\mathbf{v} \doteq \begin{pmatrix} v_1 \\ v_2 \\ \vdots \\ v_{n-1} \\ v_n \end{pmatrix} \tag{3.6}$$

where the dot over the equals sign means "is represented by."[2]

For two vectors $\mathbf{v}$ and $\mathbf{w}$ in $\mathbb{R}^n$, with entries $\{v_i\}_{i=1}^n$ and $\{w_i\}_{i=1}^n$, we define vector addition:

$$\mathbf{v} + \mathbf{w} \doteq \begin{pmatrix} v_1 + w_1 \\ v_2 + w_2 \\ \vdots \\ v_{n-1} + w_{n-1} \\ v_n + w_n \end{pmatrix}, \tag{3.7}$$

so that the components add, leaving us with another vector in $\mathbb{R}^n$ (addition of vectors with different dimensions is not defined).

Given a vector $\mathbf{v}$ with components $\{v_i\}_{i=1}^n$ and a number $\alpha \in \mathbb{R}$, we have "scalar–vector" multiplication defined by

$$\alpha \mathbf{v} \doteq \begin{pmatrix} \alpha v_1 \\ \alpha v_2 \\ \vdots \\ \alpha v_{n-1} \\ \alpha v_n \end{pmatrix} \tag{3.8}$$

where each component of the vector gets multiplied by the constant α.

---

[1] There are many ways to define the notion of dimension. This one is the most familiar, the number of numbers you need to specify an element $\mathbf{v} \in \mathbb{R}^n$.

[2] Don't ask. But, if you insist, see [18].

Finally, we can define a special kind of "vector–vector" multiplication. Given two vectors $\mathbf{v}$ and $\mathbf{w}$, both in $\mathbb{R}^n$, the "dot product" is

$$\mathbf{v} \cdot \mathbf{w} \equiv \sum_{i=1}^{n} v_i w_i. \tag{3.9}$$

Then the dot product of a vector with itself is

$$\mathbf{v} \cdot \mathbf{v} = \sum_{i=1}^{n} v_i^2 \tag{3.10}$$

which is the square of the "magnitude" or "length" of the vector (in the Pythagorean sense). We write the magnitude of the vector without the bold face, so that the length of $\mathbf{v}$ is

$$v \equiv \sqrt{\mathbf{v} \cdot \mathbf{v}}. \tag{3.11}$$

Given a collection of k vectors each in $\mathbb{R}^n$, which we'll denote $\{\mathbf{v}^j\}_{j=1}^{k \le n}$, and a set of constants $\{\alpha_j\}_{j=1}^{k}$, a general "linear combination" of the vectors can be written as

$$\mathbf{w} = \sum_{j=1}^{k} \alpha_j \mathbf{v}^j \tag{3.12}$$

where $\mathbf{w} \in \mathbb{R}^n$ is a weighted sum of the set $\{\mathbf{v}^j\}_{j=1}^{k}$.

There is a special set of vectors, called the "canonical basis vectors," $\{\mathbf{e}^j\}_{j=1}^{n}$ with the form:

$$\mathbf{e}^1 \doteq \begin{pmatrix} 1 \\ 0 \\ 0 \\ \vdots \\ 0 \\ 0 \end{pmatrix} \quad \mathbf{e}^2 \doteq \begin{pmatrix} 0 \\ 1 \\ 0 \\ \vdots \\ 0 \\ 0 \end{pmatrix} \quad \cdots \quad \mathbf{e}^n \doteq \begin{pmatrix} 0 \\ 0 \\ 0 \\ \vdots \\ 0 \\ 1 \end{pmatrix} \tag{3.13}$$

so that the jth vector, $\mathbf{e}^j$, contains all zeroes, except the jth entry which is 1. These vectors are "normalized," meaning that they have length 1. They are also "orthogonal,"[3] with $\mathbf{e}^i \cdot \mathbf{e}^j = 0$ if $i \ne j$. We can express the orthonormality using the Kronecker delta to write $\mathbf{e}^i \cdot \mathbf{e}^j = \delta_{ij}$ in general.

The collection $\{\mathbf{e}^j\}_{j=1}^{n}$ is also "complete" in the sense that *any* vector $\mathbf{v} \in \mathbb{R}^n$ can be written as a linear combination of the set $\{\mathbf{e}^j\}_{j=1}^{n}$:

$$\mathbf{v} \doteq \begin{pmatrix} v_1 \\ v_2 \\ \vdots \\ v_{n-1} \\ v_n \end{pmatrix} = \sum_{j=1}^{n} v_j \mathbf{e}^j. \tag{3.14}$$

---

[3] Orthogonality here refers to our geometric sense: the dot product of three-dimensional vectors can be written in terms of the angle between them. For vectors $\mathbf{a}$ and $\mathbf{b} \in \mathbb{R}^3$, we have $\mathbf{a} \cdot \mathbf{b} = ab \cos \theta$ (see Section 6.1). For orthogonal vectors, $\theta = \pi/2$, and $\mathbf{a} \cdot \mathbf{b} = 0$.

Such a collection of complete, orthogonal, normalized vectors is called a "basis." There are many different basis sets for $\mathbb{R}^n$, with the canonical case being the simplest.

Example in $\mathbb{R}^2$

In two dimensions, the canonical basis has elements

$$\mathbf{e}^1 \doteq \begin{pmatrix} 1 \\ 0 \end{pmatrix} \qquad \mathbf{e}^2 \doteq \begin{pmatrix} 0 \\ 1 \end{pmatrix}. \tag{3.15}$$

A generic vector in $\mathbb{R}^2$ can be written:

$$\mathbf{v} \doteq \begin{pmatrix} v_1 \\ v_2 \end{pmatrix} = v_1 \mathbf{e}^1 + v_2 \mathbf{e}^2. \tag{3.16}$$

Here is another basis, with vectors that are normalized and orthogonal:

$$\mathbf{b}^1 \doteq \frac{1}{\sqrt{2}} \begin{pmatrix} 1 \\ 1 \end{pmatrix} \qquad \mathbf{b}^2 \doteq \frac{1}{\sqrt{2}} \begin{pmatrix} 1 \\ -1 \end{pmatrix}. \tag{3.17}$$

A generic vector can also be written in this basis, although figuring out the coefficients is more difficult,

$$\begin{pmatrix} v_1 \\ v_2 \end{pmatrix} \doteq \frac{1}{\sqrt{2}} (v_1 + v_2) \, \mathbf{b}^1 + \frac{1}{\sqrt{2}} (v_1 - v_2) \, \mathbf{b}^2. \tag{3.18}$$

I obtained the coefficients in (3.18) by tedious algebra, is there a better way? Sure: we know $\mathbf{v}$ can be written as a linear combination of the basis vectors, so we start by assuming

$$\mathbf{v} = \alpha \mathbf{b}^1 + \beta \mathbf{b}^2 \tag{3.19}$$

where our target is the pair of constants $\{\alpha, \beta\}$. By orthogonality of basis vectors, we know that $\mathbf{b}^1 \cdot \mathbf{b}^2 = 0$, and the basis vectors are normalized, so that $\mathbf{b}^1 \cdot \mathbf{b}^1 = \mathbf{b}^2 \cdot \mathbf{b}^2 = 1$. Suppose we take the decomposition (3.19) and dot $\mathbf{b}^1$ into both sides

$$\mathbf{b}^1 \cdot \mathbf{v} = \alpha \tag{3.20}$$

and similarly, $\mathbf{b}^2 \cdot \mathbf{v} = \beta$, so all we need are the dot products of $\mathbf{b}^1$ and $\mathbf{b}^2$ with the vector $\mathbf{v}$,

$$\mathbf{b}^1 \cdot \mathbf{v} = \frac{1}{\sqrt{2}} (v_1 + v_2) = \alpha \qquad \mathbf{b}^2 \cdot \mathbf{v} = \frac{1}{\sqrt{2}} (v_1 - v_2) = \beta, \tag{3.21}$$

precisely the coefficients in (3.18). The dot product $\mathbf{b}^1 \cdot \mathbf{v}$ tells us "how much" $\mathbf{b}^1$ is in $\mathbf{v}$, and represents the "projection" of $\mathbf{v}$ onto $\mathbf{b}^1$. Similar language holds for the other basis vector. The whole process is pretty general, for a basis vector $\mathbf{B}^k \in \mathbb{R}^n$, the projection of any $\mathbf{v} \in \mathbb{R}^n$ onto $\mathbf{B}^k$, $\mathbf{B}^k \cdot \mathbf{v}$, gives us the coefficient multiplying $\mathbf{B}^k$ in the decomposition of $\mathbf{v}$.

Problem 3.1.1 Show that the following pair,

$$\mathbf{b}^1 \doteq \begin{pmatrix} \cos \theta \\ \sin \theta \end{pmatrix} \qquad \mathbf{b}^2 \doteq \begin{pmatrix} -\sin \theta \\ \cos \theta \end{pmatrix} \tag{3.22}$$

is a basis for $\mathbb{R}^2$ for any θ (i.e. show that these vectors are orthonormal and complete, meaning that any vector in $\mathbb{R}^2$ can be written as a linear combination of these two).

Problem 3.1.2   For the vector (written with respect to the canonical basis)

$$\mathbf{v} = \mathbf{e}^1 + 2\mathbf{e}^2 + 3\mathbf{e}^3, \tag{3.23}$$

express $\mathbf{v}$ in terms of the basis vectors

$$\mathbf{b}^1 = \frac{1}{\sqrt{6}} \begin{pmatrix} 1 \\ -2 \\ 1 \end{pmatrix} \qquad \mathbf{b}^2 = \frac{1}{\sqrt{2}} \begin{pmatrix} -1 \\ 0 \\ 1 \end{pmatrix} \qquad \mathbf{b}^3 = \frac{1}{\sqrt{3}} \begin{pmatrix} 1 \\ 1 \\ 1 \end{pmatrix}. \tag{3.24}$$

Problem 3.1.3   Given two vectors $\mathbf{v}$ and $\mathbf{w} \in \mathbb{R}^2$, show that $\mathbf{v} \cdot \mathbf{w} = vw\cos\theta$ where θ is the angle between $\mathbf{v}$ and $\mathbf{w}$ if they are drawn at a common origin (Hint: without loss of generality, orient your two-dimensional basis vectors so that one points in the direction of $\mathbf{w}$).

Problem 3.1.4   For two vectors in $\mathbb{R}^2$, $\mathbf{v}$ and $\mathbf{w}$, with $\mathbf{v} \cdot \mathbf{w} \neq 0$, there is a natural way to generate a basis: Take $\mathbf{b}^1 \equiv \mathbf{v}/v$ a unit vector pointing in the direction of $\mathbf{v}$, then let $\mathbf{a} = \mathbf{w} - (\mathbf{w} \cdot \mathbf{b}^1)\mathbf{b}^1$. Show that $\mathbf{a}$ is perpendicular to $\mathbf{b}^1$. Normalize the second basis vector, $\mathbf{b}^2 = \mathbf{a}/a$, and write $\mathbf{v}$ and $\mathbf{w}$ in the $\mathbf{b}^1$, $\mathbf{b}^2$ basis (in that decomposition, you can use the magnitudes of $\mathbf{v}$ and $\mathbf{w}$ and the dot-product $\mathbf{v} \cdot \mathbf{w}$ since those building blocks are independent of basis).

3.2 Matrices

A "matrix" is a collection of vectors, put side by side. For example, if we take a set of vectors in $\mathbb{R}^n$: $\{\mathbf{v}^j\}_{j=1}^{k \leq n}$, then we can form a matrix as follows:

$$\mathbb{M} \doteq \begin{bmatrix} \mathbf{v}^1 & \mathbf{v}^2 & \cdots & \mathbf{v}^k \end{bmatrix} \doteq \begin{pmatrix} v_1^1 & v_1^2 & \cdots & v_1^k \\ v_2^1 & v_2^2 & \cdots & v_2^k \\ \vdots & \vdots & \vdots & \vdots \\ v_n^1 & v_n^2 & \cdots & v_n^k \end{pmatrix}. \tag{3.25}$$

On the far right, we see the usual rectangular array of numbers normally associated with a matrix. There are n rows and k columns, and we typically write the indices either both up or both down. In the present case, I want to highlight the "collection of vectors" interpretation, but the more typical representation is

$$\mathbb{M} \doteq \begin{pmatrix} M_{11} & M_{12} & \cdots & M_{1k} \\ M_{21} & M_{22} & \cdots & M_{2k} \\ \vdots & \vdots & \vdots & \vdots \\ M_{n1} & M_{n2} & \cdots & M_{nk} \end{pmatrix}, \tag{3.26}$$

where M_{ij} is the number appearing in the ith row, j^{th} column, and we write (for matrices with real entries) $\mathbb{M} \in \mathbb{R}^{n \times k}$ for a matrix with n rows and k columns.

Matrices can be multiplied by numbers: given $\alpha \in \mathbb{R}$ and $\mathbb{M} \in \mathbb{R}^{n \times k}$, we have

$$\alpha \mathbb{M} \dot{\equiv} \begin{pmatrix} \alpha M_{11} & \alpha M_{12} & \cdots & \alpha M_{1k} \\ \alpha M_{21} & \alpha M_{22} & \cdots & \alpha M_{2k} \\ \vdots & \vdots & \vdots & \vdots \\ \alpha M_{n1} & \alpha M_{n2} & \cdots & \alpha M_{nk} \end{pmatrix}, \tag{3.27}$$

so that α just multiplies each entry of the matrix.

For two matrices with the same number of rows and columns, addition is defined componentwise: given $\mathbb{M}, \mathbb{W} \in \mathbb{R}^{n \times k}$

$$\mathbb{M} + \mathbb{W} \dot{\equiv} \begin{pmatrix} M_{11} + W_{11} & M_{12} + W_{12} & \cdots & M_{1k} + W_{1k} \\ M_{21} + W_{21} & M_{22} + W_{22} & \cdots & M_{2k} + W_{3k} \\ \vdots & \vdots & \vdots & \vdots \\ M_{n1} + W_{n1} & M_{n2} + W_{n2} & \cdots & M_{nk} + W_{nk} \end{pmatrix}. \tag{3.28}$$

Of more utility are the definitions of multiplication of matrices by vectors and other matrices. We can multiply a matrix $\mathbb{M} \in \mathbb{R}^{n \times k}$ by a vector $\mathbf{p} \in \mathbb{R}^k$ on the right to get a vector in $\mathbb{R}^n$:

$$\mathbb{M}\mathbf{p} \dot{\equiv} \begin{pmatrix} \sum_{j=1}^{k} M_{1j} p_j \\ \sum_{j=1}^{k} M_{2j} p_j \\ \sum_{j=1}^{k} M_{3j} p_j \\ \vdots \\ \sum_{j=1}^{k} M_{nj} p_j \end{pmatrix}. \tag{3.29}$$

If we rearrange this definition a bit, we can make contact with our original "collection of vectors" interpretation from (3.25):

$$\begin{aligned} \mathbb{M}\mathbf{p} &\dot{=} \begin{pmatrix} M_{11}p_1 + M_{12}p_2 + M_{13}p_3 + \cdots + M_{1k}p_k \\ M_{21}p_1 + M_{22}p_2 + M_{23}p_3 + \cdots + M_{2k}p_k \\ M_{31}p_1 + M_{32}p_2 + M_{33}p_3 + \cdots + M_{3k}p_k \\ \vdots \\ M_{n1}p_1 + M_{n2}p_2 + M_{n3}p_3 + \cdots + M_{nk}p_k \end{pmatrix} \\ &= \begin{pmatrix} M_{11} \\ M_{21} \\ M_{31} \\ \vdots \\ M_{n1} \end{pmatrix} p_1 + \begin{pmatrix} M_{12} \\ M_{22} \\ M_{32} \\ \vdots \\ M_{n2} \end{pmatrix} p_2 + \begin{pmatrix} M_{13} \\ M_{23} \\ M_{33} \\ \vdots \\ M_{n3} \end{pmatrix} p_3 + \cdots + \begin{pmatrix} M_{1k} \\ M_{2k} \\ M_{3k} \\ \vdots \\ M_{nk} \end{pmatrix} p_k \\ &= \sum_{j=1}^{k} p_j \mathbf{v}^j \end{aligned} \tag{3.30}$$

so that matrix–vector multiplication can be used to generate arbitrary linear combinations of vectors, just as in (3.12), with the coefficients given by the entries of the vector $\mathbf{p}$. As alternate notation,[4] if we let the vector $\mathbf{w} = \mathbb{M}\mathbf{p}$, then the entries of $\mathbf{w} \in \mathbf{R}^n$ are:

$$w_i = \sum_{j=1}^{k} v_i^j p_j = \sum_{j=1}^{k} M_{ij} p_j \text{ for } i = 1 \to n. \tag{3.31}$$

Moving on to matrix–matrix multiplication: If we have $\mathbb{M} \in \mathbf{R}^{n \times k}$ (n rows, k columns) and $\mathbb{Q} \in \mathbf{R}^{k \times \ell}$ (k rows, ℓ columns), then $\mathbb{W} = \mathbb{M}\mathbb{Q}$ is in $\mathbf{R}^{n \times \ell}$ with entries given by[5]

$$W_{ij} = \sum_{z=1}^{k} M_{iz} Q_{zj} \text{ for } i = 1 \to n, j = 1 \to \ell. \tag{3.32}$$

We can view matrix–vector multiplication as a degenerate case in which $\mathbb{Q} \in \mathbf{R}^{k \times 1}$ (k rows in 1 column, the usual representation of a column vector). Notice that the definition also gives us a way to think about vector–matrix multiplication, with a vector on the *left* – then $\mathbb{M} \in \mathbf{R}^{1 \times k}$, a "matrix" with 1 row and k columns, a "row" vector, which we sometimes think of as the "transpose"[6] of a column vector. To be concrete, if $\mathbf{v} \in \mathbf{R}^k \equiv \mathbf{R}^{k \times 1}$ is a column vector, then the transpose (denoted with a superscript "T") $\mathbf{v}^T \in \mathbf{R}^{1 \times k}$ is a row vector. For a matrix $\mathbb{Q} \in \mathbf{R}^{k \times \ell}$, we can left-multiply by $\mathbf{v}^T$,

$$\mathbf{w} = \mathbf{v}^T \mathbb{Q} \text{ has } w_i = \sum_{j=1}^{k} v_j Q_{ji} \text{ for } i = 1 \to \ell. \tag{3.33}$$

We can also use the transpose of a column vector to represent the usual dot product. For column vectors $\mathbf{v}, \mathbf{w} \in \mathbf{R}^n$,

$$\mathbf{v} \cdot \mathbf{w} \equiv \sum_{j=1}^{n} v_j w_j = \mathbf{v}^T \mathbf{w} = \mathbf{w}^T \mathbf{v}. \tag{3.34}$$

Finally, we define the inverse of a matrix $\mathbb{M} \in \mathbf{R}^{n \times n}$ to be the matrix that, when multiplied by $\mathbb{M}$, returns the "identity" matrix $\mathbb{I} \in \mathbf{R}^{n \times n}$ with ones along the diagonal, zero everywhere else. We denote the inverse as $\mathbb{M}^{-1}$, with $\mathbb{M}^{-1}\mathbb{M} = \mathbb{I}$.

Matrix multiplication (for both vectors and matrices) is the real workhorse of the notation in expressing (and allowing us to solve) linear algebraic relationships. For example, suppose we have the set of coupled, linear equations:

$$\begin{aligned} ax + by &= c \\ dx + ey &= f \end{aligned} \tag{3.35}$$

for (provided) constants $\{a, b, d, e\}$ on the left, and $\{c, f\}$, on the right. We want to find the values of $\{x, y\}$ that make both equations true (or establish that no set of values allows

---

[4] There are a lot of different ways of *talking* about linear algebra, and you should get used to as many as possible.

[5] Matrix–matrix multiplication is basically just matrix–vector multiplication applied to each column vector of the matrix on the right.

[6] The matrix operation "transpose" takes rows to columns and vice-versa. If you have a matrix $\mathbb{M}$ with M_{ij} the entry associated with the ith row, j^{th} column, then the transpose of $\mathbb{M}$, $\mathbb{M}^T$, has $\left(\mathbb{M}^T\right)_{ij} = M_{ji}$.

both equations to hold). We can write the set of equations in terms of matrix–vector multiplication:

$$\underbrace{\begin{pmatrix} a & b \\ d & e \end{pmatrix}}_{\equiv \mathbb{A}} \underbrace{\begin{pmatrix} x \\ y \end{pmatrix}}_{\equiv \mathbf{x}} = \underbrace{\begin{pmatrix} c \\ f \end{pmatrix}}_{\equiv \mathbf{b}}. \tag{3.36}$$

Then the formal solution to $\mathbb{A}\mathbf{x} = \mathbf{b}$ is, in terms of the "inverse" of $\mathbb{A}$, $\mathbf{x} = \mathbb{A}^{-1}\mathbf{b}$, provided the inverse exists.[7]

The solution for $\{x, y\}$, obtained by algebraic manipulation, is

$$\begin{aligned} x &= \frac{ce - bf}{ae - bd} \\ y &= \frac{af - cd}{ae - bd}. \end{aligned} \tag{3.37}$$

It is clear from this general solution that if $ae - bd = 0$, the solution in the form of (3.37), does not exist. Otherwise, we have the most general solution to the pair of equations in (3.35). Aside from the solution, we would like a test for the solution's existence, and the denominator in (3.37) provides a provocative clue.

Determinant

The determinant of a matrix is a number we assign to a (square, although the notion can be generalized) matrix. For a two-by-two matrix $\mathbb{M} \in \mathbb{R}^{2 \times 2}$, the determinant is *defined* to be

$$\det \mathbb{M} \equiv M_{11}M_{22} - M_{12}M_{21}. \tag{3.38}$$

For $\mathbb{M} \in \mathbb{R}^{3 \times 3}$, the determinant is defined recursively, according to

$$\det \mathbb{M} = M_{11} \det \tilde{\mathbb{M}}_{11} - M_{12} \det \tilde{\mathbb{M}}_{12} + M_{13} \det \tilde{\mathbb{M}}_{13} \tag{3.39}$$

where $\tilde{\mathbb{M}}_{ij}$ is the 2×2 matrix obtained by striking out the ith row and j^{th} column of $\mathbb{M}$. We can continue in this manner to higher and higher dimensional matrices. The recursive structure allows us to write the determinant, for $\mathbb{M} \in \mathbb{R}^{n \times n}$ as

$$\det \mathbb{M} = \sum_{j=1}^{n} (-1)^{j+1} M_{1j} \det \tilde{\mathbb{M}}_{1j}. \tag{3.40}$$

For the defining case of two dimensions (and from its extension), the determinant gives us a way of answering the question: "does a solution to (3.35) exist?" For the matrix $\mathbb{A}$ defined in (3.36), the determinant is

$$\det \mathbb{A} = ae - bd. \tag{3.41}$$

That is precisely the quantity that "determines" whether or not a solution to (3.35) exists (as is evidenced by the denominators in (3.37)). If a matrix (in any dimension) has a

---

[7] *Finding* the inverse in general is another whole problem, and one which we will ignore (you can find relevant methods in [7, 11]) – most numerical packages include a routine to produce the inverse if it exists.

determinant that vanishes, then it is *not* invertible. The determinant is the bellwether of invertibility. Prior to solving a set of linear algebraic equations, one should calculate the determinant of the associated matrix to ensure that a solution exists.

An important property of the determinant, as you will establish in Problem 3.2.2, is "the determinant of the product of two matrices $\mathbb{A}$ and $\mathbb{B}$ is the product of the determinants (of $\mathbb{A}$ and $\mathbb{B}$),"

$$\det \mathbb{AB} = \det \mathbb{A} \det \mathbb{B}. \tag{3.42}$$

Problem 3.2.1   Given two matrices:

$$\mathbb{M} \doteq \begin{pmatrix} 1 & 5 \\ 2 & -1 \\ 3 & 0 \end{pmatrix} \qquad \mathbb{W} \doteq \begin{pmatrix} 4 & 2 \\ 0 & 1 \end{pmatrix},$$

what products can you form out of the matrices themselves and their transposes? For example, is $\mathbb{MW}$ a valid matrix product? How about $\mathbb{WM}$? For the ones that are valid, evaluate the product.

Problem 3.2.2   Show, for 2×2 matrices, that $\det(\mathbb{AB}) = \det(\mathbb{A})\det(\mathbb{B})$. As a corollary, show that $\det(\mathbb{A}^{-1}) = 1/\det(\mathbb{A})$.

Problem 3.2.3   What is the relation between $\det(\mathbb{A})$ and $\det(\mathbb{A}^T)$? (work with $\mathbb{A} \in \mathbb{R}^{2 \times 2}$ for simplicity, the result holds for all dimensions.)

Problem 3.2.4   For a square matrix $\mathbb{M}$ with inverse $\mathbb{M}^{-1}$, we have $\mathbb{M}^{-1}\mathbb{M} = \mathbb{I}$. Show that $\mathbb{MM}^{-1} = \mathbb{I}$ (establishing that $\mathbb{M}$ is the matrix inverse of $\mathbb{M}^{-1}$).

Problem 3.2.5   For matrices $\mathbb{A}, \mathbb{B} \in \mathbb{R}^{n \times n}$, show that $(\mathbb{AB})^{-1} = \mathbb{B}^{-1}\mathbb{A}^{-1}$.

Problem 3.2.6   Show that for matrices $\mathbb{A}$ and $\mathbb{B}$ both in $\mathbb{R}^{n \times n}$, $(\mathbb{AB})^T = \mathbb{B}^T\mathbb{A}^T$.

Problem 3.2.7   For a matrix $\mathbb{M} \in \mathbb{R}^{n \times n}$, evaluate the determinant of $\mathbb{M}^T\mathbb{M}$ in terms of the determinant of $\mathbb{M}$.

3.3 Linear Transformations

Given a vector $\mathbf{v} \in \mathbb{R}^n$, the most general linear "transformation" that can act on $\mathbf{v}$ is represented by a matrix–vector multiplication

$$\mathbf{w} = \mathbb{M}\mathbf{v} \tag{3.43}$$

for some[8] $\mathbb{M} \in \mathbb{R}^{n \times n}$.

From one point of view, this operation takes the columns of $\mathbb{M}$ and forms a linear combination with the weighting given by the elements of $\mathbf{v}$ as in (3.30). But if we think of the action of $\mathbb{M}$ *on* $\mathbf{v}$, we are taking a vector $\mathbf{v}$ and mapping it to a new vector $\mathbf{w}$ by multiplying $\mathbf{v}$ by $\mathbb{M}$. There are two important ways that $\mathbf{v}$ can be changed into $\mathbf{w}$: rotation

---

[8]   We could have $\mathbb{M} \in \mathbb{R}^{n \times \ell}$ but then $\mathbf{w}$ and $\mathbf{v}$ live in different spaces.

and stretching.[9] Under a rotation, the length of the vector is unchanged but the direction changes. Under a stretch, the direction of the vector is unchanged, while its length increases or decreases.

3.3.1 Rotation

For a pure rotation, turning $\mathbf{v}$ into $\mathbf{w}$, we can enforce length preservation, $\mathbf{w} \cdot \mathbf{w} = \mathbf{v} \cdot \mathbf{v}$ by putting a constraint on the matrix $\mathbb{M}$ representing the rotation:

$$\mathbf{w} \cdot \mathbf{w} = (\mathbb{M}\mathbf{v})^T \mathbb{M}\mathbf{v} = \mathbf{v}^T \mathbb{M}^T \mathbb{M}\mathbf{v} = \mathbf{v} \cdot \mathbf{v} \tag{3.44}$$

so that we must have $\mathbb{M}^T \mathbb{M} = \mathbb{I}$, the identity matrix, in order for the vectors $\mathbf{w}$ and $\mathbf{v}$ to have the same length. A matrix that satisfies $\mathbb{M}^T \mathbb{M} = \mathbb{I}$ is called an "orthogonal" matrix, and we see that in this special case, the inverse of $\mathbb{M}$ is just the transpose of $\mathbb{M}$, i.e. $\mathbb{M}^{-1} = \mathbb{M}^T$.

If you think of the matrix $\mathbb{M}$ in terms of its column vectors,

$$\mathbb{M} \doteq \begin{bmatrix} \mathbf{v}^1 & \mathbf{v}^2 & \cdots & \mathbf{v}^n \end{bmatrix}, \tag{3.45}$$

then its transpose has these column vectors as its rows

$$\mathbb{M}^T \doteq \begin{bmatrix} \left(\mathbf{v}^1\right)^T \\ \left(\mathbf{v}^2\right)^T \\ \vdots \\ \left(\mathbf{v}^n\right)^T \end{bmatrix}. \tag{3.46}$$

The matrix–matrix product can be viewed as a series of dot-products

$$\mathbb{M}^T \mathbb{M} \doteq \begin{pmatrix} \mathbf{v}^1 \cdot \mathbf{v}^1 & \mathbf{v}^1 \cdot \mathbf{v}^2 & \cdots & \mathbf{v}^1 \cdot \mathbf{v}^n \\ \mathbf{v}^2 \cdot \mathbf{v}^1 & \mathbf{v}^2 \cdot \mathbf{v}^2 & \mathbf{v}^2 \cdot \mathbf{v}^3 & \cdots \\ \vdots & \vdots & \ddots & \cdots \\ \mathbf{v}^n \cdot \mathbf{v}^1 & \mathbf{v}^n \cdot \mathbf{v}^2 & \cdots & \mathbf{v}^n \cdot \mathbf{v}^n \end{pmatrix}. \tag{3.47}$$

For an orthogonal matrix, this product is the identity, meaning that $\mathbf{v}^i \cdot \mathbf{v}^j = \delta_{ij}$, the column vectors of $\mathbb{M}$ are orthogonal and normalized to one, the properties that give an orthogonal matrix its name. These vectors also form a natural basis for the vector space $\mathbb{R}^n$.

As a simple example of a rotation matrix, consider the following 2×2:

$$\mathbb{M} \doteq \begin{pmatrix} \cos\theta & \sin\theta \\ -\sin\theta & \cos\theta \end{pmatrix}. \tag{3.48}$$

If you take a point in the two-dimensional plane, with coordinates x and y, and form the vector

$$\mathbf{v} \doteq \begin{pmatrix} x \\ y \end{pmatrix}, \tag{3.49}$$

then the new vector $\mathbf{w} \equiv \mathbb{M}\mathbf{v}$ is rotated clockwise through an angle θ as shown in Figure 3.2.

---

[9] How about axis inversion?

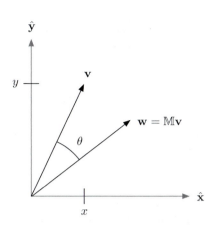

Fig. 3.2 A vector **v** and the rotated vector **w** = $\mathbb{M}$**v**, where $\mathbb{M}$ is given by (3.48). The vector **v** is rotated clockwise through an angle θ to get **w**.

3.3.2 Eigenvectors

A pure stretch relating **w** to **v** (again through **w** = $\mathbb{M}$**v**) means that there is a number λ such that

$$\mathbf{w} = \mathbb{M}\mathbf{v} = \lambda\mathbf{v}. \tag{3.50}$$

In this case, the vector **w** points in the same direction as **v** (okay, parallel or anti-parallel, depending on the sign of λ), and the magnitude has changed.

As an example, the matrix

$$\mathbb{M} \doteq \lambda \begin{pmatrix} 1 & 0 \\ 0 & 1 \end{pmatrix} \tag{3.51}$$

takes any vector **v** and scales it by λ: $\mathbb{M}\mathbf{v} = \lambda\mathbf{v}$. But this matrix is very special in its universality. For most matrices $\mathbb{M}$, there are some vectors that are stretched, and others that are not purely stretched (they may be rotated as well). As an example, take the matrix

$$\mathbb{M} \doteq \begin{pmatrix} \alpha & 0 \\ 0 & \beta \end{pmatrix} \tag{3.52}$$

for constants $\alpha \neq \beta$. Again thinking of a point in two dimensions with coordinates x and y placed into a vector **v** as in (3.49), if we multiply by the matrix $\mathbb{M}$, we get

$$\mathbf{w} = \mathbb{M}\mathbf{v} \doteq \begin{pmatrix} \alpha x \\ \beta y \end{pmatrix}, \tag{3.53}$$

and this **w** is not parallel to the original **v** (unless $\alpha = \beta$, which we already covered). So we turn the problem around a little, and ask for the vectors that are stretched (and only stretched) under multiplication by $\mathbb{M}$. In this case, the vectors are:

$$\mathbf{v}^1 \doteq \begin{pmatrix} 1 \\ 0 \end{pmatrix} \qquad \mathbf{v}^2 \doteq \begin{pmatrix} 0 \\ 1 \end{pmatrix}, \tag{3.54}$$

with $\mathbb{M}\mathbf{v}^1 = \alpha\mathbf{v}^1$ and $\mathbb{M}\mathbf{v}^2 = \beta\mathbf{v}^2$. These vectors, $\mathbf{v}^1$ and $\mathbf{v}^2$ are special, and adapted to the matrix $\mathbb{M}$, they are called "eigenvectors of $\mathbb{M}$." Notice that the scaling coefficients, α and β here, also come from $\mathbb{M}$ itself. These coefficients are known as the "eigenvalues of $\mathbb{M}$."

The "eigenvalue problem" comes from the observation that not all vectors behave in the same way under multiplication by $\mathbb{M}$ and is defined as follows: given a matrix $\mathbb{M} \in \mathbb{R}^{n\times n}$, find the collection of vectors $\{\mathbf{v}^j\}_{j=1}^n$ that are only scaled upon multiplication by $\mathbb{M}$. That is, we want to find the set of vectors $\{\mathbf{v}^j\}_{j=1}^n$ (the "eigenvectors") and scaling factors $\{\lambda_j\}_{j=1}^n$ (the "eigenvalues") such that

$$\mathbb{M}\mathbf{v}^j = \lambda_j\mathbf{v}^j. \tag{3.55}$$

The numerical values found in the eigenvectors and eigenvalues are set by the matrix (linear transformation) $\mathbb{M}$.

Solving for the eigenvalues and eigenvectors is not easy. The idea is to take the defining equation: $\mathbb{M}\mathbf{v}^j = \lambda_j\mathbf{v}^j$ and rewrite it with the unknown eigenvalue on the left:

$$(\mathbb{M} - \lambda_j\mathbb{I})\,\mathbf{v}^j = 0, \tag{3.56}$$

where $\mathbb{I}$ is again the $n \times n$ identity matrix. If the matrix $\mathbb{M} - \lambda_j\mathbb{I}$ is invertible, then the only way this equation can be satisfied is by $\mathbf{v}^j = 0$, giving us no eigenvectors (this is the situation for most values λ_j could take). We are interested in values of λ_j that make $\mathbb{M}-\lambda_j\mathbb{I}$ un-invertible. In that case, there will be nontrivial vectors that satisfy $(\mathbb{M} - \lambda_j\mathbb{I})\mathbf{v}^j = 0$. So we want the determinant of $\mathbb{M} - \lambda_j\mathbb{I}$ to be zero, and we have an equation involving only λ_j (and the entries of $\mathbb{M}$):

$$\det(\mathbb{M} - \lambda_j\mathbb{I}) = 0. \tag{3.57}$$

Given $\mathbb{M}$, this is a polynomial of degree n in λ_j and so has n roots, there are n values that λ_j can take on to satisfy (3.57). We find each of those, somehow,[10] and then plug each back in to $\mathbb{M}\mathbf{v}^j = \lambda_j\mathbf{v}^j$ to find the vector $\mathbf{v}^j$.

As an example, take

$$\mathbb{M} \doteq \begin{pmatrix} a & b \\ c & d \end{pmatrix}, \tag{3.58}$$

then

$$\det(\mathbb{M} - \lambda_j\mathbb{I}) = \lambda_j^2 - \lambda_j(a + d) + (ad - bc) = 0 \tag{3.59}$$

with two solutions (from the quadratic formula),

$$\lambda_j = \frac{1}{2}\left(a + d \pm \sqrt{a^2 + 4bc - 2ad + d^2}\right). \tag{3.60}$$

Taking each of these in turn, we solve for x and y in

$$\begin{pmatrix} a & b \\ c & d \end{pmatrix}\begin{pmatrix} x \\ y \end{pmatrix} = \lambda_j\begin{pmatrix} x \\ y \end{pmatrix} \tag{3.61}$$

[10] This can be done using a numerical root-finding procedure as described in Section 8.1, but an extension of the procedure sketched in Section 8.5.3 is more commonly used.

to get $\mathbf{v}^1$ and $\mathbf{v}^2$. Notice that the eigenvalue equation in (3.55) puts no constraint on the *length* of the $\mathbf{v}^j$ (multiply both sides of (3.55) by a constant α and let $\alpha\mathbf{v}^j \to \mathbf{v}^j$, scaling the eigenvector does not change the eigenvalue), and it is typical to normalize the eigenvectors to have unit length.

We can write equation (3.55) as a matrix–vector equation, for $j = 1 \to n$, by making a matrix $\mathbb{V}$ whose columns are the individual vectors $\mathbf{v}^j$. That matrix will be[11] $\mathbb{V} \in \mathbb{C}^{n \times n}$ (an $n \times n$ matrix with complex entries),

$$\mathbb{V} \doteq \begin{bmatrix} \mathbf{v}^1 & \mathbf{v}^2 & \cdots & \mathbf{v}^n \end{bmatrix}. \tag{3.62}$$

Then the product $\mathbb{M}\mathbb{V}$ is a matrix, and we can write it in terms of $\mathbb{V}$ itself by defining the diagonal matrix $\mathbb{L} \in \mathbb{R}^{n \times n}$, with the eigenvalues along its diagonal, zeroes elsewhere,

$$\mathbb{L} \doteq \begin{pmatrix} \lambda_1 & 0 & \cdots & 0 \\ 0 & \lambda_2 & 0 & \cdots \\ 0 & 0 & \lambda_3 & \ddots \\ \vdots & \vdots & 0 & \ddots \end{pmatrix}. \tag{3.63}$$

We can, finally, write (3.55) as a collection of equations, expressed as a single matrix equation:

$$\mathbb{M}\mathbb{V} = \mathbb{V}\mathbb{L}. \tag{3.64}$$

If $\mathbb{V}$ is invertible, we can write:

$$\mathbb{V}^{-1}\mathbb{M}\mathbb{V} = \mathbb{L} \tag{3.65}$$

and we say that we have "diagonalized" the matrix $\mathbb{M}$ (we can also turn the equation around to write $\mathbb{M} = \mathbb{V}\mathbb{L}\mathbb{V}^{-1}$).

3.3.3 Symmetric Matrices

We'll end our discussion of matrices by proving two special properties of "symmetric" matrices. A symmetric matrix is one whose transpose is equal to itself: $\mathbb{M}^T = \mathbb{M}$. What we will show is that if a matrix is symmetric, then all of its eigenvalues are real, and its eigenvectors are orthogonal: $\mathbf{v}^j \cdot \mathbf{v}^k = \delta_{jk}$.

Suppose $\mathbb{M}$ is a square symmetric matrix with real entries. Then if we take the eigenvalue equation and dot $(\mathbf{v}^j)^*$ (the complex conjugate of the jth eigenvector) into both sides, we get:

$$\left(\left(\mathbf{v}^j\right)^*\right)^T \left(\mathbb{M}\mathbf{v}^j\right) = \lambda_j \left(\mathbf{v}^j\right)^* \cdot \mathbf{v}^j \tag{3.66}$$

where on the left we are using the fact that $\left(\mathbf{v}^j\right)^* \cdot \left(\mathbb{M}\mathbf{v}^j\right) = \left(\left(\mathbf{v}^j\right)^*\right)^T \left(\mathbb{M}\mathbf{v}^j\right)$ by the definition of dot product (see (3.34)). On the right, it is clear that $\left(\mathbf{v}^j\right)^* \cdot \mathbf{v}^j$ is real, since this is just the

---

[11] Even for matrices with real entries, the eigenvalues and eigenvectors could be complex (that's where the fundamental theorem of algebra, ensuring n solutions to $\det(\mathbb{M} - \lambda\mathbb{I}) = 0$, lives), so we have to expand our target space.

length (squared) of a complex vector. When dealing with complex vectors and matrices, we often end up conjugating and transposing, and the combination is given its own symbol:

$$\left(\mathbf{v}^j\right)^\dagger \equiv \left(\left(\mathbf{v}^j\right)^*\right)^T, \tag{3.67}$$

so that we would write (3.66) as

$$\left(\mathbf{v}^j\right)^\dagger \mathbb{M}\mathbf{v}^j = \lambda_j\left(\mathbf{v}^j\right)^\dagger \mathbf{v}^j. \tag{3.68}$$

If we take the transpose and conjugate of (3.55), then

$$\left(\mathbf{v}^j\right)^\dagger \mathbb{M}^T = \lambda_j^*\left(\mathbf{v}^j\right)^\dagger \tag{3.69}$$

and we can multiply by $\mathbf{v}^j$ on the right to get:

$$\left(\mathbf{v}^j\right)^\dagger \mathbb{M}^T\mathbf{v}^j = \lambda_j^*\left(\mathbf{v}^j\right)^\dagger \mathbf{v}^j. \tag{3.70}$$

Subtracting (3.70) from (3.68) gives

$$\left(\mathbf{v}^j\right)^\dagger \left(\mathbb{M} - \mathbb{M}^T\right)\mathbf{v}^j = \left(\lambda_j - \lambda_j^*\right)\left(\mathbf{v}^j\right)^\dagger \mathbf{v}^j. \tag{3.71}$$

For symmetric matrices with $\mathbb{M}^T = \mathbb{M}$, the left-hand side is zero. Assuming $\mathbf{v}^j \neq 0$, the only way the right-hand side can be zero is if $\lambda_j^* = \lambda_j$, so that the eigenvalue is a real number.

Moving on to orthogonality, we use a similar approach. This time, we'll start with two different eigenvectors, $\mathbf{v}^j$ and $\mathbf{v}^k$, with

$$\begin{aligned}\mathbb{M}\mathbf{v}^j &= \lambda_j\mathbf{v}^j\\ \mathbb{M}\mathbf{v}^k &= \lambda_k\mathbf{v}^k,\end{aligned} \tag{3.72}$$

and assume $\lambda_j \neq \lambda_k$. Multiply the top equation by $\left(\mathbf{v}^k\right)^\dagger$ on the left, take the conjugate transpose of the equation for $\mathbf{v}^k$ (noting the already established reality of the eigenvalue) and multiply by $\mathbf{v}^j$ on the right to get the two equations

$$\begin{aligned}\left(\mathbf{v}^k\right)^\dagger \mathbb{M}\mathbf{v}^j &= \lambda_j\left(\mathbf{v}^k\right)^\dagger \mathbf{v}^j\\ \left(\mathbf{v}^k\right)^\dagger \mathbb{M}^T\mathbf{v}^j &= \lambda_k\left(\mathbf{v}^k\right)^\dagger \mathbf{v}^j.\end{aligned} \tag{3.73}$$

Subtracting bottom from top, we get an equation similar to (3.71)

$$\left(\mathbf{v}^k\right)^\dagger \left(\mathbb{M} - \mathbb{M}^T\right)\mathbf{v}^j = \left(\lambda_j - \lambda_k\right)\left(\mathbf{v}^k\right)^\dagger \mathbf{v}^j, \tag{3.74}$$

with the left-hand side vanishing since $\mathbb{M} = \mathbb{M}^T$. For the right-hand-side to vanish, we either need $\lambda_j = \lambda_k$ which we have already excluded, or[12]

$$\left(\mathbf{v}^k\right)^\dagger \mathbf{v}^j = \mathbf{v}^k \cdot \mathbf{v}^j = 0 \tag{3.75}$$

so that we can conclude that the eigenvectors are orthogonal.[13] Since eigenvectors can be normalized to unity, we have

$$\mathbf{v}^j \cdot \mathbf{v}^k = \delta_{jk}, \tag{3.76}$$

---

[12] Since the entries of $\mathbb{M}$ are real, and the λ_j are real, the eigenvectors will have real entries.
[13] Even for $\lambda_k = \lambda_j$, we can pick orthogonal eigenvectors, see Problem 3.3.10.

and we have a collection of n orthogonal unit vectors. These can be used as a "natural" basis for $\mathbb{R}^n$. Evidently, symmetric matrices have a built-in preferred basis. It is this property that we will use to solve our coupled oscillator problem.

When we have a symmetric matrix, the matrix form of the eigenvector problem from (3.64) has $\mathbb{V}$ with columns that are orthogonal vectors. Then it is clear that $\mathbb{V}^T\mathbb{V} = \mathbb{I}$ and the transpose of $\mathbb{V}$ is the inverse of $\mathbb{V}$ (whence $\mathbb{V}$ is an orthogonal matrix). That makes it easy to factor $\mathbb{M}$ into a product of orthogonal and diagonal matrices,

$$\mathbb{M} = \mathbb{V}\mathbb{L}\mathbb{V}^T. \tag{3.77}$$

Example: Image Compression

The factorization of a symmetric matrix as in (3.77) motivates a typical type of compression for matrices. Roughly, the importance of an eigenvector in the decomposition of $\mathbb{M}$ is related to the size of the associated eigenvalue magnitude. The larger the eigenvalue, the more its eigenvector contributes to $\mathbb{M}$ (see Problem 3.3.17). Assume that the eigenvalues in $\mathbb{L}$ have been ordered so that $|\lambda_1| > |\lambda_2| > \cdots > |\lambda_n|$, and the matrix $\mathbb{V}$ respects that ordering (so that $\mathbf{v}^1$, the first column, goes along with λ_1, etc.).

Let the truncated matrix $\mathbb{V}_k \in \mathbb{R}^{n \times k}$ be defined to contain, as its columns, the first k eigenvectors of $\mathbb{M}$. Similarly, take $\mathbb{L}_k \in \mathbb{R}^{k \times k}$ to be the square matrix with $\lambda_1 \to \lambda_k$ as its diagonal entries, all else are zero. Then we can define the "truncated" $\mathbb{M}_k \equiv \mathbb{V}_k\mathbb{L}_k\mathbb{V}_k^T$, and it is this matrix that we hope will approximate $\mathbb{M}$ for some value of k. Obviously, if $k = n$, we recover $\mathbb{M}$, we can write $\mathbb{M}_n = \mathbb{M}$. As it turns out (see, for example, [7]), the error that is made is bounded by the sum of the unapproximated eigenvalues:

$$\|\mathbb{M} - \mathbb{M}_k\|^2 \leq \sum_{j=k+1}^{n} |\lambda_j|^2. \tag{3.78}$$

How does compression factor in here? If you had to store the entries of $\mathbb{M}$, you'd need to hold onto $\sim n^2/2$ values, or roughly n^2 pieces of data. If you just take the first k eigenvectors, you have to store $\sim n \times k$ values plus the eigenvalues themselves, so you end up with $\sim n \times k + k \sim n \times k$. If $k \ll n$, then you've traded memory for reconstruction effort (you actually have to perform the multiplications necessary to generate $\mathbb{M}_k = \mathbb{V}_k\mathbb{L}_k\mathbb{V}_k^T$).

Finally, the connection to images. A matrix that has values from $0 \to 1$ can be interpreted as a grayscale image (with 0 associated with black, 1 with white). You can visually analyze the error of truncation on a particular image. As an example, in Figure 3.3, we can see the "exact" image on the right, with approximate truncations at $k = 10$, 50, and 100 (out of 10,000 total eigenvectors) from left to right.

Problem 3.3.1   Find the eigenvalues of the matrix

$$\mathbb{A} \doteq \begin{pmatrix} 1 & -10 \\ 1 & -1 \end{pmatrix}.$$

The lesson here is that just because a matrix has real entries doesn't mean its eigenvalues are real.

Fig. 3.3 From left to right, the image views of the grayscale matrices M_{10}, M_{50}, M_{100}, and M_{10000} (the "exact" image). Image by Hugh Hochman.

Problem 3.3.2 Using the two-dimensional rotation matrix from (3.48), find the mapping of the points $x = 1$, $y = 0$ and $x = 0$, $y = 1$ under the rotation with $\theta = \pi/2$. Do they make sense?

Problem 3.3.3 Verify, for the rotation matrix (3.48), that $\mathbb{M}^T\mathbb{M} = \mathbb{I}$. What is the geometrical significance of the matrix $\mathbb{M}^T$ when viewed as a transformation (i.e. what is the relation between $\mathbf{v}$ and $\mathbf{w} = \mathbb{M}^T\mathbf{v}$)?

Problem 3.3.4 Find the eigenvectors for the matrix $\mathbb{M}$ in (3.58) (don't worry about normalization).

Problem 3.3.5 The "trace" of a matrix is the sum of its diagonal entries. Show, for the 2×2 matrix in (3.58) that

$$\text{tr}(\mathbb{M}) \equiv \sum_{j=1}^{2} M_{jj} = \sum_{j=1}^{2} \lambda_j$$

where λ_j is the jth eigenvalue. This property, that the trace of a matrix is equal to the sum of its eigenvalues, is true in any dimension.

Problem 3.3.6 Suppose you have decomposed a matrix $\mathbb{M} \in \mathbb{R}^{n \times n}$ into its eigenvalues and eigenvectors as in (3.65) which can also be written as $\mathbb{M} = \mathbb{V}\mathbb{L}\mathbb{V}^{-1}$. Show that

$$\det \mathbb{M} = \prod_{j=1}^{n} L_{jj},$$

so that the determinant of a matrix is equal to the product of its eigenvalues.

Problem 3.3.7 Show that if $\mathbf{v}$ is an eigenvector of a matrix $\mathbb{A} \in \mathbb{R}^{n \times n}$, with associated eigenvalue λ, then $\mathbf{w} \equiv \alpha\mathbf{v}$ (for constant, real $\alpha \neq 0$) is also an eigenvector of $\mathbb{A}$. What is the eigenvalue of $\alpha\mathbf{v}$?

Problem 3.3.8 Show that if two eigenvectors $\mathbf{v}$ and $\mathbf{w}$ of the matrix $\mathbb{A} \in \mathbb{R}^{n \times n}$ have the same eigenvalue λ, then the linear combination $\mathbf{z} = a\mathbf{v} + b\mathbf{w}$ (for real constants a and b) is also an eigenvector with eigenvalue λ. What about $\mathbf{z}$ if $\mathbf{v}$ and $\mathbf{w}$ are eigenvectors with *different* eigenvalues, is $\mathbf{z}$ an eigenvector in that case?

Problem 3.3.9 Find the eigenvalues and eigenvectors of the matrix

$$\mathbb{A} \doteq -\kappa^2 \begin{pmatrix} -1 & 1 \\ 1 & -1 \end{pmatrix},$$

normalize the eigenvectors, so that each has unit length.

Problem 3.3.10 Suppose you have two unit vectors $\mathbf{v}$ and $\mathbf{w}$ which are both eigenvectors of a symmetric matrix $\mathbb{A} \in \mathbb{R}^{n \times n}$ with the same eigenvalue λ, so that: $\mathbb{A}\mathbf{v} = \lambda\mathbf{v}$ and $\mathbb{A}\mathbf{w} = \lambda\mathbf{w}$, and further assume that $\mathbf{v} \neq \pm\mathbf{w}$. Show that I can pick two (unit) eigenvectors, $\mathbf{p}$ and $\mathbf{q}$ that both have eigenvalue λ, but that are also orthogonal: $\mathbf{p} \cdot \mathbf{q} = 0$. This observation, together with the notion that all eigenvectors with eigenvalue not equal to λ are orthogonal to both $\mathbf{v}$ and $\mathbf{w}$ (and hence to $\mathbf{p}$ and $\mathbf{q}$) ensures that we can form a complete orthonormal basis out of the eigenvectors of $\mathbb{A}$ even if some of the eigenvectors share the same eigenvalue.

Problem 3.3.11 There is a continuous form of the eigenvalue problem, where the linear operators (represented by matrices for the vector eigenvalue problem) are now differential operators. The simplest case is:

$$\left[\frac{d^2}{dx^2}\right] f(x) = -\omega^2 f(x),$$

where $\frac{d^2}{dx^2}$ is the linear operator, $f(x)$ is the "eigenfunction", and $-\omega^2$ is its associated eigenvalue. Find $f(x)$ and $-\omega^2$ given the boundary conditions: $f(0) = 0$ and $f(L) = 0$ for constant L. Since this is a continuous problem, we expect there to be an infinite number of eigenvalues and eigenfunctions (i.e. make sure you get solutions other than $f(x) = 0$).

Problem 3.3.12 For a matrix $\mathbb{A} \in \mathbb{R}^{n \times n}$ with $\mathbb{A} = \mathbb{A}^T$, suppose we construct a matrix with the unit (i.e. length 1) eigenvectors of $\mathbb{A}$ as its columns, $\mathbb{V} \doteq \begin{bmatrix} \mathbf{v}^1 & \mathbf{v}^2 & \cdots & \mathbf{v}^n \end{bmatrix}$. What is the determinant (up to sign) of $\mathbb{V} \in \mathbb{R}^{n \times n}$?

Problem 3.3.13 For a symmetric matrix $\mathbb{A} \in \mathbb{R}^{n \times n}$ what is the relationship between the eigenvalues of $\mathbb{A}$ and the eigenvalues of $\alpha\mathbb{A}$ for constant (real) α?

Problem 3.3.14 For $\mathbb{M} \in \mathbb{R}^{n \times n}$, show that if $\mathbb{M}$ is antisymmetric, $\mathbb{M}^T = -\mathbb{M}$, its eigenvalues have no real part. Show that the eigenvectors have the property: $\mathbf{v}^T\mathbf{v} = 0$ – note the lack of conjugation in this product – these vectors do have $\mathbf{v}^\dagger\mathbf{v} = 1$, so they are normalized in the appropriate complex sense.

Problem 3.3.15 Show that any matrix $\mathbb{M} \in \mathbb{R}^{n \times n}$ can be written as a symmetric matrix plus an antisymmetric matrix.

Problem 3.3.16 Given two matrices $\mathbb{A}$ and $\mathbb{B} \in \mathbb{R}^{n \times n}$, suppose there is a vector $\mathbf{v} \in \mathbb{R}^n$ (with $\mathbf{v} \neq 0$) that is an eigenvector of both matrices, but with different eigenvalues: $\mathbb{A}\mathbf{v} = \lambda\mathbf{v}$, $\mathbb{B}\mathbf{v} = \sigma\mathbf{v}$ with $\sigma \neq \lambda$. Show that it must be the case that $\det(\mathbb{A}\mathbb{B} - \mathbb{B}\mathbb{A}) = 0$.

Problem 3.3.17 In this problem, we will establish (3.78) for diagonal matrices. Let $\mathbb{D}$ be a diagonal matrix in $\mathbb{R}^n$ with nonzero entries $D_{ii} = d_i$ for $i = 1 \rightarrow n$. Then $\mathbb{D}_k$ is the diagonal matrix with all entries above the kth one set to zero. Let the elements of $\mathbf{r}$ be $r_i \equiv (\mathbb{D} - \mathbb{D}_k)_{ii}$ for $i = 1 \rightarrow n$ (i.e. the diagonal elements of the matrix $\mathbb{D} - \mathbb{D}_k$), show that

$$r^2 \leq \sum_{j=k+1}^{n} r_j^2 = \sum_{j=k+1}^{n} \lambda_j^2 \tag{3.79}$$

where λ_j is the jth eigenvalue of $\mathbb{D}$.

3.4 Free Oscillator Chain

Suppose we have a set of n masses connected by springs all in a row, a few are shown in Figure 3.4. Each spring has equilibrium length a and spring constant k and we'll take all the masses to have the same mass m. If we denote the position of the jth mass by $x_j(t)$, then the equations of motion look like

$$m\ddot{x}_1(t) = k(x_2(t) - x_1(t) - a)$$
$$m\ddot{x}_2(t) = -k(x_2(t) - x_1(t) - a) + k(x_3(t) - x_2(t) - a) = k(x_1(t) - 2x_2(t) + x_3(t))$$
$$\vdots$$
$$m\ddot{x}_j(t) = k(x_{j-1}(t) - 2x_j(t) + x_{j+1}(t))$$
$$\vdots$$
$$m\ddot{x}_n(t) = -k(x_n(t) - x_{n-1}(t) - a).$$

(3.80)

It is interesting that only the first and last masses have equations of motion that depend on the equilibrium spacing. Defining $\kappa^2 \equiv k/m$, we can write these equations as

$$\ddot{x}_1(t) = \kappa^2(x_2(t) - x_1(t) - a)$$
$$\ddot{x}_j(t) = \kappa^2(x_{j-1}(t) - 2x_j(t) + x_{j+1}(t)) \text{ for } j = 2 \to n-1$$
$$\ddot{x}_n(t) = \kappa^2(-x_n(t) + x_{n-1}(t) + a).$$

(3.81)

Now we get to the point: The equations of motion can be expressed in terms of matrices and vectors. Define $\mathbf{X}(t) \in \mathbb{R}^n$ to be the vector with the unknown $x_j(t)$ as its jth entry, then the set (3.81) can be written

$$\ddot{\mathbf{X}}(t) = -\mathbb{Q}(\mathbf{X}(t) - \mathbf{A})$$

(3.82)

where the matrix $\mathbb{Q} \in \mathbb{R}^{n \times n}$ has the form

$$\mathbb{Q} \doteq -\kappa^2 \begin{pmatrix} -1 & 1 & 0 & 0 & 0 & \cdots \\ 1 & -2 & 1 & 0 & 0 & \cdots \\ 0 & 1 & -2 & 1 & 0 & \cdots \\ 0 & & \ddots & \ddots & \ddots & 0 \\ 0 & \cdots & 0 & 1 & -2 & 1 \\ 0 & \cdots & 0 & 0 & 1 & -1 \end{pmatrix},$$

(3.83)

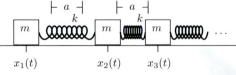

Fig. 3.4 A chain of identical masses connected by identical springs. The location of the jth mass at time t is $x_j(t)$.

and the vector $\mathbb{Q}\mathbf{A} \in \mathbb{R}^n$ has two nonzero entries, its first, $-\kappa^2 a$ and last, $\kappa^2 a$. Notice that the matrix $\mathbb{Q}$ is symmetric, so we know it has real eigenvalues and orthogonal eigenvectors, and we can decompose it into orthogonal matrices multiplying a diagonal one: $\mathbb{Q} = \mathbb{V}\mathbb{L}\mathbb{V}^T$, where the columns of $\mathbb{V}$ are the eigenvectors of $\mathbb{Q}$ and the diagonal entries of $\mathbb{L}$ the eigenvalues, $\{\lambda_j\}_{j=1}^n$.

Following our single, one-dimensional spring solution from Section 1.1, let $\mathbf{Y}(t) \equiv \mathbf{X}(t) - \mathbf{A}$, so that the equations of motion are

$$\ddot{\mathbf{Y}}(t) = -\mathbb{Q}\mathbf{Y}(t) \tag{3.84}$$

and using the decomposition of $\mathbb{Q}$, we have

$$\frac{d^2}{dt^2}\mathbf{Y}(t) = -\mathbb{V}\mathbb{L}\mathbb{V}^T\mathbf{Y}(t) \longrightarrow \frac{d^2}{dt^2}\left(\mathbb{V}^T\mathbf{Y}(t)\right) = -\mathbb{L}\left(\mathbb{V}^T\mathbf{Y}(t)\right) \tag{3.85}$$

where we multiplied both sides by $\mathbb{V}^T$ (the inverse of $\mathbb{V}$) and noted that $\mathbb{V}$ is time-independent (so we could slip it through the temporal derivatives). Now let $\mathbf{Z}(t) \equiv \mathbb{V}^T\mathbf{Y}(t)$, and the equation of motion is $\ddot{\mathbf{Z}}(t) = -\mathbb{L}\mathbf{Z}(t)$. Remember that $\mathbb{L}$ is diagonal, so that we have effectively "decoupled" the equations of motion, each one has the form

$$\ddot{z}_j(t) = -\lambda_j z_j(t) \tag{3.86}$$

for $j = 1 \rightarrow n$. We can now solve each equation of motion separately:

$$z_j(t) = p_j \cos(\sqrt{\lambda_j}t) + q_j \sin(\sqrt{\lambda_j}t) \tag{3.87}$$

for constants p_j and q_j. With $\mathbf{Z}(t)$ in hand, we can work our way backwards

$$\mathbf{Z}(t) = \mathbb{V}^T\mathbf{Y}(t) = \mathbb{V}^T(\mathbf{X}(t) - \mathbf{A}) \longrightarrow \mathbf{X}(t) = \mathbb{V}\mathbf{Z}(t) + \mathbf{A} \tag{3.88}$$

and from here, we can set the constants $\{p_j\}_{j=1}^n$ and $\{q_j\}_{j=1}^n$ using the initial position and velocity of each mass.

That's it, in theory – we construct $\mathbb{Q}$, find the eigenvalues and eigenvectors, and use those to take the de-coupled solution for $\mathbf{Z}(t)$ back to $\mathbf{X}(t)$ setting constants of integration as we go. We'll work out the case for the pair of masses that we solved explicitly at the start of the chapter using this new formulation to see how it all works. That process will also provide some insight into the physical meaning of the eigenvectors.

3.4.1 Two Masses

For two masses and one spring, the matrix $\mathbb{Q}$ is

$$\mathbb{Q} = -\kappa^2 \begin{pmatrix} -1 & 1 \\ 1 & -1 \end{pmatrix}, \tag{3.89}$$

with eigenvalues $\lambda_1 = 2\kappa^2 = 2k/m$ and $\lambda_2 = 0$ from Problem 3.3.9. The first eigenvalue is precisely the ω^2 we had in (3.4) for equal masses, so we associate this eigenvalue with the oscillatory motion. The second eigenvalue is zero, and taken as a frequency this would correspond to motion that has an infinite period, code for nonoscillatory motion. Recall from our previous solution to this problem that the center of mass motion is not periodic, so we expect it to be associated with this eigenvalue.

The eigenvectors that go along with these eigenvalues are

$$\mathbf{v}^1 \doteq \frac{1}{\sqrt{2}} \begin{pmatrix} -1 \\ 1 \end{pmatrix} \qquad \mathbf{v}^2 \doteq \frac{1}{\sqrt{2}} \begin{pmatrix} 1 \\ 1 \end{pmatrix}. \tag{3.90}$$

Suppose we had set up our initial conditions so that the solution for $\mathbf{X}(t)$ was parallel to $\mathbf{v}^1$ – there would be an oscillatory portion out front that had frequency governed by $\sqrt{2k/m}$ and the two masses would be moving in opposite directions relative to one another. For $\mathbf{X}(t)$ parallel to $\mathbf{v}^2$, we see that the motion will always be in the same direction (if the first mass moves to the right with unit magnitude, so does the second), and this represents the center of mass motion. These two different types of motion are orthogonal, and the general solution can be made up of a linear combination of the pair (with appropriate coefficients out front). In the end, we reproduce the solution from (3.5).

3.4.2 Three Masses

Things get more interesting if we introduce a third mass. Now the matrix $\mathbb{Q}$ looks like

$$\mathbb{Q} \doteq -\kappa^2 \begin{pmatrix} -1 & 1 & 0 \\ 1 & -2 & 1 \\ 0 & 1 & -1 \end{pmatrix} \tag{3.91}$$

with eigenvalues

$$\lambda_1 = 3\kappa^2 \qquad \lambda_2 = \kappa^2 \qquad \lambda_3 = 0 \tag{3.92}$$

and associated (normalized) eigenvectors

$$\mathbf{v}^1 \doteq \frac{1}{\sqrt{6}} \begin{pmatrix} 1 \\ -2 \\ 1 \end{pmatrix} \qquad \mathbf{v}^2 \doteq \frac{1}{\sqrt{2}} \begin{pmatrix} -1 \\ 0 \\ 1 \end{pmatrix} \qquad \mathbf{v}^3 \doteq \frac{1}{\sqrt{3}} \begin{pmatrix} 1 \\ 1 \\ 1 \end{pmatrix}. \tag{3.93}$$

Once again, the eigenvectors describe the relative motion of the masses, and to each is associated the frequency $\sqrt{\lambda_j}$. For the first eigenvector, we have the end masses moving one way while the middle mass moves in the opposite direction, this occurs with frequency $\omega = \sqrt{3}\kappa$. For the second eigenvector, with frequency κ, the outer masses move opposite one another (similar to the first eigenvector in (3.90) for the two-mass problem). The third eigenvector is again the center of mass one, with no frequency. Notice that the set $\{\mathbf{v}^1, \mathbf{v}^2, \mathbf{v}^3\}$ form a basis for $\mathbb{R}^3$ (this is to be expected since $\mathbb{Q}$ is symmetric), so *any* relative motion between the masses can be expressed as a linear combination of the basis motions shown in Figure 3.5.

Setting the initial conditions for the masses sets the admixture of the eigenvectors with connected frequencies in the solution. The eigenvectors that describe the characteristic relative motion are called "normal modes," and finding them amounts to solving the problem of motion.

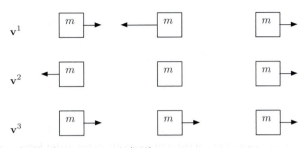

Fig. 3.5　The relative motion described by the eigenvectors in (3.93).

3.4.3 Normal Modes

Putting a system into one of its normal modes can be done by appropriate choice of initial conditions. The advantage of running an oscillator chain in a particular normal mode is that only one frequency of oscillation will govern the motion. There are a variety of ways to set up the excitation of a single mode, and we'll take the simplest, in which the system starts from rest with tuned initial displacements, a configuration that is easy to establish in a classroom demonstration.

For a system of n masses, and the usual description of motion given by $\mathbf{X}(t)$, we'll set $\dot{\mathbf{X}}(0) = 0$, the masses start from rest. In the diagonalized space of motions represented by the vector $\mathbf{Z}(t)$ with entries from (3.87), starting from rest means setting $q_j = 0$ for all j. The real-space positions can be recovered from (3.88), $\mathbf{X}(t) = \mathbb{V}\mathbf{Z}(t) + \mathbf{A}$, which can be written as the sum of the columns of $\mathbb{V}$ weighted by the entries of $\mathbf{Z}$,

$$\mathbf{X}(t) = \mathbf{A} + \sum_{j=1}^{n} z_j(t)\mathbf{v}^j = \mathbf{A} + \sum_{j=1}^{n} p_j \cos\left(\sqrt{\lambda_j}t\right)\mathbf{v}^j. \tag{3.94}$$

In terms of the initial values, if we define the vector $\mathbf{P} \doteq \begin{pmatrix} p_1 & p_2 & \cdots & p_n \end{pmatrix}^T$, then we have

$$\mathbf{X}(0) = \mathbf{A} + \mathbb{V}\mathbf{P}. \tag{3.95}$$

If our goal is to start in the kth normal mode, then we should set $\mathbf{P} = \alpha \mathbf{e}^k$ where $\mathbf{e}^k \in \mathbb{R}^n$ is the canonical basis vector (with zeroes in all positions except the kth which has a 1) and α is a constant setting the size of the initial extension. So we can isolate this normal mode motion by starting with $\mathbf{X}(0) = \mathbf{X}_0 = \mathbf{A} + \mathbb{V}\mathbf{e}^k$ (setting $\alpha \rightarrow 1$).

How can we set up such an initial configuration physically (see [9] for some good examples)? Well, if you just set $\mathbf{X}(0) = \mathbf{A}$, the masses start, and remain, in equilibrium. For $\mathbf{X}(0) = \mathbf{A} + \mathbb{V}\mathbf{e}^k$, the initial extension beyond equilibrium is given by the $\mathbb{V}\mathbf{e}^k = \mathbf{v}^k$ term. You can realize this experimentally by setting the initial displacement according to $\mathbf{v}^k$. Going back to the case of three masses, if you wanted to set up motion in the first mode ($\mathbf{v}^1$ in (3.93)), you'd start with the masses in their equilibrium positions, and move the outer two masses left (say), and the inner mass twice as far to the right, then release from rest. This type of solution can be set up using an air track and masses attached by springs. With some practice, you can achieve a visually clear normal mode motion.

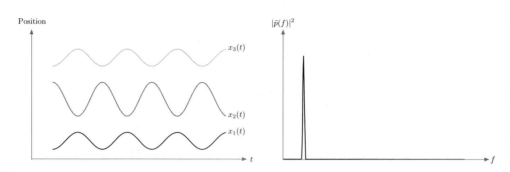

Fig. 3.6 The plot on the left shows the position of three masses that began from rest in the first normal mode. The plot on the right is the power spectrum of the motion, with a single peak, indicating that the motion occurs with one and only one frequency.

If started correctly using extension from $\mathbf{v}^k$, you end up with a solution that looks like

$$\mathbf{X}(t) = \mathbf{A} + \alpha \cos(\sqrt{\lambda_k} t)\mathbf{v}^k. \tag{3.96}$$

The defining property of the normal mode is its single-frequency of oscillation applied to a collective motion of the masses. This means that if you looked at any mass in the chain, you would see the frequency $f = \sqrt{\lambda_k}/(2\pi)$, or no motion at all, since there are normal modes in which a mass remains stationary, like the second mass in $\mathbf{v}^2$ from (3.93). In Figure 3.6, we see the motion in time of all three masses (left) started off in the first normal mode, and on the right, the power spectrum associated with the motion of the first mass. The outer pair of masses are in phase and have the same magnitude, while the central mass is exactly out of phase with twice the magnitude of the outer pairs as indicated by the eigenvector $\mathbf{v}^1$ in (3.93). Note that the peak in the power spectrum always occurs at the same spot regardless of which mass we use to make the power spectrum, but the peak height can change based on which mass we use.

If we combine two normal modes by taking $\mathbf{P} = \alpha \mathbf{v}^k + \beta \mathbf{v}^j$ for constants α and β, and use it to set $\mathbf{X}_0 = \mathbf{A} + \mathbb{V}\mathbf{P}$, then we get motion that is a linear combination of the two modes. This time, there will be two frequencies present in the spectrum of most of the masses. Using $\alpha = \beta = 1/2$ and taking $k = 1, j = 2$, we can again plot the motion of the three masses, and the power spectrum associated with the first mass (note that the second mass has frequency due only to $\mathbf{v}^1$ since $\mathbf{v}^2$ does not move the second mass at all). The motion and spectrum for this case are shown in Figure 3.7. In the power spectrum, the first peak is the new one associated with $\mathbf{v}^2$. The second peak occurs at the same location as the peak in the power spectrum from Figure 3.6, where it is caused by the $\mathbf{v}^1$ motion.

The point is that if you have a time series of the motion of a single mass, you can actually determine quite a bit about the system by looking at the power spectrum of the motion. That spectrum gives the characteristic normal mode frequencies from the physical configuration (although again, the heights of the various peaks depend on the particular mass that is sampled).

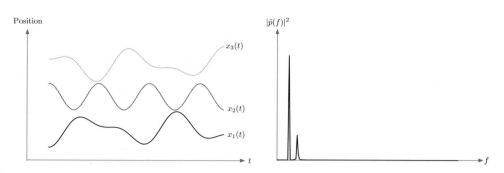

Fig. 3.7 The plot on the left shows the position of three masses that began from rest in a mixture of the first and second normal modes. The plot on the right is the power spectrum of the motion, with two peaks this time, the second peak associated with the first normal mode, the first peak with the second normal mode.

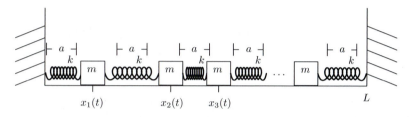

Fig. 3.8 A chain of masses connected to walls separated by a distance L.

Problem 3.4.1 We know how to set up the chain so that it starts from rest with relative displacements from one of the normal modes. But we can also select a normal mode by setting the chain to be in equilibrium initially with a tuned initial velocity. Find the starting velocity $\dot{\mathbf{X}}(0)$ that excites oscillation in the kth normal mode. This approach to setting up a particular normal mode oscillation is harder to realize as a demonstration.

3.5 Fixed Oscillator Chain

Suppose we connect the first and last masses of an oscillator chain to a "wall" by a spring (with constant k and equilibrium length a) as depicted in Figure 3.8. We'll imagine that the left edge of the wall is at the origin, and the right edge is at a distance L. All that changes in the story in the last section is an extra term for the first and last masses, so that their equations of motion become

$$m\ddot{x}_1(t) = k(x_2(t) - x_1(t) - a) - k(x_1(t) - a) = k(-2x_1(t) + x_2(t))$$
$$m\ddot{x}_n(t) = -k(x_n(t) - x_{n-1}(t) - a) + k(L - x_n(t) - a) = k(-2x_n(t) + x_{n-1}(t) + L).$$
$$(3.97)$$

Going through the same steps as before, with $\kappa^2 \equiv k/m$, only the definitions of $\mathbb{Q}$ and $\mathbf{A}$ change. $\mathbb{Q}$ has a slightly nicer form,

$$
\mathbb{Q} \doteq -\kappa^2 \begin{pmatrix}
-2 & 1 & 0 & 0 & 0 & \cdots \\
1 & -2 & 1 & 0 & 0 & \cdots \\
0 & 1 & -2 & 1 & 0 & \cdots \\
0 & & \ddots & \ddots & \ddots & 0 \\
0 & \cdots & 0 & 1 & -2 & 1 \\
0 & \cdots & 0 & 0 & 1 & -2
\end{pmatrix},
\tag{3.98}
$$

and for $\mathbf{A}$, we have

$$
\mathbb{Q}\mathbf{A} \doteq \begin{pmatrix}
0 \\
0 \\
\vdots \\
0 \\
\kappa^2 L
\end{pmatrix}.
\tag{3.99}
$$

The matrix $\mathbb{Q}$ in (3.98) has eigenvalues and eigenvectors that we can calculate analytically, as it turns out. We want to solve for the jth eigenvalue/vector pair

$$
\mathbb{Q}\mathbf{v}^j = \lambda_j \mathbf{v}^j.
\tag{3.100}
$$

Think of the pth row of this vector equation,

$$
-\kappa^2 \left(v_{p-1}^j - 2v_p^j + v_{p+1}^j \right) = \lambda_j v_p^j.
\tag{3.101}
$$

Suppose we assume that

$$
v_k^j = c_j \sin\left(\frac{\pi j k}{n+1} \right)
\tag{3.102}
$$

for constants $\{c_j\}_{j=1}^n$. Then inserting this form into (3.101) gives

$$
2\kappa^2 \left(1 - \cos\left(\frac{\pi j}{n+1} \right) \right) = \lambda_j.
\tag{3.103}
$$

For $n = 2$, the eigenvalues, according to this formula, are $\lambda_1 = \kappa^2$ and $\lambda_2 = 3\kappa^2$ (try using the 2×2 matrix to find the eigenvalues directly). If we take $n = 3$, then the formula gives $\lambda_1 = (2 - \sqrt{2})\kappa^2$, $\lambda_2 = 2\kappa^2$, and $\lambda_3 = (2 + \sqrt{2})\kappa^2$. Notice that in both of these cases, we have lost the $\lambda = 0$ solution, meaning that there is no drift of the center of mass. That makes sense, the masses are all attached to each other or a wall now, so the center of mass, if it moves, must move in an oscillatory fashion.

Problem 3.5.1   For a matrix $\mathbb{A} \in \mathbb{R}^{n \times n}$, with nonzero entries in the kth row (for $1 < k < n$): $A_{kk-1} = a$, $A_{kk} = b$, and $A_{kk+1} = a$ (so that the matrix is symmetric) find the n eigenvalues $\{\lambda_j\}_{j=1}^n$ and eigenvectors $\{\mathbf{v}^j\}_{j=1}^n$ analytically (Hint: start from a guess like the one in (3.102)).

Problem 3.5.2   For the following signal $p(t)$, sketch the power spectrum (no explicit calculation allowed), $|\tilde{p}(f)|^2$ vs. f using appropriate units and values for the frequency axis, and get the correct relative heights, but don't worry about the height values themselves.

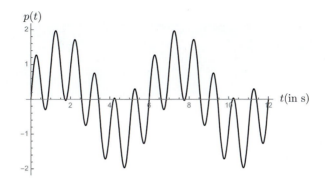

 **Fig. 3.9** Signal for Problem 3.5.2.

Problem 3.5.3   Suppose you have n masses (each with mass m) coupled by identical springs (with spring constant k) and attached to walls separated a distance L. The chain is immersed in a viscous medium with coefficient b, so that the jth mass (where $j \neq 0$ or n) has equation of motion (let $\omega^2 \equiv k/m$)

$$\ddot{x}_j(t) = -\omega^2(-x_{j-1}(t) + 2x_j(t) - x_{j+1}(t)) - 2b\dot{x}_j(t).$$

Find the decoupled equation of motion for the kth normal mode, i.e. what is $\ddot{z}_k(t) =$? (write in terms of the eigenvalues/vectors of the matrix $\mathbb{Q}$, our usual tri-diagonal one for masses connected by springs).

4   The Wave Equation

We can take our n balls and springs from the last chapter and ask: What happens if we take $n \to \infty$? We'll do that carefully, of course, otherwise the answer is meaningless. In particular, our model will be n balls of identical mass m connected by springs with spring constant k and equilibrium spacing a. The total mass of the string of balls is $M = mn$, the total length at equilibrium is $L = (n - 1)a$. These bulk physical properties we will leave fixed as we send $n \to \infty$. Then it is clear that we will also have to send $a \to 0$, the equilibrium spacing must shrink as well. In addition, the mass of each individual ball must go to zero like $1/n$ to keep M fixed.

Spring Constant

There is one more bulk quantity to consider, the "net" spring constant of our chain of balls and springs. We can think of the springs connected to one another in series, and try to find the effective spring constant of the chain. We will require that that be constant as we increase the number of balls (while shrinking down the mass of the individual balls).

Suppose we have a spring with constant K, and we stretch it out to a distance Δx (with respect to its equilbrium length). The energy stored in the spring is

$$U = \frac{1}{2}K(\Delta x)^2. \tag{4.1}$$

Where is the energy stored? The spring is the same everywhere, so it is reasonable to assume the energy is spread uniformly throughout the extended spring. If you think of two halves of the spring, half the energy should be stored on the left, half on the right. So the energy in the left-half of the spring is $U_\ell = U/2$. Call the effective spring constant on the left k (same as on the right), then we know that

$$U_\ell = \frac{1}{2}k\left(\frac{\Delta x}{2}\right)^2 \tag{4.2}$$

is the energy stored in the left-half spring. Now we can equate this with half of the total energy, $U/2$, to find an expression for K in terms of k:

$$\frac{1}{8}k\Delta x^2 = \frac{1}{4}K\Delta x^2 \longrightarrow K = \frac{k}{2}. \tag{4.3}$$

If we divide up the extended spring into $n - 1$ intervals, we expect energy $U/(n-1)$ to be stored in each interval. Again let k be the effective spring constant of each of the intervals.

Then

$$\frac{1}{2}k\left(\frac{\Delta x}{n-1}\right)^2 = \frac{\frac{1}{2}K\Delta x^2}{n-1} \longrightarrow K = \frac{k}{n-1}. \tag{4.4}$$

Turning it around: the chain of n balls connected by $n-1$ springs each with constant k has an effective bulk spring constant of $K = k/(n-1)$. As we send $n \to \infty$, then, we must take $k \to \infty$ so as to keep K fixed.

4.1 Continuum Limit

The equation of motion for the jth mass in a chain of masses-connected-by-springs is, from (3.80),

$$m\ddot{x}_j(t) = k(x_{j-1}(t) - 2x_j(t) + x_{j+1}(t)), \tag{4.5}$$

and we can express this in terms of the fixed M and K

$$\ddot{x}_j(t) = n(n-1)\frac{K}{M}(x_{j-1}(t) - 2x_j(t) + x_{j+1}(t)). \tag{4.6}$$

We will assume that the motion of the individual masses is small. The jth mass has equilibrium location $e_j \equiv ja$, and we will take $x_j(t) \approx e_j$ for all times of interest. This is part of our continuum assumption – we don't expect the mass that is at equilibrium at e_j to be found far to the right of the mass at equilibrium at e_{j+1} unless we allow the individual masses to pass through each other without consequence (not usually allowed physically). To exploit this "close-to-equilibrium" assumption, define the new relative separation variable $\phi(e_j, t) \equiv x_j(t) - e_j$ so that $x_j(t) = \phi(e_j, t) + e_j$. Putting this into (4.6) gives[1]

$$\frac{\partial^2}{\partial t^2}\phi(e_j, t) = n(n-1)\frac{K}{M}(\phi(e_{j-1}, t) - 2\phi(e_j, t) + \phi(e_{j+1}, t))$$

$$= n(n-1)\frac{K}{M}(\phi(e_j - a, t) - 2\phi(e_j, t) + \phi(e_j + a, t)). \tag{4.7}$$

Now $e_j = ja$ will eventually be a continuous variable. As $n \to \infty$, a gets smaller, and the values of ja come closer and closer to describing any point along the x axis (ja becomes a finer and finer gradation of points along the axis). Call[2] $x \equiv e_j = ja$, the point of equilibrium for the jth mass. We are being general – in the continuum limit, *every* point x along the x-axis has an associated massless ball that is in equilibrium at it. So every point has one of the infinite number of balls lying at it when in equilibrium. This allows us to write $\phi(e_j, t) = \phi(x, t)$ and provides a language for the Taylor expansion we are about to perform. With this notation in place, we can write the equation of motion for the jth ball, now the "ball at equilibrium at x" as

---

[1] The function $\phi(e_j, t)$ depends on both e_j and t, so we'll turn the dots appearing in (4.6) into partial time derivatives.

[2] Technically, $x \equiv \lim_{a \to 0} e_j = ja$, of course.

$$\frac{\partial^2}{\partial t^2}\phi(x,t) = n(n-1)\frac{K}{M}(\phi(x-a,t) - 2\phi(x,t) + \phi(x+a,t)) \tag{4.8}$$

and we are finally ready to Taylor expand for small a (remember that $L = (n-1)a$ and L is fixed). Using O-notation,[3]

$$\begin{aligned}
\frac{\partial^2\phi(x,t)}{\partial t^2} &= n(n-1)\frac{K}{M}\left(\frac{\partial^2\phi(x,t)}{\partial x^2}a^2 + O(a^4)\right) \\
&= \frac{n}{n-1}\frac{KL^2}{M}\left(\frac{\partial^2\phi(x,t)}{\partial x^2} + O\left(\frac{1}{(n-1)^2}\right)\right).
\end{aligned} \tag{4.9}$$

And now, finally, for the limit as $n \to \infty$, the term out front goes to 1, and $O((n-1)^{-2}) \to 0$ leaving

$$\frac{\partial^2\phi(x,t)}{\partial t^2} = \frac{KL^2}{M}\frac{\partial^2\phi(x,t)}{\partial x^2}. \tag{4.10}$$

We have a continuous "spring" with spring constant K, mass M distributed uniformly along it, and equilibrium length L. Almost any material can be described in this manner, at least to good approximation. Examples include The Slinky® (obviously), and rods of material (in which context the stiffness K is known as "Young's modulus" and can be measured by stretching the rod and seeing with what force it responds). But more diffuse materials also have displacements described by (4.10), the density of air in a one-dimensional column behaves approximately according to (4.10).

The constant KL^2/M has the dimension of velocity-squared. Define the speed v by $v^2 \equiv KL^2/M$ so that (4.10) becomes

$$-\frac{\partial^2\phi(x,t)}{\partial t^2} + v^2\frac{\partial^2\phi(x,t)}{\partial x^2} = 0, \tag{4.11}$$

and in this form, we have the "wave equation." The wave equation shows up all over the place in physics, with v being set by physically relevant parameters.[4] For materials, it is the mass, length, and Young's modulus that determines v. For density in air, it is the temperature and pressure of the ambient environment that determines v. Whatever the mechanism, this equation governs, at least in approximation, far more of physics than it has any right to.

Partial Derivatives

I'll take this opportunity to review the definition and manipulations associated with partial derivatives. A partial derivative just refers to the derivative of a multivariable function with respect to one of its variables. As an example, take a function of two variables, $f(x,y)$. The partial derivative of this function with respect to x is

---

[3] Without being overly formal, an expression like $O(a^p)$ refers to any function whose leading-order contribution is a^p, with higher-order contributions omitted. As an example, take $\sin\theta$ for θ near zero, then we say that $\sin\theta = O(\theta)$ since the first term in the Taylor expansion is θ. For $\cos\theta$ with $\theta \approx 0$, we have $\cos\theta = 1 + O(\theta^2)$.

[4] That is, not all instantiations of the wave equation can be associated with masses connected by springs in some sort of limit. It arises from totally different physical configurations in, for example, Maxwell's theory of electromagnetism, in which case the "speed" v is set by the geometry of space–time. See Chapter 7 for more examples and extensions.

$$\frac{\partial f(x,y)}{\partial x} \equiv \lim_{\epsilon \to 0} \frac{f(x+\epsilon,y) - f(x,y)}{\epsilon}, \tag{4.12}$$

and the partial derivative with respect to y is

$$\frac{\partial f(x,y)}{\partial y} \equiv \lim_{\epsilon \to 0} \frac{f(x,y+\epsilon) - f(x,y)}{\epsilon}. \tag{4.13}$$

For a warmup, take the function $f(x,y) = \alpha x + \beta y$ for constants α and β, then the two different partial derivatives are

$$\begin{aligned}
\frac{\partial f(x,y)}{\partial x} &= \lim_{\epsilon \to 0} \frac{(\alpha(x+\epsilon) + \beta y) - (\alpha x + \beta y)}{\epsilon} = \alpha \\
\frac{\partial f(x,y)}{\partial y} &= \lim_{\epsilon \to 0} \frac{(\alpha x + \beta(y+\epsilon)) - (\alpha x + \beta y)}{\epsilon} = \beta.
\end{aligned} \tag{4.14}$$

These derivatives tell us, for example, that $\frac{\partial x}{\partial y} = 0$ (take $\alpha = 1$, $\beta = 0$ in our test $f(x,y) = \alpha x + \beta y$), a reminder here that x does not explicitly depend on y. You *can* imagine a relationship between x and y, that would define a curve in one dimension, and in that setting, the ordinary derivative is appropriate. We use partials when we assume the coordinates are independent, and you can really treat the partials as "regular" derivatives with respect to one or the other variable, with the remaining one(s) as constant(s) for the purposes of differentiation.

We'll look at a more complicated case, let $f(x,y) = \alpha x \sin(\beta y)$ (for constant α and β again). To take the x-partial, we imagine $\sin(\beta y)$ is a fixed constant, and just take the x derivative of the resulting "ordinary" expression:

$$\frac{\partial f(x,y)}{\partial x} = \alpha \sin(\beta y). \tag{4.15}$$

Similarly, treating x as a constant for the y-derivative, we get

$$\frac{\partial f(x,y)}{\partial y} = \alpha x \beta \cos(\beta y). \tag{4.16}$$

Additional derivatives function similarly, so we can define the second partial derivatives

$$\frac{\partial^2 f(x,y)}{\partial x^2} \equiv \lim_{\epsilon \to 0} \frac{1}{\epsilon} \left(\left. \frac{\partial f(x,y)}{\partial x} \right|_{x+\epsilon,y} - \left. \frac{\partial f(x,y)}{\partial x} \right|_{x,y} \right) \tag{4.17}$$

and similarly for the second derivative with respect to y. Again, you just treat everything except the derivative variable as constant and proceed to differentiate.

There are also "mixed" partials to consider. We can take the x-derivative of the y-derivative of $f(x,y)$:

$$\frac{\partial^2 f(x,y)}{\partial x \partial y} \equiv \frac{\partial}{\partial x} \left(\frac{\partial f(x,y)}{\partial y} \right), \tag{4.18}$$

and vice-versa, the y-derivative of the x-derivative is

$$\frac{\partial^2 f(x,y)}{\partial y \partial x} \equiv \frac{\partial}{\partial y} \left(\frac{\partial f(x,y)}{\partial x} \right). \tag{4.19}$$

Fortunately, these two are equal (as you will show in Problem 4.1.3), so we can be sloppy about which we have in mind, they are numerically identical.

Problem 4.1.1   For the function $f(x,y) = A(x^2 \sin(y/L) - xy)$ (with constants A and L), evaluate the following partial derivatives:

$$\frac{\partial f(x,y)}{\partial x} \qquad \frac{\partial^2 f(x,y)}{\partial y^2} \qquad \frac{\partial^3 f(x,y)}{\partial x^3} \qquad \frac{\partial^2 f(x,y)}{\partial x \partial y}$$

Problem 4.1.2   In three dimensions, with coordinates x, y, and z, let $q_1 \equiv x$, $q_2 \equiv y$, and $q_3 \equiv z$. Define a matrix $\mathbb{M}$ by its entries,

$$M_{ij} \equiv \frac{\partial q_i}{\partial q_j}$$

for $i = 1 \to 3$ and $j = 1 \to 3$. Write out the entries of this matrix of partial derivatives.

Problem 4.1.3   Show that mixed partials satisfy "cross-derivative" equality:

$$\frac{\partial^2 f(x,y)}{\partial x \partial y} = \frac{\partial^2 f(x,y)}{\partial y \partial x}$$

so that the order in which you take the derivatives doesn't matter. Check this result explicitly using the function $f(x,y) = xy \sin(x^2 y)$.

Problem 4.1.4   You can generate the series spring addition formula by considering a mass M attached by a spring with constant k_1 to a mass m which is attached by a spring with constant k_2 to the wall. Taking $m \to 0$ eliminates the intermediate mass, leaving us with an effective force acting on M. To extract that effective force and the effective spring constant it implies, write the equations of motion for M and $m \neq 0$, then take $m \to 0$ and see what the equation of motion for M must become – the constant in front of the distance to M defines the spring constant. Do you recover (4.3) for $k_1 = k_2 \equiv k$?

Problem 4.1.5   The relative separation $\phi(x,t)$ tells us, at time t, "the distance, relative to x, of the mass that is at equilibrium at x." For a continuous spring system extending from $x = 0$ to L, suppose you were given $\phi(x,0) = 0$, describe the configuration of masses. Describe the system if $\phi(x,0) = \alpha$ for positive, constant α. If $\phi(x,0) = x/2$, where, relative to $x = 0$, is the mass that is at equilibrium at $1/4$ m? (the m here stands for "meter").

Problem 4.1.6   Show that the wave equation (4.11) supports "superposition," so that if $\phi_1(x,t)$ and $\phi_2(x,t)$ both satisfy (4.11) so does the sum $\phi_1(x,t) + \phi_2(x,t)$.

Problem 4.1.7   Which of the following PDEs supports "superposition" (see the previous problem):

$$\frac{\partial^2 f(x,y)}{\partial x^2} + \frac{\partial^2 f(x,y)}{\partial y^2} = 0$$

$$\frac{\partial^2 f(x,y)}{\partial x^2} + \frac{\partial^2 f(x,y)}{\partial y^2} = \rho(x,y) f(x,y)$$

$$i\hbar \frac{\partial \Psi(x,t)}{\partial t} + \frac{\hbar^2}{2m} \frac{\partial^2 \Psi(x,t)}{\partial x^2} - V(x,t) \Psi(x,t) = 0$$

$$\frac{\partial^2 u(x,t)}{\partial x^2} - \frac{u(x,t)}{\alpha} \frac{\partial u(x,t)}{\partial t} = 0.$$

Problem 4.1.8   For a discrete chain of masses connected by springs, with $x_j(t)$ giving the location of the jth mass at time t, the total energy of the chain is

$$E = \frac{1}{2}m\sum_{j=1}^{n}\dot{x}_j(t)^2 + \frac{1}{2}k\sum_{j=1}^{n-1}(x_{j+1}(t) - x_j(t) - a)^2 .$$

Find the continuum limit of the energy following the procedure in this section: start by letting $e_j \equiv ja$ and switch to the difference variable $\phi(e_j, t) \equiv x_j(t) - e_j$, replace the constants k and m using their relation to K and M, and take the limit as the number of masses goes to infinity (with fixed total length at equilibrium L). The sums will become integrals ($\sum\_\_a \to \int\_\_dx$), and the integrand is itself the "energy density" (energy per unit length here) of the continuous system. You should end up with an energy density that is

$$u(x, t) = \frac{1}{2}\frac{M}{L}\left(\left(\frac{\partial\phi(x,t)}{\partial t}\right)^2 + v^2\left(\frac{\partial\phi(x,t)}{\partial x}\right)^2\right).$$

4.2 Wave Equation for Strings

Another physical system in which the wave equation appears is the taut string. Suppose you have a string with constant mass density μ (mass per unit length), and we think about a segment of string extending from x to $x + dx$ as shown in Figure 4.1. The tension in the string exerts a force on the left and right ends of the segment. If we demand that the string be "inextensible," so that the pieces of the string do not move left or right (opposite our

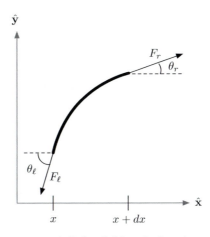

Fig. 4.1   A portion of string extending from $x \to x + dx$. Each end of the string has a force on it due to the tension in the string. The magnitude of the force on the left is F_ℓ and it makes an angle of θ_ℓ with respect to horizontal. On the right, the force magnitude is F_r and the angle is θ_r.

spring system from the previous section), then the forces in the horizontal direction must cancel:

$$F_\ell \cos \theta_\ell = F_r \cos \theta_r. \tag{4.20}$$

We'll make two further assumptions (in Section 7.3, we will see what happens if you relax these): (1) the tension in the string is constant throughout, so that $F_\ell = F_r \equiv T$, and (2) the angles θ_ℓ and θ_r are small. With these two assumptions, the inextensibility requirement (4.20) is automatically satisfied, since $\cos \theta_\ell \approx 1$ and $\cos \theta_r \approx 1$ for small angles.

Moving on to the y component of Newton's second law, let $y(x, t)$ be the height of the string at location x and time t. The portion of the string extending from x to $x + dx$ has mass $m = \mu dx$, and the net vertical force is just $T(\sin \theta_r - \sin \theta_\ell)$ (from Figure 4.1 again), so that we have

$$m\frac{\partial^2 y(x, t)}{\partial t^2} = T(\sin \theta_r - \sin \theta_\ell) \longrightarrow \mu dx \frac{\partial^2 y(x, t)}{\partial t^2} \approx T(\tan \theta_r - \tan \theta_\ell) \tag{4.21}$$

where we have used the small angle approximation $\sin \theta \approx \theta \approx \tan \theta$. But those tangent terms just represent the slopes at the left and right ends of the string segment. If we let $y'(x, t) \equiv \frac{\partial y(x,t)}{\partial x}$, then we can write Newton's second law as

$$\mu dx \frac{\partial^2 y(x, t)}{\partial t^2} \approx T(y'(x + dx, t) - y'(x, t)), \tag{4.22}$$

and Taylor expanding on the right, we get

$$\mu dx \frac{\partial^2 y(x, t)}{\partial t^2} \approx Ty''(x, t)dx \longrightarrow -\frac{\partial^2 y(x, t)}{\partial t^2} + \frac{T}{\mu}\frac{\partial^2 y(x, t)}{\partial x^2} = 0 \tag{4.23}$$

which is the wave equation again. This time, the characteristic speed is set by $v^2 = T/\mu$.

Problem 4.2.1 A hundred yard ball of string has mass .5 kg. One meter of the string has one end attached to a wall, and the other attached to a 1 lb weight and hung over a pulley. Assuming constant tension, what is the speed with which waves will travel between the wall and the pulley?

4.3 Solving the Wave Equation

We know how to solve ordinary differential equations of various sorts. The wave equation is our first example of a "partial" differential equation (PDE). How should we solve it? There are a number of approaches, and we'll think about three specific ones that reduce the PDE to an ODE (or set of ODEs) that we can then solve.

4.3.1 First-Order Form: Method of Characteristics

From (4.11), we can factor the derivatives:

$$-\frac{\partial^2 \phi(x,t)}{\partial t^2} + v^2 \frac{\partial^2 \phi(x,t)}{\partial x^2} = \left(v\frac{\partial}{\partial x} - \frac{\partial}{\partial t} \right)\left(v\frac{\partial}{\partial x} + \frac{\partial}{\partial t} \right)\phi(x,t) = 0 \qquad (4.24)$$

and then it is clear that one way to satisfy the wave equation is to have

$$\left(v\frac{\partial}{\partial x} + \frac{\partial}{\partial t} \right)\phi(x,t) = 0. \qquad (4.25)$$

We have reduced a second-order partial differential equation to a first-order one (actually a *pair* of first-order ones, as you will see). Suppose we were given the values of $\phi(x,t)$ at $t = 0$: $\phi(x,0) = u(x)$ for some provided $u(x)$. Then we can ask: Are there curves, $x(t)$, along which the value of $\phi(x,t)$ solving (4.25) remains constant? Then we'd just pick up our known values of $u(x)$ at $t = 0$ and move them along the curve $x(t)$. That's an easy way to find solutions for $\phi(x,t)$. Operationally, we take the curves $x(t)$ and evaluate $\phi(x(t),t)$ along them, then the total time derivative of $\phi(x(t),t)$ along the curve is

$$\frac{d\phi(x(t),t)}{dt} = \frac{\partial \phi(x,t)}{\partial x}\frac{dx(t)}{dt} + \frac{\partial \phi(x,t)}{\partial t}. \qquad (4.26)$$

Now if we pick $\frac{dx(t)}{dt} = v$, then (4.25) applies and $\frac{d\phi}{dt} = 0$. So for curves of the form $x(t) = x_0 + vt$, the value of ϕ is the same all along the curve. These curves are straight lines with slope v (or slope $1/v$ if we plot t versus x). Then we know that the solution for $\phi(x,t)$ at x is $\phi(x,t) = u(x_0) = u(x - vt)$. You can easily check that this form solves the original wave equation, and of course the factored form in (4.25), by taking derivatives. The curves that we generated are called "characteristic curves," and this solution was obtained by the "method of characteristics" where we evaluate the total time derivative of $\phi(x(t),t)$ and use that to define the curves $x(t)$.

This approach has a pictorial interpretation. The initial curve $u(x)$ has values that get "picked up" and moved along the straight lines of slope $1/v$ emanating from the x axis. So the "picture" of $u(x)$ at $t = 0$ gets shifted to the *right* in time, with constant speed v. An example of such curves and an initial function that is moved along them is shown in Figure 4.2.

All of this right-traveling initial data is for $\phi(x,t) = u(x - vt)$, but of course if we factored the wave equation in the other direction, then we'd have

$$\left(v\frac{\partial}{\partial x} + \frac{\partial}{\partial t} \right)\left(v\frac{\partial}{\partial x} - \frac{\partial}{\partial t} \right)\phi(x,t) = 0 \qquad (4.27)$$

and the first-order form that solves this is

$$\left(v\frac{\partial}{\partial x} - \frac{\partial}{\partial t} \right)\phi(x,t) = 0. \qquad (4.28)$$

Here, the solutions are related to the initial values $u(x)$ by $\phi(x,t) = u(x + vt)$, and the solution moves to the *left* with constant speed v.

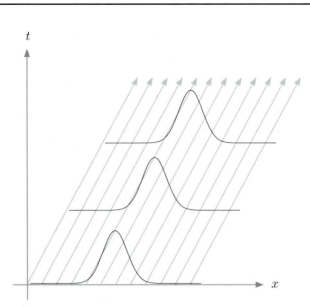

Fig. 4.2 An example of some characteristic curves, defined by $t = (x - x_0)/v$ for initial position x_0. There is a reference Gaussian for $u(x)$ plotted at $t = 0$ and some other times to demonstrate how the constant values of the initial data along the characteristic curves can be viewed as a right-traveling waveform.

In general, the solution to the full wave equation is given by

$$\phi(x, t) = f(x - vt) + g(x + vt) \tag{4.29}$$

for single-variable functions $f(x), g(x)$ which would be used to satisfy the initial conditions. For example, suppose we are given functions $u(x)$ and $w(x)$ such that:

$$\phi(x, 0) = u(x) \qquad \left.\frac{\partial \phi}{\partial t}\right|_{t=0} = w'(x), \tag{4.30}$$

then we would take

$$f(x) = \frac{1}{2}\left(u(x) - \frac{1}{v}w(x)\right) \qquad g(x) = \frac{1}{2}\left(u(x) + \frac{1}{v}w(x)\right) \tag{4.31}$$

and the full solution would be

$$\phi(x, t) = \frac{1}{2}\left(u(x - vt) - \frac{1}{v}w(x - vt) + u(x + vt) + \frac{1}{v}w(x + vt)\right). \tag{4.32}$$

4.3.2 Additive Separation of Variables

The other way of turning a PDE into an ODE (possibly many) is to make some assumptions about the form of the solution. As an example, let's take $\phi(x, t) = X(x) + T(t)$, where $X(x)$ and $T(t)$ are unknown functions of x and t respectively, so that we are looking for solutions that depend on x and t separately in additive combination. Running this through the full wave equation from (4.11) gives

$$\frac{d^2 T(t)}{dt^2} = v^2 \frac{d^2 X(x)}{dx^2}. \tag{4.33}$$

Now the logic of separation of variables (SOV) is as follows: The left-hand side of the equation depends only on t, while the right-hand side depends only on x. The only way that the equation can hold, *for all values of x and t* is if both sides are constant. So we demand, calling the shared constant α,

$$\begin{aligned}
\frac{d^2 T(t)}{dt^2} &= \alpha \longrightarrow T(t) = \frac{1}{2}\alpha t^2 + bt + c \\
v^2 \frac{d^2 X(x)}{dx^2} &= \alpha \longrightarrow X(x) = \frac{1}{2}\frac{\alpha}{v^2}x^2 + dx + e
\end{aligned} \tag{4.34}$$

and "the" solution is

$$\phi(x,t) = \frac{1}{2}\alpha\left(t^2 + \frac{x^2}{v^2}\right) + bt + dx + (c + e). \tag{4.35}$$

There are four parameters here: α, b, d, and $c + e \equiv f$. That's the right number for a PDE that is second order in both time and space. Yet it is clear that we will not be able to satisfy many different types of boundary conditions with the solution in (4.35). There is a notion of superposition here (the idea that sums of solutions are solutions), so we could add together solutions with different values of the four independent constants. But there is another type of separation of variables that comes up a lot, and we will think about issues of superposition and boundary values in that setting.

4.3.3 Multiplicative Separation of Variables

Another way to achieve the separation found in, for example, (4.33) is to start with the "multiplicative" form: $\phi(x,t) = X(x)T(t)$. We still have independent functions that depend on x and t only, but this time they are multiplied together instead of added. Running this ansatz through (4.11), we have

$$X(x)\frac{d^2 T(t)}{dt^2} = v^2 \frac{d^2 X(x)}{dx^2}T(t). \tag{4.36}$$

This is not obviously in the form of (4.33), since there is x dependence on the left, t dependence on the right. Yet if we divide both sides by ϕ itself, we get

$$\frac{1}{T(t)}\frac{d^2 T(t)}{dt^2} = \frac{v^2}{X(x)}\frac{d^2 X(x)}{dx^2}, \tag{4.37}$$

and now we can make the separation argument: The left-hand side depends only on t and the right depends only on x, so each must be separately equal to a constant. Call that constant, for reasons that will become clear later on, $-\alpha^2$. Then we have to solve the pair of ODEs

$$\frac{d^2 T(t)}{dt^2} = -\alpha^2 T(t) \longrightarrow T(t) = A\cos(\alpha t) + B\sin(\alpha t)$$

$$\frac{d^2 X(x)}{dx^2} = -\frac{\alpha^2}{v^2} X(x) \longrightarrow X(x) = C\cos\left(\frac{\alpha x}{v}\right) + D\sin\left(\frac{\alpha x}{v}\right)$$

$$(4.38)$$

for constants A, B, C, and D.

Combining the two factors, the solution is

$$\phi(x,t) = \left(C\cos\left(\frac{\alpha x}{v}\right) + D\sin\left(\frac{\alpha x}{v}\right)\right)(A\cos(\alpha t) + B\sin(\alpha t)). \qquad (4.39)$$

We can use the constants to satisfy boundary conditions. For example, suppose we require that at $x = 0$ and $x = \ell$, the function $\phi(x,t)$ vanish at all times: $\phi(0,t) = \phi(\ell,t) = 0$. Physically, we are requiring that the mass at the endpoints of the chain be at fixed locations, for example. Then we have

$$\phi(0,t) = C(A\cos(\alpha t) + B\sin(\alpha t)) = 0 \longrightarrow C = 0 \qquad (4.40)$$

and we are left with (absorbing the constant D into A and B)

$$\phi(x,t) = (A\cos(\alpha t) + B\sin(\alpha t))\sin\left(\frac{\alpha x}{v}\right). \qquad (4.41)$$

For the other end, we have

$$\phi(\ell,t) = (A\cos(\alpha t) + B\sin(\alpha t))\sin\left(\frac{\alpha \ell}{v}\right) = 0 \qquad (4.42)$$

and we could satisfy this equation, for all times t, by taking $A = B = 0$. This choice leaves us with $\phi(x,t) = 0$, and we have no hope of satisfying initial conditions. To retain a nontrivial solution, we instead set $\sin(\alpha \ell/v) = 0$, giving us a set of values for α,

$$\frac{\alpha \ell}{v} = n\pi \longrightarrow \alpha = \frac{n\pi v}{\ell} \text{ for integer } n. \qquad (4.43)$$

There is still an infinite family of solutions here, one for each integer n.

The wave equation is linear in $\phi(x,t)$, and so superposition holds: If you have two solutions to the wave equation, $\phi_1(x,t)$ and $\phi_2(x,t)$, then the sum is also a solution. For our integer-indexed $\phi(x,t)$, we can form an infinite sum that is still a solution and satisfies the boundary conditions term-by-term,

$$\phi(x,t) = \sum_{n=1}^{\infty} \left[A_n \cos\left(\frac{n\pi v}{\ell}t\right) + B_n \sin\left(\frac{n\pi v}{\ell}t\right)\right] \sin\left(\frac{n\pi}{\ell}x\right). \qquad (4.44)$$

We could write the sine term in position as exponentials to make the connection with the Fourier series we studied earlier in Section 2.3. Then it is clear that the coefficients $\{A_n, B_n\}_{n=1}^{\infty}$ are associated with the decomposition of the initial conditions. Remember that we must be given $\phi(x,0) = u(x)$ and the time-derivative at $t = 0$, $\dot{\phi}(x,0) = w'(x)$. From the solution, we have

$$\phi(x,0) = \sum_{n=1}^{\infty} A_n \sin\left(\frac{n\pi}{\ell}x\right) = u(x)$$

$$\left.\frac{\partial \phi(x,t)}{\partial t}\right|_{t=0} = \sum_{n=1}^{\infty} B_n \left(\frac{n\pi v}{\ell}\right) \sin\left(\frac{n\pi}{\ell}x\right) = w'(x). \tag{4.45}$$

Using the orthogonality of sine:

$$\int_0^\ell \sin\left(\frac{n\pi x}{\ell}\right) \sin\left(\frac{k\pi x}{\ell}\right) dx = \frac{\ell}{2}\delta_{nk}, \tag{4.46}$$

we can multiply both sides of the equations in (4.45) by $\sin(k\pi x/\ell)$ and integrate to isolate the coefficients

$$A_k = \frac{2}{\ell}\int_0^\ell u(x)\sin\left(\frac{k\pi x}{\ell}\right) dx$$

$$B_k = \frac{2}{k\pi v}\int_0^\ell w'(x)\sin\left(\frac{k\pi x}{\ell}\right) dx. \tag{4.47}$$

4.3.4 Series Solution

The general solution in (4.44) has the basic form of a (spatial) Fourier sine series like (2.55) but with time-varying coefficients. We could have started with a solution ansatz of the form

$$\phi(x,t) = \sum_{j=1}^{\infty} a_j(t)\sin\left(\frac{j\pi x}{\ell}\right). \tag{4.48}$$

We know, for appropriate choice of $\{a_j(t)\}_{j=1}^{\infty}$, that "any" function can be written in this form, and because the coefficients can change in time, so can the function we are representing in its sine series decomposition. So a solution to the wave equation, with the boundary conditions $\phi(0,t) = \phi(\ell,t) = 0$ can be expressed as (4.48) for a particular set of $\{a_j(t)\}_{j=1}^{\infty}$.

This observation leads us to turn around and use (4.48) as the starting point in generating a solution to the wave equation. If we take $\phi(x,t)$ from (4.48) and run it through the wave equation, we get

$$\sum_{j=1}^{\infty} \left[\ddot{a}_j(t) + v^2\left(\frac{j\pi}{\ell}\right)^2 a_j(t)\right] \sin\left(\frac{j\pi x}{\ell}\right) = 0. \tag{4.49}$$

Now we use the orthogonality of sine to demand that each term in the sum vanish separately.[5] Then we get an infinite family of ordinary differential equations in time,

$$\ddot{a}_j(t) = -v^2\left(\frac{j\pi}{\ell}\right)^2 a_j(t) \tag{4.50}$$

---

[5] We could do this carefully, as we have before, by multiplying (4.49) by $\sin(k\pi x/\ell)$ and integrating $x: 0 \to \ell$, this would eliminate all terms in the sum with $j \neq k$.

an equation that we recognize, and whose solutions we know all too well:

$$a_j(t) = A_j \cos\left(\frac{j\pi v}{\ell}t\right) + B_j \sin\left(\frac{j\pi v}{\ell}t\right) \tag{4.51}$$

giving precisely the term in front of $\sin(j\pi x/\ell)$ in (4.44). The utility of the approach, in this case, comes from applying it to PDEs other than the wave equation where we already know quite a bit about the solutions. You can explore some examples in the problems.

Problem 4.3.1 Suppose we had a factored (i.e. first-order) wave equation of the form

$$v_0 \frac{x}{\ell} \frac{\partial \phi(x,t)}{\partial x} + \frac{\partial \phi(x,t)}{\partial t} = 0$$

for constant v_0 and ℓ and with $\phi(x,0) = u(x)$ given. Find the solution to this equation using the method of characteristics.

Problem 4.3.2 The Laplace equation in two dimensions is

$$\frac{\partial^2 f(x,y)}{\partial x^2} + \frac{\partial^2 f(x,y)}{\partial y^2} = 0,$$

for a function $f(x,y)$ of the two Cartesian variables x and y, find the solution that has $f(0,y) = 0, f(L,y) = 0$ (for constant L) with $f(x,0) = f_0 \sin(2\pi x/L)$ and $f(x,L) = 0$. We will study this type of equation and its solutions in Chapter 6.

Problem 4.3.3 The "heat equation" in one spatial dimension is

$$\frac{\partial^2 u(x,t)}{\partial x^2} - \frac{1}{\alpha}\frac{\partial u(x,t)}{\partial t} = 0, \tag{4.52}$$

find $u(x,t)$ given $u(0,t) = 0$, $u(L,t) = 0$, and $u(x,0) = u_0 \sin(\pi x/L)$. Sketch the solution at different times indicating its behavior as t goes from $0 \to \infty$. What are the units of α?

Problem 4.3.4 For the partial differential equation:

$$-\frac{\partial^2 u(x,t)}{\partial t^2} + v^2\frac{\partial^2 u(x,t)}{\partial x^2} + m^2 u(x,t) = 0,$$

with constant m (some real number[6]), solve for $u(x,t)$ given the boundary conditions $u(0,t) = 0$, $u(L,t) = 0$ with $\frac{\partial u(x,t)}{\partial t}\big|_{t=0} = 0$ and $u(x,0) = u_0 \sin(q\pi x/L)$ for integer q. For what values of q (relative to m, say) is your solution bounded (does not become infinite) for all time $t \geq 0$.

Problem 4.3.5 The wave equation governing the height of a rope is

$$-\frac{\partial^2 y(x,t)}{\partial t^2} + v^2\frac{\partial^2 y(x,t)}{\partial x^2} = 0,$$

where $y(x,t)$ is the height of the piece of rope with horizontal location x at time t. We can write this equation in terms of the Fourier transform of $y(x,t)$ *in time* (one can also Fourier transform the spatial coordinate, but we'll leave x alone here):

$$\tilde{y}(x,f) = \int_{-\infty}^{\infty} y(x,t)e^{i2\pi ft}\,dt.$$

---

[6] This equation is what you would get from Maxwell's electricity and magnetism if light had mass.

Take the inverse transform version of $y(x, t)$,

$$y(x, t) = \int_{-\infty}^{\infty} \tilde{y}(x, f) e^{-i2\pi ft} \, df,$$

and insert it into the wave equation. Solve the resulting ODE (in x) for $\tilde{y}(x, f)$ (you should end up with two undetermined constants of integration, which, for once, we'll leave as undetermined constants).

Problem 4.3.6   A solution to the wave equation,

$$p(x, t) = p_0 \left[\cos(2\pi(x - vt)g_1) + \cos(2\pi(x - vt)g_2)\right] \tag{4.53}$$

(for constants g_1 and g_2) is observed by you at your location, $x = 0$. Show that you can write that signal as

$$p(0, t) = F \cos(2\pi f_1 t) \cos(2\pi f_2 t) \tag{4.54}$$

for constant frequencies f_1 and f_2 and magnitude F. What are f_1 and f_2 in terms of g_1 and g_2? If the original signal was made up of $g_1 v = 441$ Hz and $g_2 v = 440$ Hz, what frequencies appear in (4.54)? Can you hear both of those frequencies? The expression (4.54) is an example of the "beats" that the intereference of two signals can make – you perceive a new frequency that has time-varying amplitude (that's how (4.54) is interpreted by your ears and brain).

Problem 4.3.7   **a.**   A ball of mass m moving with constant speed v bounces elastically off of walls separated by a distance a (no gravity here). We want to describe the probability-per-unit-length of finding the ball in the vicinity of $x \in [0, a]$, we'll call this quantity $\rho(x)$ and the probability of finding the ball within dx of x is $dP = \rho(x)dx$ so that the probability of finding the ball between two locations, x_1 and $x_2 > x_1$ (with both in $[0, a]$) is

$$P(x_1, x_2) = \int_{x_1}^{x_2} \rho(x) \, dx.$$

Construct $\rho(x)$ for the ball – think about the following two observations: (1) given two locations x and y (both in between 0 and a), at which location is the ball more likely to be found? and (2) What is the total probability of finding the ball between $x = 0$ and a?

    b.   When you study quantum mechanics (a great resource is [13]), you will learn that the fundamental object is a complex function $\Psi(x, t)$. The quantum mechanical probability density, $\rho(x, t)$ (defined similarly to a.) is given by: $\rho(x, t) = \Psi^*(x, t)\Psi(x, t)$. For a ball of mass m bouncing back and forth between elastic walls, the equation governing $\Psi(x, t)$, called the "Schrödinger wave equation," is:

$$-\frac{\hbar^2}{2m} \frac{\partial^2 \Psi(x, t)}{\partial x^2} = i\hbar \frac{\partial \Psi(x, t)}{\partial t}, \tag{4.55}$$

for constant $\hbar$ (with what units?). The boundary conditions are: $\Psi(0, t) = 0$ and $\Psi(a, t) = 0$. Solve this equation for $\Psi(x, 0) = \psi_0 \sin(\pi x/a)$. Find the value of the constant ψ_0 (Hint: what is the total probability of finding the ball between 0 and a?).

Sketch the resulting probability density $\rho(x, t)$. Do the same for the initial condition $\Psi(x, 0) = \psi_0 \sin(100\pi x/a)$. Which probability density $\rho(x, t)$ looks more like your result from part a.?

Problem 4.3.8   Use the approach from Section 4.3.4 to write the general, time-dependent solution to Schrödinger's equation (4.55) subject to the boundary conditions $\Psi(0, t) = 0 = \Psi(a, t)$.

Problem 4.3.9   Using a series ansatz, solve the heat equation (4.52) with boundary conditions $u(0, t) = 0 = u(L, t)$.

Problem 4.3.10   Find "a" (by hook or crook, with any boundary/initial conditions that you like) solution to the wave equation with "driving" term on the right-hand side

$$-\frac{\partial^2 y(x, t)}{\partial t^2} + v^2 \frac{\partial^2 y(x, t)}{\partial x^2} = F_0 \sin(2\pi f t).$$

Problem 4.3.11   Solve the wave equation with $\phi(0, t) = \phi(\ell, t) = 0$ and initial conditions $\phi(x, 0) = u_0 x(x - \ell)/\ell^2$ and $\frac{\partial \phi(x,t)}{\partial t}|_{t=0} = 0$.

Problem 4.3.12   Solve the wave equation with $\phi(0, t) = \phi(\ell, t) = 0$ and initial conditions $\phi(x, 0) = 0$ and $\frac{\partial \phi(x,t)}{\partial t}|_{t=0} = w_0 x(x - \ell)/\ell^2$.

4.4 Standing Waves

The individual solutions that we got from separation of variables are already interesting on their own. Think of

$$\phi(x, t) = \phi_0 \sin\left(\frac{j\pi v t}{\ell}\right) \sin\left(\frac{j\pi x}{\ell}\right) \tag{4.56}$$

for integer j. This solution can be thought of as a time-varying amplitude for the spatial sine function selected by j. That is why these solutions are called "standing" waves. All that changes in time is the magnitude of the solution, the spatial form always looks the same (although when the coefficient out front is negative, the spatial function has flipped upside down, and of course, it also goes through zero as it changes sign). An example, for $j = 4$, is shown in Figure 4.3 where the plot of $\phi(x, t)$ is shown for a few different times. The points at which the solution goes through the x axis are fixed in time, and are called "nodes." The maxima and minima that occur in between the nodes are called "antinodes."

The oscillatory pieces of $\phi(x, t)$ in space and time determine a characteristic length and time for the solution. The "wavelength," usually denoted λ, is the spatial length of a single cycle of sine. If you took a snapshot of $\phi(x, t)$ at time t_0 and measured the distance between peaks as in Figure 4.4, that distance is the wavelength, and it is related to j and ℓ by

$$\frac{j\pi\lambda}{\ell} = 2\pi \longrightarrow \lambda = \frac{2\ell}{j}. \tag{4.57}$$

The temporal oscillation defines the "period" of the wave. If you stood at a single location x_0 and measured the wave's magnitude at that location, the time it takes for one

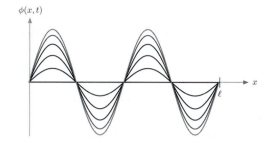

The standing wave solution from (4.56) with $j = 4$ shown for six different times (from dark to light). The zero crossings on the x axis are "nodes," and the max/min that occur in between are "antinodes."

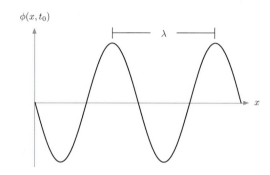

Given a snapshot of $\phi(x, t)$ at time t_0, the distance from one peak to the next defines the wavelength λ.

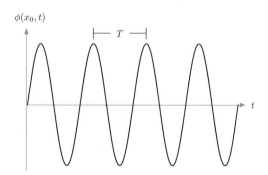

Plotting the wave $\phi(x_0, t)$ in time at a particular location x_0. The period of the wave, T, is the time from one peak to the next.

full cycle of the wave to pass you (assuming it is not a node) is the period, usually denoted T – this is shown in Figure 4.5. In terms of the solution parameters, the period is given by

$$\frac{j\pi v T}{\ell} = 2\pi \longrightarrow T = \frac{2\ell}{jv}. \tag{4.58}$$

From the period, we also define the "frequency": $f \equiv 1/T = jv/(2\ell)$, and the "angular frequency" $\omega \equiv 2\pi f = j\pi v/\ell$.

We can relate the wavelength and period, $T = \lambda/v$, so that λ is also interpretable as the distance travelled by the wave in one full cycle (the wave, remember, travels with characteristic speed v). That's strange since these standing waves aren't really traveling left and right in the usual way. If we were thinking of waves on a string, the standing waves would describe pieces of string moving up and down. Is anything moving left or right? How should we think about standing waves in terms of the general solution to the wave equation in (4.29)?

As it turns out, standing waves are the sum of carefully tuned left- and right-traveling solutions. We can write the standing wave solution using angle addition:

$$\cos\left(\frac{j\pi}{\ell}(x \pm vt)\right) = \cos\left(\frac{j\pi v}{\ell}t\right)\cos\left(\frac{j\pi}{\ell}x\right) \mp \sin\left(\frac{j\pi v}{\ell}t\right)\sin\left(\frac{j\pi}{\ell}x\right) \tag{4.59}$$

so that

$$\phi(x,t) = \frac{\phi_0}{2}\left[\cos\left(\frac{j\pi}{\ell}(x-vt)\right) - \cos\left(\frac{j\pi}{\ell}(x+vt)\right)\right], \tag{4.60}$$

where the first term represents the right-traveling piece, the second term the left-traveling one. So there are still traveling waves here, it's just the superposition that makes it *look* as if we have a horizontally static waveform with time-varying vertical height.

The standing wave solution can also be used to connect with the oscillator chain from Section 3.5. Suppose we discretized space into a grid with fixed spacing $\Delta x = \ell/(n+1)$ for integer n determining the number of grid points. The grid points are at $x_k = k\Delta x$ for $k = 1 \to n$, and then the evaluation of $\phi(x,t)$ at those grid points is

$$\phi(x_k,t) = \phi_0 \sin\left(\frac{j\pi vt}{\ell}\right)\sin\left(\frac{j\pi x_k}{\ell}\right) = \phi_0 \sin\left(\frac{j\pi vt}{\ell}\right)\sin\left(\frac{\pi jk}{n+1}\right). \tag{4.61}$$

The spatial portion of this function gives precisely the entries of v_k^j from (3.102). The oscillatory angular frequency is (for $\kappa^2 \equiv Kn(n-1)/M$ and using $\ell = L$ to match the previous definition)

$$\omega_j = \frac{j\pi v}{L} = \frac{j\pi}{L}\sqrt{\frac{KL^2}{M}} = j\pi\sqrt{\frac{\kappa^2}{n(n-1)}} \approx \frac{j\pi\kappa}{n} \tag{4.62}$$

for large n. Meanwhile, if we expand the expression for the eigenvalue λ_j in (3.103) for large n, noting that the angular frequency there is $\omega^2 \sim \lambda_j$, we get

$$\omega \approx \left(\frac{j\pi\kappa}{n}\right), \tag{4.63}$$

so that ω_j is the same for the continuous solution and its (large n) discrete approximation. Of course, you must immediately turn around and ask the question: What (continuous?!) eigenvalue problem does $\phi(x,t)$ satisfy? Why does this correspondence exist at all?

Problem 4.4.1   Find standing wave solutions to the wave equation for a string,

$$-\frac{\partial^2 y(x,t)}{\partial t^2} + v^2\frac{\partial^2 y(x,t)}{\partial x^2} = 0,$$

with boundary conditions $y(0,t) = 0$ and $\frac{\partial y(x,t)}{\partial x}\big|_{x=L} = 0$.

Problem 4.4.2   Does the heat equation (4.52) support standing wave solutions for $u(0,t) = u(L,t) = 0$?

4.5 Plane Waves

Standing waves are made up of a linear combination of left- and right-traveling waves as we saw explicitly in (4.60). Those individual traveling waves are themselves oscillatory with a single frequency and wavelength. Take a solution to the wave equation

$$\phi(x,t) = A\cos(k(x - vt)) \tag{4.64}$$

with k a constant playing the role of $j\pi/\ell$ from (4.60). The solution at time $t = 0$ is $\phi(x,0) = A\cos(kx)$ and this waveform moves to the right with speed v, a special case of (4.29). Because of its purely oscillatory form, we can identify a wavelength and frequency just as we did in Section 4.4. From the $t = 0$ solution, the wavelength is $k\lambda = 2\pi \rightarrow \lambda = 2\pi/k$. Now taking $x = 0$, we have $\phi(0,t) = A\cos(kvt)$ and the period is $kvT = 2\pi \rightarrow T = 2\pi/(kv) = \lambda/v$ so that the frequency is $f = 1/T = v/\lambda$. We can write the solution to highlight the roles of wavelength and frequency,

$$\phi(x,t) = A\cos\left(2\pi\left[\frac{x}{\lambda} - ft\right]\right), \tag{4.65}$$

and if we had a left-traveling oscillatory wave solution, of the form $\cos(2\pi(x/\lambda + ft)) = \cos(2\pi(-x/\lambda - ft))$, we could write a sum of left and right travelers,

$$\phi(x,t) = A\cos\left(2\pi\left[\frac{x}{\lambda} - ft\right]\right) + B\cos\left(2\pi\left[-\frac{x}{\lambda} - ft\right]\right). \tag{4.66}$$

Finally, we can take the view from Chapter 2, and think of $\phi(x,t)$ as a complex function of exponentials,

$$\phi(x,t) = Ae^{i2\pi(-ft+x/\lambda)} + Be^{i2\pi(-ft-x/\lambda)} = e^{-i2\pi ft}\left(Ae^{i2\pi x/\lambda} + Be^{-i2\pi x/\lambda}\right)$$
$$f = \frac{v}{\lambda}. \tag{4.67}$$

If you want to recover (4.66), just take the real part of the exponential. You can also get a sinusoidal version of the solution by taking the imaginary portion of the complex solution. Individual solutions of the form (4.67) are called "plane waves." They have well-defined frequency and wavelength, and are written to highlight those, but the auxiliary condition relating them, $f = v/\lambda$, must be kept in mind, otherwise the function $\phi(x,t)$ in (4.67) does not satisfy the wave equation (i.e. the values of f and λ are not independent). Plane waves with different frequencies and wavelengths can be added together with the sum itself a solution to the wave equation by superposition. Indeed, the spatial exponentials in the second expression for $\phi(x,t)$ in (4.67) look a lot like the terms in a spatial Fourier series decomposition. Because of the shared temporal pre-factor, we sometimes focus on the spatial piece of $\phi(x,t)$ with the understanding that the temporal piece can be put back in when necessary.

An example of a cartoon that focuses attention on the spatial piece is shown in Figure 4.6, where we identify the term $e^{i2\pi x/\lambda}$ with a right-traveling sinusoidal wave, and $e^{-i2\pi x/\lambda}$ with a left-traveling one. Those identifications are impossible to justify without referring to the full solution, complete with temporal dependence, in (4.67), and that reference is often implicit.

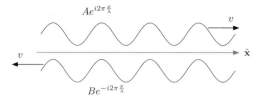

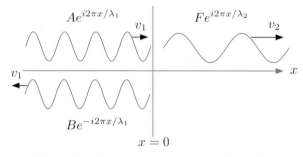

Fig. 4.6 A cartoon of the spatial piece of the solution in (4.67). The real part of the left and right traveling pieces are shown separately, identified by their spatial dependence.

Fig. 4.7 A string has speed v_1 on the left ($x < 0$) and is attached to another string with speed v_2 on the right. An "incident" wave coming from the left gives rise to a "transmitted" wave on the right and a "reflected" wave on the left.

We can use this picture to set up the physically interesting case of a wave that moves from one medium to another. The hallmark of that change in medium will be a change in the speed of the wave. Referring to the wave equation for a string (4.23), for example, we see that the physical parameters of the string are the tension T and the mass density μ. But these always appear together in $v^2 = T/\mu$, so regardless of which you want to change, the relevant parameter describing the wave equation in each medium will be the speed v.

Staying with the string example, suppose we took two strings with different mass densities, leading to different speeds v_1 and v_2. We combine the strings at $x = 0$, and let sinusoidal waves travel from the left to the right, from the region $x < 0$ to $x > 0$, encountering the change in speed at the join point. If we think of a plane wave coming in from the left, traveling to the right, then it is natural to have, on the left, a solution with spatial dependence $Ae^{i2\pi x/\lambda_1}$ where λ_1 is the wavelength of the wave on the left string. When the wave encounters the new string at $x = 0$, we expect a piece of the wave to propagate onto the new string, so on the right, $x > 0$, we expect a right-traveling plane wave with new wavelength λ_2, $Fe^{i2\pi x/\lambda_2}$. Meanwhile, we could also have reflection at the join point,[7] a left-traveling wave occurring on the left, $x < 0$, which we denote $Be^{-i2\pi x/\lambda_1}$. A representation of this setup is shown in Figure 4.7. The two strings naturally partition space into $x < 0$ and $x > 0$, and the assumed spatial form of the solutions on either side is

$$\phi_1(x) = Ae^{i2\pi x/\lambda_1} + Be^{-i2\pi x/\lambda_1} \text{ for } x < 0$$
$$\phi_2(x) = Fe^{i2\pi x/\lambda_2} \text{ for } x > 0. \tag{4.68}$$

---

[7] Imagine that the second string has an infinite mass density, then the incoming wave will not enter the new section of string and must return along the first string.

How should we relate these two solutions? Since the wave equation is second order in space, we expect its solutions to be continuous and derivative-continuous, similar to the continuity argument we made in Section 1.7.1 (and see Problem 2.0.2). Those two continuity conditions give

$$\phi_1(0, t) = \phi_2(0, t) \qquad \frac{\partial \phi_1(x, t)}{\partial x}\bigg|_{x=0} = \frac{\partial \phi_2(x, t)}{\partial x}\bigg|_{x=0}, \tag{4.69}$$

which here read

$$A + B = F \qquad \frac{i2\pi}{\lambda_1}(A - B) = \frac{i2\pi}{\lambda_2}F. \tag{4.70}$$

Adding these two equations eliminates B, and we get a relation between F and A:

$$F = \frac{2A}{1 + \frac{\lambda_1}{\lambda_2}}. \tag{4.71}$$

Now $\lambda_1 = v_1/f$ and $\lambda_2 = v_2/f$ so that $\lambda_1/\lambda_2 = v_1/v_2$. Think about what we just did: we propagated the change in speed to a change in wavelength while assuming the frequency is the same on the left and right. Does that make sense? Sure, the frequency is fixed by whatever mechanism generated the plane wave on the left, some person moving the left edge of the string up and down rhythmically with frequency f, for example. So the frequency is fixed by the production of the waves, then the only thing that can change in the relation $v = \lambda f$ from (4.67) is the wavelength. The upshot is that we can write

$$F = \frac{2A}{1 + \frac{v_1}{v_2}} \tag{4.72}$$

to directly relate F to A and the characteristic speeds on the left and right. Going back to (4.70), we can subtract to isolate B

$$2B = \left(1 - \frac{\lambda_1}{\lambda_2}\right)F = \frac{2A\left(1 - \frac{\lambda_1}{\lambda_2}\right)}{\left(1 + \frac{\lambda_1}{\lambda_2}\right)} \longrightarrow B = \frac{A\left(1 - \frac{v_1}{v_2}\right)}{1 + \frac{v_1}{v_2}}. \tag{4.73}$$

This case of two strings is the simplest example of a wave equation with speed that has spatial dependence, and we shall see more of this type of spatially-varying wave speed in Section 4.8 and Chapter 7.

Problem 4.5.1   What happens to the F and B coefficients in the case $v_2 \to 0$? Does this make sense?

Problem 4.5.2   For a wavelength λ_1 on the left, is the wavelength on the right larger or smaller than λ_1 if $v_1 < v_2$?

Problem 4.5.3   The "reflection" coefficient is defined to be $R \equiv |B/A|^2$ and the "transmission" coefficient is $T \equiv v_1/v_2|F/A|^2$. These coefficients tell us "how much" of the incident wave is reflected, and how much passes through to the other string. Then we must have $R + T = 1$, show that this is indeed the case.

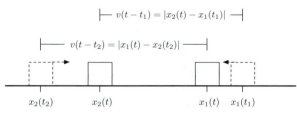

The distances involved in the relations (4.75) and (4.76). The lengths themselves are defined in terms of the location of the masses at different times, and the equations enforce the idea that "something" is traveling from one location to the other at constant speed.

4.6 Delays

One important feature of the wave equation is the finite speed v exhibited by its solutions. That speed is set by the physical environment in which the wave equation appears. But once set, all wave propagation occurs at this speed. That means that there are fundamental speed limits in place, and information of whatever sort (sound, electromagnetic radiation, pulses in a game of tug-of-war) arrive with a lag relative to their production.

If we think about a pair of masses connected by a spring, the force exerted by the spring depends on the relative locations of the masses, but it takes some time for that information to be transmitted from one particle to the other. We started with (3.1) for a pair of masses, but in order to capture the idea that signals travel at finite speed, we should have

$$m_1\ddot{x}_1(t) = k(x_2(t_2) - x_1(t) - a)$$
$$m_2\ddot{x}_2(t) = -k(x_2(t) - x_1(t_1) - a). \tag{4.74}$$

Looking at the equation for x_1, the time t_2 refers to the time at which the signal from the second mass left, arriving at the first mass at time t. The signal from the second mass travels along the spring with constant speed v (assuming a continuum spring model), so we get an implicit equation governing t_2:

$$v(t - t_2) = |x_1(t) - x_2(t_2)| \tag{4.75}$$

which we need to solve for t_2. Similarly, t_1 in (4.74) is the time at which a signal leaves the first mass, arriving at the second mass at t, so that

$$v(t - t_1) = |x_1(t_1) - x_2(t)|. \tag{4.76}$$

These algebraic relations are shown in Figure 4.8 where it is easy to see the geometry of the delay. The force acting on the first particle depends on the location of the second particle at time t_2, while the force acting on the second particle depends on the location of the first particle at time t_1 – there is no guarantee that, for example, those forces (at time t) are equal and opposite.[8]

There is no obvious analytical way to solve (4.74) together with (4.75) and (4.76), and even thinking about the solution is difficult. For example, if we specify the position and

---

[8] Indeed, Newton's third law is problematic in settings where these delays occur, which is almost all.

velocity of each particle at $t = 0$, we cannot solve the equations of motion since the "forces" depend on the motion of the masses prior to $t = 0$. You can sidestep the problem of the (extreme) increase in "initial" data that is required by imagining the masses at rest in a fixed position for all times $t < 0$. Then there will be some time $t^* = (x_2(0) - x_1(0))/v$ at which the particles begin interacting, and you still have to figure out, once the masses are moving, at what time each influences the other.

Let's simplify the problem to see the issues in a concrete setting. Take a single mass attached to a wall with a spring, and suppose that the delay is *constant*. For a time constant τ, we'll take the equation of motion to be

$$m\ddot{x}(t) = -kx(t - \tau) \tag{4.77}$$

where we have set the equilibrium position of the spring to be at zero. Before we solve this, we can look at the limiting behavior. Suppose τ is small (compared to some time scale of interest), then we can expand the equation of motion

$$m\ddot{x}(t) \approx -k(x(t) - \tau\dot{x}(t)) = -kx(t) + k\tau\dot{x}(t). \tag{4.78}$$

This looks like a damped harmonic oscillator, but with the wrong sign for the damping. The solution to this equation will involve exponential *growth*, which is problematic since we do not observe that type of behavior in nature.

Going back to the full equation of motion (4.77), we can insert the usual guess: $x(t) \sim e^{\alpha t}$ and try to find α. Taking $\omega^2 \equiv k/m$ as always, we get

$$m\alpha^2 e^{\alpha t} = -ke^{\alpha t}e^{-\alpha\tau} \longrightarrow \alpha^2 = -\omega^2 e^{\alpha\tau}, \tag{4.79}$$

a transcendental equation. We'll look at numerical solutions to these in Section 8.1.

Problem 4.6.1 A charge moves according to $\mathbf{w}(t) = vt\,\hat{\mathbf{x}} + d\,\hat{\mathbf{y}}$. Sketch the trajectory of the particle (i.e. sketch the path it traces out in time). You are at the origin, $x = y = 0$. You measure the electric field of the charge – where was the charge when it emitted the electric field that you receive at $t = 0$ (electric field information travels at the speed of light, c)?

Problem 4.6.2 For the delay differential equation

$$\frac{dx(t)}{dt} = -x(t - \tau), \tag{4.80}$$

write the approximate ODE for τ small (using Taylor expansion). Solve this ODE with $x(0) = x_0$.

Problem 4.6.3 Find an exact solution to the delay differential equation (4.80) for constant $\tau = 1$, and with "initial" value $x(0) = 1$. (Hint: look up the definition of the "product log").

Problem 4.6.4 A pair of equal and opposite charges (with the same mass) undergoes uniform circular motion via their electrostatic interaction. Assuming no delay, draw the charges at an instant in time and indicate the force directions on each. Now if there is a delay (which there is, light takes a finite amount of time to get from one charge to the other), sketch the forces on each charge at an instant in time – from your force diagram, can you have uniform circular motion in a system with delays?

4.7 Shocks

The time-of-flight required for a "signal" to go from one place to another leads to some interesting physics, even in the simplified case where the signal travels with constant speed. If you clap your hands at time $t = 0$, making a sonic "point source," the sound propagates in air at roughly constant speed. As time goes on, the signal propagates further from the source, making a sphere of influence that has radius $R(t) = vt$ centered on the source.

Now suppose that the source is moving at constant speed v_{src}, and the hand claps occur at equally spaced times, so that the center of the source changes as time goes on as shown on the left in Figure 4.9. The "x" marks the location of the source at different times, and the circles centered on the "x"'s show the sphere of influence for the hand clap (meaning, the furthest point that could hear the clap at time t). The circles get smaller moving from left to right since less time has elapsed for the points to the right of the starting position. In Figure 4.9, the source is moving faster than the sound speed, $v_{src} > v$. On the right, we see the situation in the rest frame of the source (the source is fixed, and I just shifted the centers of the circles as if they moved with constant speed $v_{src} - v$). Notice that the rest frame of the source lies *outside* all of the spheres of influence, so that the source itself hears nothing. Meanwhile, the spheres "pile up" behind the source forming a "cone" that has a lot of different fronts superimposed. This is the "sonic boom" that you hear when an airplane goes faster than the speed of sound.

Depending on the ratio of the source and sound speeds, that cone can open at different angles. In Figure 4.10, we have a source moving with constant speed that is greater than the sound speed. At time $t = 0$ the source emits a signal, then at time t the source's location together with the largest sphere of influence can be used to determine the "Mach angle," θ_m,

$$\tan \theta_m = \frac{v}{v_{src}}. \tag{4.81}$$

Problem 4.7.1 Some jets can go faster than the speed of sound. What is the angle of the Mach cone for a jet traveling at five times the speed of sound in air?

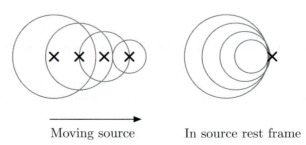

Moving source   In source rest frame

Fig. 4.9 On the left, a moving source makes a sound (hand clap) at each "x" location. The circles show how far the sound has travelled since it was generated. On the right, the same picture but in the rest frame of the source.

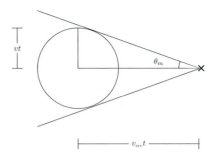

Fig. 4.10 A source moves faster than the sound speed. A "Mach cone" forms behind the source, making an angle θ_m that is determined by the ratio of the source and sound speeds.

4.8 Wave Equation with Varying Propagation Speed

Going back to the first-order wave equation developed in Section 4.3.1,

$$\frac{\partial \phi(x,t)}{\partial t} = -v\frac{\partial \phi(x,t)}{\partial x}, \tag{4.82}$$

we assumed that v, a speed, was fixed. But that speed comes to us as part of the continuum limit of identical balls connected by identical springs. If some of the springs had different spring constants, or the mass of the balls changed, we could end up with a speed v that itself depended on position (and potentially time as well). This is not an unreasonable situation to consider – in air, the speed with which acoustic disturbances propagate can change based on the local temperature, pressure, etc. How must the wave equation be modified to incorporate a spatially varying $v(x)$? (We'll return to questions like this in more detail in Chapter 7.)

We could go back to the ball and spring model, but we can also develop (4.82) by appealing to conservation of mass. Suppose we interpret $\phi(x,t)$ as a mass-per-unit-length,[9] so that in an interval between x and $x + dx$, the mass contained at time t is $\phi(x,t)dx$ for infinitesimal dx. Now in order for the mass contained in the interval to change to some new value at time $t + dt$, it must be the case that some mass entered the interval from the left (say), and some mass exited the interval on the right. That's the statement of conservation: no mass is created or destroyed in the interval. Let $v(x,t)$ be the velocity of the mass at location x, time t, then the amount of mass coming in from the left is $\phi(x,t)v(x,t)dt$ since vdt is the length of material that enters from the left over the time interval dt. The amount of mass exiting on the right over that same interval is $\phi(x + dx,t)v(x + dx,t)dt$. A sketch of the situation is shown in Figure 4.11.

The difference between the mass in the interval at time $t + dt$ and the mass contained at time t is accounted for by the contributions entering and leaving:

$$\phi(x,t+dt)dx - \phi(x,t)dx = \phi(x,t)v(x,t)dt - \phi(x+dx,t)v(x+dx,t)dt. \tag{4.83}$$

---

[9] That's just a dimensional change from its usual interpretation as "displacement of material that has equilibrium location x."

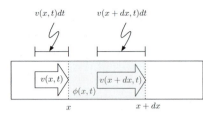

Fig. 4.11 The mass in the region $x \rightarrow x + dx$ changes because mass flows in from the left with speed $v(x, t)$ and out to the right with speed $v(x + dx, t)$.

If we Taylor expand in dt on the left, and in dx on the right, we get

$$\frac{\partial \phi(x, t)}{\partial t} dxdt = -\frac{\partial}{\partial x} (\phi(x, t)v(x, t)) \, dtdx \longrightarrow \frac{\partial \phi(x, t)}{\partial t} = -\frac{\partial}{\partial x} (\phi(x, t)v(x, t)) \,. \quad (4.84)$$

This becomes (4.82) when $v(x, t)$ is just a constant, and tells us how to account for local changes in speed.

We can again use the method of characteristics to make some progress here. Assume that the local speed, $v(x)$, is constant in time, but varies spatially. Then the conservative wave equation reads

$$\frac{\partial \phi(x, t)}{\partial t} + v(x)\frac{\partial \phi(x, t)}{\partial x} = -\phi(x, t)\frac{\partial v(x)}{\partial x} \,. \quad (4.85)$$

Suppose we consider curves $x(t)$ with $\dot{x}(t) = v(x(t))$ and $x(0) = x_0$. Along those curves,

$$\frac{d\phi(x(t), t)}{dt} = \frac{\partial \phi(x, t)}{\partial x}\bigg|_{x=x(t)} \dot{x}(t) + \frac{\partial \phi(x(t), t)}{\partial t} = -\phi(x, t)\frac{dv(x)}{dx} \,. \quad (4.86)$$

You find the curves $x(t)$ by solving a first-order ODE, $\dot{x}(t) = v(x(t))$, then you can find the value of $\phi(x, t)$ along those curves by solving the ODE (4.86) with $\phi(x(0), 0)$ given (see Section 4.8.2). This is not necessarily an easy problem to solve analytically, but it can be done numerically.

4.8.1 Conservation Laws

An equation of the form (4.84) is known as a "conservation law." These show up a lot in physics because there are many things that are not created or destroyed, and conservation laws can be used to keep track of those quantities, like charge and mass, that change because of a flow of the relevant quantity through a boundary. In the differential form of (4.84), it is hard to see this association, so we switch to the "integral" form of the conservation law.

Consider a portion of the x axis, from $a \rightarrow b$ with $a < b$. We can integrate both sides of (4.84) from a to b to get:

$$\frac{d}{dt} \int_a^b \phi(x, t) \, dx = -(\phi(b, t)v(b, t) - \phi(a, t)v(a, t)) \,. \quad (4.87)$$

On the left, we have the temporal variation of the total amount of $\phi(x, t)dx$ (the total amount of "stuff," mass in the setup from the previous section) contained between a and b. On the right, we evaluate the product ϕv at the left and right boundary. The reason the total amount of "stuff" in the interval changes is because of stuff entering on the left at a, and stuff leaving on the right at b (the signs are based on positive v pointing from a to b), as expressed by the right-hand side. Notice that if nothing is flowing, so that $v(x, t) = 0$, then nothing enters or leaves and the total amount of $\phi(x, t)$ found between a and b is just a constant.

4.8.2 Traffic Flow

There is another class of varying $v(x, t)$ that can be used to generate interesting nonlinear wave equations. Suppose $v(x, t)$ depended on $\phi(x, t)$ itself in some manner. This happens in many places, notably fluid dynamics, where the speed with which sound propagates depends on things like the density of air, which is itself changing due to wave-like motion that depends on the local speed. As a fun example (described formally in [14]), if we take $\phi(x, t)$ to be the density of cars along a road, then conservation of "number of cars" gives

$$\frac{\partial \phi(x, t)}{\partial t} + \frac{\partial}{\partial x}(v(x, t)\phi(x, t)) = 0. \tag{4.88}$$

To make a model for the local car speed $v(x, t)$, we could assume that the speed is inversely proportional to the car density itself. If there are no cars on the road, you travel at some maximum speed (the speed limit) v_{max}. If cars are bumper-to-bumper with some maximum density ϕ_{max}, the speed is zero. The simplest (linear) relationship that captures these assumptions is

$$v(x, t) = v_{max}\left(1 - \frac{\phi(x, t)}{\phi_{max}}\right), \tag{4.89}$$

but now the wave equation itself has become nonlinear (in $\phi(x, t)$)

$$\frac{\partial \phi(x, t)}{\partial t} + \frac{\partial}{\partial x}\left(v_{max}\left(1 - \frac{\phi(x, t)}{\phi_{max}}\right)\phi(x, t)\right) = 0. \tag{4.90}$$

One can still use the method of characteristics to solve this, up to a point. If we expand the derivatives, this "traffic flow" equation becomes

$$\frac{\partial \phi(x, t)}{\partial t} + v_{max}\left(1 - 2\frac{\phi(x, t)}{\phi_{max}}\right)\frac{\partial \phi(x, t)}{\partial x} = 0, \tag{4.91}$$

and we can define characteristic curves $x(t)$ that have slope

$$\dot{x}(t) = v_{max}\left(1 - 2\frac{\phi(x, t)}{\phi_{max}}\right) \tag{4.92}$$

so that (4.91) is satisfied. Think of the initial situation: we are given $\phi(x, 0)$, and that defines a local slope, for the curves governed by (4.92), that varies from one location to another. For $t \approx 0$, the characteristic curve emanating from x-location x_0 has slope

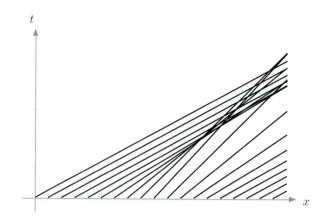

Fig. 4.12 Some characteristic curves generated by the initial data, with slope given by the inverse of (4.93) (inverse because we are plotting t vs. x here). The characteristic curves, along which $\phi(x(t), t)$ is constant, cross at some finite time. But then the solution is double-valued, and not physically relevant.

$$v_{\max} \left(1 - 2 \frac{\phi(x_0, 0)}{\phi_{\max}} \right). \tag{4.93}$$

The slope depends on both the initial values $\phi(x, 0)$ and on where the curve starts. It is possible to get slopes such that characteristic curves that start close to one another end up crossing, giving multiple values for $\phi(x, t)$ at certain points as shown in Figure 4.12. That is not allowable (you can't have both five cars per mile and two cars per mile at the same location), and the solution is to allow a discontinuity in density to form. This discontinuity is called a "shock," and represents the same sort of discontinuous behavior as sound waves piling up behind a fast-moving jet. Many nonlinear partial differential equations share this behavior. Shocks typically form when physical variables in the problem travel faster than the local speed in the wave equation.

4.8.3 Riemann Problem

One way to capture discontinuous shock-like behavior is to build it in from the start. The "Riemann problem" is defined by an initial condition that has one constant value for $x < 0$ and another constant value for $x > 0$. The initial data, then, is

$$\phi(x, 0) = \begin{cases} \phi_\ell & x < 0 \\ \phi_r & x > 0 \end{cases} = \phi_\ell + (\phi_r - \phi_\ell)\,\theta(x) \tag{4.94}$$

for constants ϕ_ℓ and ϕ_r (and step function, $\theta(x)$, defined in (2.133)).

Take the traffic flow equation in the form (4.91) as our example PDE. It is clear that constant functions satisfy the PDE, and this motivates us to guess the following piecewise constant "solution,"

$$\phi(x, t) = \phi_\ell + (\phi_r - \phi_\ell)\,\theta(x - D(t)). \tag{4.95}$$

This tentative solution takes on constant values to the left and right of some location $D(t)$. We want to find $D(t)$ such that this $\phi(x,t)$ solves (4.91). Clearly the initial condition (4.94) is satisfied as long as $D(0) = 0$. What else constrains the function separating the two sides? Think of the spatial and temporal derivatives here, using the derivative of the step function from Problem 2.6.8:

$$\frac{\partial \phi(x,t)}{\partial x} = (\phi_r - \phi_\ell)\,\delta(x - D(t)) \qquad \frac{\partial \phi(x,t)}{\partial t} = -(\phi_r - \phi_\ell)\,\delta(x - D(t))\dot{D}(t), \quad (4.96)$$

and we can relate the two derivatives

$$\frac{\partial \phi(x,t)}{\partial t} = -\dot{D}(t)\frac{\partial \phi(x,t)}{\partial x}. \tag{4.97}$$

Using this relation in (4.91), we have

$$-\dot{D}(t)\frac{\partial \phi(x,t)}{\partial x} + v_{\max}\left(1 - 2\frac{\phi(x,t)}{\phi_{\max}}\right)\frac{\partial \phi(x,t)}{\partial x} = 0, \tag{4.98}$$

or written out,

$$\dot{D}(t)(\phi_r - \phi_\ell)\,\delta(x - D(t)) = v_{\max}\left(1 - 2\frac{\phi(x,t)}{\phi_{\max}}\right)(\phi_r - \phi_\ell)\,\delta(x - D(t)). \tag{4.99}$$

If we integrate both sides of this equation in x from $x = -\infty \to \infty$, then the delta functions go away,

$$\dot{D}(t) = v_{\max}\left(1 - 2\frac{\phi(D(t),t)}{\phi_{\max}}\right). \tag{4.100}$$

We will use this equation to isolate $D(t)$, but it is worth noting that our $\phi(x,t)$ solves the integrated form of (4.91) rather than the PDE itself. Such a solution is called a "weak" solution, and these are relevant when there is a discontinuity built-in to the ansatz – the PDE itself has trouble with functions $\phi(x,t)$ that have ill-defined derivatives (here, the derivatives of $\phi(x,t)$ involve delta functions).

Back to (4.100), we have to evaluate $\phi(D(t),t) = \phi_\ell + 1/2(\phi_r - \phi_\ell)$ where the $1/2$ comes from evaluating the step function at the discontinuity.[10] Putting this in, we get a constant (in time) right-hand side

$$\dot{D}(t) = v_{\max}\left(1 - \frac{\phi_\ell + \phi_r}{\phi_{\max}}\right) \longrightarrow D(t) = v_{\max}\left(1 - \frac{\phi_\ell + \phi_r}{\phi_{\max}}\right)t \tag{4.101}$$

with $D(0) = 0$ required by the initial conditions.

The function $D(t)$ is itself a line with constant slope, and it separates the left and right values of the weak solution $\phi(x,t)$. The "shock" discontinuity can travel to the left (negative slope) or right (positive slope) depending on the relative constants ϕ_ℓ and ϕ_r (see $\dot{D}(t)$ in (4.101)). In Figure 4.13, we have the maximum density on the right, so that "cars" have to stop as they get to the more dense side, the discontinuity, then, moves *left*. The characteristic curves emanating from initial points with density ϕ_ℓ on the left and ϕ_r on the right have slope given by

---

[10] See Problem 2.6.8.

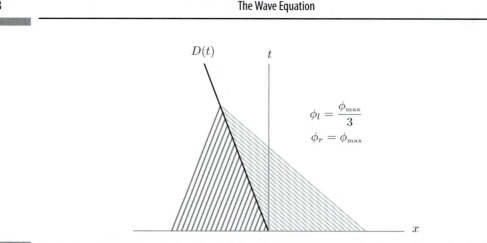

Fig. 4.13   The density on the right is maximal, so the shock discontinuity, $D(t)$, has a negative slope, and proceeds to the left. The characteristic curves, along which the value of $\phi(x, t)$ is constant, are shown for both left and right. These run right into the discontinuity curve.

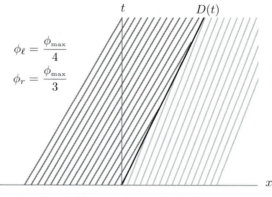

Fig. 4.14   Here, the constant densities on the left and right are tuned so that the shock discontinuity moves to the right ($D(t)$ has a positive slope). Again, the characteristic curves from the left and right run into the discontinuity.

$$v_\ell = v_{\max}\left(1 - 2\frac{\phi_\ell}{\phi_{\max}}\right)$$

$$v_r = v_{\max}\left(1 - 2\frac{\phi_r}{\phi_{\max}}\right)$$

(4.102)

from (4.93). We can also get a shock that travels to the right, and the characteristic curves and shock curve for $\phi_\ell = \phi_{\max}/4$ with $\phi_r = \phi_{\max}/3$ are shown in Figure 4.14.

There is another type of behavior we can get from this weak piecewise solution. Suppose the characteristic curves coming from densities on the left have negative slope, while characteristic curves coming from densities on the right have positive slope. Referring to (4.102), this could happen if $\phi_\ell > \phi_{\max}/2$ and $\phi_r < \phi_{\max}/2$. In Figure 4.15, we see that there is no clash of values (as happens when characteristic curves run into each other,

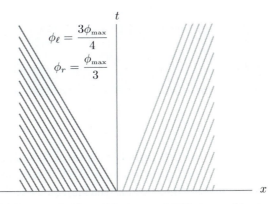

Fig. 4.15   Characteristic curves on the left have negative slope, while those on the right have positive slope. The curves do not intersect, but they do leave a "void" of values in the center.

suggesting that a given point has two different values for density, precisely the situation that is resolved by the weak solution with traveling discontinuity) but there is a growing (in time) gap in which there is no information about the density.

Problem 4.8.1   Draw the solution to the Riemann problem for traffic flow with initial data $\phi_\ell = \phi_{max}$ and $\phi_r = \phi_{max}/4$.

5 Integration

In this chapter, we will study harmonic oscillators in a setting where we need to explicitly integrate various different functions. We'll start by developing a general integral solution to the one-dimensional damped driven harmonic oscillator. Then we'll move on to problems for which a closed form trajectory is not immediately attainable, and show how to make progress characterizing motion by, for example, computing its period.

5.1 First-Order ODEs

We start with a version of integration that takes us back to our work in Section 2.1. Any ODE can be written as a set of first-order ODEs. As an example, take the harmonic oscillator problem,

$$\ddot{x}(t) = -\omega^2 x(t) \text{ with } x(0) = x_0 \text{ and } \dot{x}(0) = v_0 \text{ given.} \tag{5.1}$$

This equation can be written in terms of the vector

$$\mathbf{X}(t) \dot{\equiv} \begin{pmatrix} x(t) \\ \dot{x}(t) \end{pmatrix} \tag{5.2}$$

as

$$\frac{d}{dt} \begin{pmatrix} x(t) \\ \dot{x}(t) \end{pmatrix} = \underbrace{\begin{pmatrix} 0 & 1 \\ -\omega^2 & 0 \end{pmatrix}}_{\equiv \mathbb{M}} \begin{pmatrix} x(t) \\ \dot{x}(t) \end{pmatrix}, \tag{5.3}$$

so that we have

$$\frac{d\mathbf{X}(t)}{dt} = \mathbb{M}\mathbf{X}(t) \text{ with } \mathbf{X}(0) = \begin{pmatrix} x_0 \\ v_0 \end{pmatrix}. \tag{5.4}$$

This is, in a sense, just like $\dot{x}(t) = mx(t)$ with solution $x(t) = e^{mt}x_0$. The only difference is that $\mathbf{X}(t)$ is a vector and $\mathbb{M}$ is a matrix. Still, it is tempting to take

$$\mathbf{X}(t) = e^{\mathbb{M}t}\mathbf{X}(0) \tag{5.5}$$

and try to appropriately *define* the exponential of a matrix. The most natural definition relies on the series expansion from (1.51),

$$e^{\mathbb{M}t} = \sum_{j=0}^{\infty} \frac{t^j}{j!} \mathbb{M}^j. \tag{5.6}$$

The matrix multiplication makes it clear that the exponential of a matrix is itself a matrix of the same (square) size as $\mathbb{M}$ itself. Suppose, further, that the eigenvalues and eigenvectors of the matrix $\mathbb{M}$ are known. If the eigenvectors are the columns of a matrix $\mathbb{V}$, and the eigenvalues appear on the diagonal of the (diagonal) matrix $\mathbb{L}$, then

$$\mathbb{M} = \mathbb{V}\mathbb{L}\mathbb{V}^{-1} \tag{5.7}$$

from Section 3.3. The powers of $\mathbb{M}$ can be written in terms of $\mathbb{V}$ and $\mathbb{L}$. Let's start with $\mathbb{M}^2$ to see the pattern,

$$\mathbb{M}^2 = \mathbb{V}\mathbb{L}\mathbb{V}^{-1}\mathbb{V}\mathbb{L}\mathbb{V}^{-1} = \mathbb{V}\mathbb{L}^2\mathbb{V}^{-1}. \tag{5.8}$$

The diagonal matrix $\mathbb{L}$ has powers $\mathbb{L}^p$ that are themselves diagonal matrices with entries that are the entries of $\mathbb{L}$ raised to the pth power,

$$\mathbb{L}^p = \begin{pmatrix} \lambda_1^p & 0 & 0 & \cdots \\ 0 & \lambda_2^p & 0 & \cdots \\ \vdots & \vdots & \ddots & \vdots \\ 0 & 0 & \cdots & \lambda_n^p \end{pmatrix}. \tag{5.9}$$

Since we always pair a $\mathbb{V}$ with a $\mathbb{V}^{-1}$ (except the outermost pair), the pth power of $\mathbb{M}$ is

$$\mathbb{M}^p = \mathbb{V}\mathbb{L}^p\mathbb{V}^{-1} \tag{5.10}$$

and we can use this in (5.6)

$$e^{\mathbb{M}t} = \sum_{j=0}^{\infty} \frac{t^j}{j!} \mathbb{V}\mathbb{L}^j\mathbb{V}^{-1} = \mathbb{V}\left[\sum_{j=0}^{\infty} \frac{t^j}{j!}\mathbb{L}^j\right]\mathbb{V}^{-1}. \tag{5.11}$$

Since the matrix $\mathbb{L}^j$ is diagonal, the expression in brackets is just a set of exponentials arrayed along the diagonal of a matrix $e^{\mathbb{L}t}$, and we can simplify further,

$$e^{\mathbb{M}t} = \mathbb{V}\underbrace{\begin{pmatrix} e^{\lambda_1 t} & 0 & 0 & \cdots \\ 0 & e^{\lambda_2 t} & 0 & \cdots \\ \vdots & \vdots & \ddots & \vdots \\ 0 & 0 & \cdots & e^{\lambda_n t} \end{pmatrix}}_{=e^{\mathbb{L}t}}\mathbb{V}^{-1}. \tag{5.12}$$

The solution for $\mathbf{X}(t)$ is

$$\mathbf{X}(t) = \mathbb{V}e^{\mathbb{L}t}\mathbb{V}^{-1}\mathbf{X}(0). \tag{5.13}$$

For the matrix $\mathbb{M}$ in (5.3), we have eigenvalues $\lambda_1 = -i\omega$, $\lambda_2 = i\omega$ with un-normalized eigenvectors

$$\mathbf{v}^1 \doteq \begin{pmatrix} \frac{i}{\omega} \\ 1 \end{pmatrix} \qquad \mathbf{v}^2 \doteq \begin{pmatrix} -\frac{i}{\omega} \\ 1 \end{pmatrix}. \tag{5.14}$$

The matrices of interest for constructing $\mathbf{X}(t)$ are

$$\mathbb{V} \doteq \begin{pmatrix} \frac{i}{\omega} & -\frac{i}{\omega} \\ 1 & 1 \end{pmatrix} \qquad e^{\mathbb{L}t} \doteq \begin{pmatrix} e^{-i\omega t} & 0 \\ 0 & e^{i\omega t} \end{pmatrix} \tag{5.15}$$

and putting these into (5.13),

$$\mathbf{X}(t) = \begin{pmatrix} x_0 \cos(\omega t) + \frac{v_0}{\omega} \sin(\omega t) \\ v_0 \cos(\omega t) - \omega x_0 \sin(\omega t) \end{pmatrix} \tag{5.16}$$

as always.

To include damping, we would just take the equation of motion

$$\ddot{x}(t) = -\omega^2 x(t) - 2b\dot{x}(t) \tag{5.17}$$

and find the new matrix $\mathbb{M}$ that comes from writing (5.17) as a pair of first-order ODEs:

$$\frac{d}{dt} \begin{pmatrix} x(t) \\ \dot{x}(t) \end{pmatrix} = \underbrace{\begin{pmatrix} 0 & 1 \\ -\omega^2 & -2b \end{pmatrix}}_{\equiv \mathbb{M}} \begin{pmatrix} x(t) \\ \dot{x}(t) \end{pmatrix}, \tag{5.18}$$

which modifies the eigenvalues and eigenvectors. The two eigenvalues are $-b \pm \sqrt{b^2 - \omega^2}$, which we recognize from our last pass at the damped harmonic oscillator in Section 2.1.

5.1.1 Driven Harmonic Oscillator

To include a driving term, we start with

$$\ddot{x}(t) = -\omega^2 x(t) + f(t), \qquad x(0) = x_0 \qquad \dot{x}(0) = v_0, \tag{5.19}$$

for driving force $F(t) \equiv mf(t)$, and again write in vector form to obtain a first-order ODE

$$\frac{d}{dt} \begin{pmatrix} x(t) \\ \dot{x}(t) \end{pmatrix} = \underbrace{\begin{pmatrix} 0 & 1 \\ -\omega^2 & 0 \end{pmatrix}}_{\equiv \mathbb{M}} \begin{pmatrix} x(t) \\ \dot{x}(t) \end{pmatrix} + \underbrace{\begin{pmatrix} 0 \\ f(t) \end{pmatrix}}_{\equiv \mathbf{A}}. \tag{5.20}$$

This looks like a linear, first-order ODE with an "offset" $\mathbf{A}$. Consider a solution $\mathbf{X}(t) = e^{\mathbb{M}t}(\mathbf{X}(0) + \mathbf{Y}(t))$ for some new vector $\mathbf{Y}(t)$, then the derivative of $\mathbf{X}(t)$ is related to the derivative of $\mathbf{Y}(t)$ by

$$\dot{\mathbf{X}}(t) = \mathbb{M}\mathbf{X}(t) + e^{\mathbb{M}t}\dot{\mathbf{Y}}(t), \tag{5.21}$$

and if we put this into $\dot{\mathbf{X}}(t) - \mathbb{M}\mathbf{X}(t) = \mathbf{A}$,

$$e^{\mathbb{M}t}\dot{\mathbf{Y}} = \mathbf{A}, \tag{5.22}$$

so that we can set[1]

$$\mathbf{Y}(t) = \int_0^t e^{-\mathbb{M}\bar{t}}\mathbf{A}(\bar{t})\, d\bar{t} \tag{5.23}$$

---

[1] Note that $e^{-\mathbb{M}t}$ is the matrix inverse of $e^{\mathbb{M}t}$. The matrix $e^{\mathbb{M}t} = \mathbb{V}e^{\mathbb{L}t}\mathbb{V}^{-1}$ has inverse $e^{-\mathbb{M}t} = \mathbb{V}e^{-\mathbb{L}t}\mathbb{V}^{-1}$.

and the solution is

$$\mathbf{X}(t) = e^{\mathbb{M}t}\mathbf{X}(0) + e^{\mathbb{M}t}\left(\int_0^t e^{-\mathbb{M}\bar{t}}\mathbf{A}(\bar{t})\,d\bar{t}\right). \tag{5.24}$$

Written in terms of the eigenvalue decomposition of $\mathbb{M}$, we have

$$\mathbf{X}(t) = \mathbb{V}e^{\mathbb{L}t}\mathbb{V}^{-1}\mathbf{X}(0) + \mathbb{V}e^{\mathbb{L}t}\int_0^t e^{-\mathbb{L}\bar{t}}\mathbb{V}^{-1}\mathbf{A}(\bar{t})\,d\bar{t}. \tag{5.25}$$

Using the $\mathbb{M}$ for the harmonic oscillator (without damping) from (5.3), we can write out the solution in terms of the integral of $f(t)$

$$\mathbf{X}(t) = \begin{pmatrix} x_0\cos(\omega t) + \frac{v_0}{\omega}\sin(\omega t) \\ v_0\cos(\omega t) - \omega x_0\sin(\omega t) \end{pmatrix} + \begin{pmatrix} \frac{i}{2\omega}\left[e^{-i\omega t}\int_0^t e^{i\omega\bar{t}}f(\bar{t})\,d\bar{t} - e^{i\omega t}\int_0^t e^{-i\omega\bar{t}}f(\bar{t})\,d\bar{t}\right] \\ \frac{1}{2}\left[e^{-i\omega t}\int_0^t e^{i\omega\bar{t}}f(\bar{t})\,d\bar{t} + e^{i\omega t}\int_0^t e^{-i\omega\bar{t}}f(\bar{t})\,d\bar{t}\right] \end{pmatrix}. \tag{5.26}$$

In the language of our previous ODE solutions, the first term here is the homogeneous piece, $h(t)$, and the term with integrals is the sourced piece of the solution, denoted $\bar{x}(t)$ previously. Recall that you got the top equation using a variation of parameters approach in Problem 1.7.3.

5.1.2 Damped Driven Harmonic Oscillator

We can carry out the same setup for a damped, driven harmonic oscillator. Start with the equation of motion from (5.17) with matrix form in (5.18). You will calculate the matrix exponential in Problem 5.1.4, and get

$$e^{\mathbb{M}t} = e^{-bt}\begin{pmatrix} \cos(\bar{\omega}t) + \frac{b}{\bar{\omega}}\sin(\bar{\omega}t) & \frac{1}{\bar{\omega}}\sin(\bar{\omega}t) \\ -\frac{\omega^2}{\bar{\omega}}\sin(\bar{\omega}t) & \cos(\bar{\omega}t) - \frac{b}{\bar{\omega}}\sin(\bar{\omega}t) \end{pmatrix} \tag{5.27}$$

$$\bar{\omega} \equiv \sqrt{\omega^2 - b^2}.$$

Using this matrix exponential in (5.24), we can write out the integral solution to the damped, driven harmonic oscillator problem for an arbitrary forcing function. To keep things simple, let's just record the position as a function of time,

$$x(t) = e^{-bt}\bigg[x_0\left(\cos(\bar{\omega}t) + \frac{b}{\bar{\omega}}\sin(\bar{\omega}t)\right) + \frac{v_0}{\bar{\omega}}\sin(\bar{\omega}t)$$

$$+ \frac{\sin(\bar{\omega}t)}{\bar{\omega}}\int_0^t \left(\cos(\bar{\omega}\bar{t}) + \frac{b}{\bar{\omega}}\sin(\bar{\omega}\bar{t})\right)e^{b\bar{t}}f(\bar{t})\,d\bar{t} \tag{5.28}$$

$$- \frac{1}{\bar{\omega}}\left(\cos(\bar{\omega}t) + \frac{b}{\bar{\omega}}\sin(\bar{\omega}t)\right)\int_0^t \sin(\bar{\omega}\bar{t})e^{b\bar{t}}f(\bar{t})\,d\bar{t}\bigg].$$

We can try the familiar case of a single oscillatory driving frequency, as we did back in Section 2.5. This time, we are getting the solution directly by integrating a bunch of exponentials (with single-frequency driving, the driving force itself can be written in exponential form). An example is shown in Figure 5.1 where we can see both $x(t)$

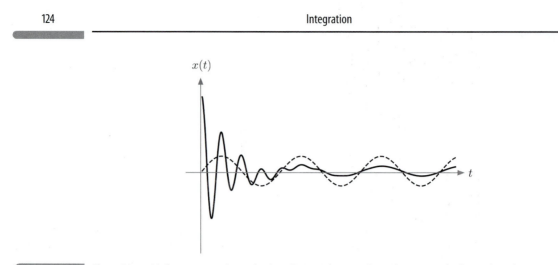

$x(t)$

Fig. 5.1   The position, $x(t)$, for a mass moving under the influence of a spring force, damping, and a driving force that is sinusoidal with frequency that is not associated with the natural spring frequency. The driving force is shown as the dashed curve so that you can see the transient solution decay, leaving an oscillatory piece with frequency that matches the driving frequency.

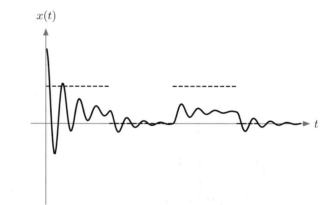

$x(t)$

Fig. 5.2   A square wave driving force (shown as dashed lines) and the position as a function of time of a damped oscillator moving under the influence of the square driver.

for some set of ω, b and initial conditions, with the driving force overlaid. In that case, the transient piece, associated with the initial conditions (i.e. the homogenous solution) dies out giving back the driven part of the solution which shares the driving frequency.

As another example of interest, we can take a square wave driving force, and apply it together with the damped oscillator forces to get $x(t)$. An example here would be an RLC circuit that is driven by a square wave voltage from a function generator. In Figure 5.2 we can see the curve of $x(t)$ together with the square wave driving function. There is, again, a notion of transient decay. As time goes on, the damped oscillator responds only to the square wave force that periodically "pings" the oscillator which then decays due to damping until another square wave pulse hits it.

Problem 5.1.1 Show, from its definition, that the matrix exponential of a matrix $\mathbb{A} \in \mathbb{R}^{n \times n}$ has:

$$\frac{d}{dt} e^{\mathbb{A}t} = \mathbb{A}e^{\mathbb{A}t} = e^{\mathbb{A}t}\mathbb{A}.$$

Problem 5.1.2 Given a function $h(\bar{t})$, define the integral:

$$I(t) = \int_a^t h(\bar{t}) \, d\bar{t}$$

for constant $a < t$. Evaluate the derivative: $\frac{dI(t)}{dt}$.

Problem 5.1.3 We are most interested in $x(t)$ in a solution like (5.26), and since $\dot{x}(t)$ is recoverable from $x(t)$, there is no need to record the second entry in $\mathbf{X}(t)$. Show that $\dot{x}(t)$ from (5.26) is, indeed, the time derivative of the $x(t)$ term.

Problem 5.1.4 For the matrix:

$$\mathbb{A} \doteq \begin{pmatrix} 0 & 1 \\ -\omega^2 & -2b \end{pmatrix}$$

what is $e^{\mathbb{A}t}$ (write the 2×2 matrix that you get, involving exponents and the constants ω, b)?

Problem 5.1.5 A charged particle of mass m moves under the influence of an external force $F(t)$. Newton's second law reads:

$$m\ddot{x}(t) = F(t) + m\tau\dddot{x}(t) \tag{5.29}$$

where τ is a constant. Write this third-order differential equation as three first-order ones.

Problem 5.1.6 For (5.29), find the general integral solution analogous to (5.26) given $F(t)$. For initial conditions, use $x(0) = x_0$, $\dot{x}(0) = v_0$, and $\ddot{x}(0) = a_0$.

Problem 5.1.7 Two masses (m) are connected by (identical) springs (spring constant k, equilibrium length a). They move on a frictionless table under the influence of both the springs and a constant force F_0 pointing to the right as shown in Figure 5.3.

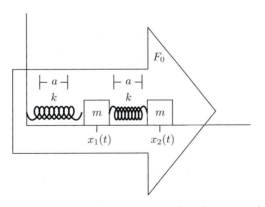

Fig. 5.3 Masses connected by identical springs, a constant force F_0 pointing to the right acts on each mass (for Problem 5.1.7).

Write down the equations of motion for the two masses. Convert this pair of second-order ODEs into four first-order ODEs. Write your four equations in the form: $\frac{d}{dt}\mathbf{Z}(t) = \mathbb{A}\mathbf{Z}(t) + \mathbf{B}$ for $\mathbf{Z}(t) \in \mathbb{R}^4$, $\mathbb{A} \in \mathbb{R}^{4\times4}$, and $\mathbf{B} \in \mathbb{R}^4$.

Problem 5.1.8 Suppose we had a particle acted on by the force $F(t) = mf_0T\delta(t)$, a "kick" delivered at time $t = 0$. Using Newton's second law, $m\ddot{x}(t) = F(t)$, find $x(t)$ for a particle that starts from rest at $x(0) = 0$. Note the discontinuity here, since the force violates the finiteness assumption we made in Problem 2.0.2.

Problem 5.1.9 Take the force $F(t) = mf_0T\delta(t - t')$ for constants m, f_0, T, and $t' > 0$, in (5.19). Find the position as a function of time, $x(t)$, for this impulsive driving force with $x(0) = 0$ and $\dot{x}(0) = 0$.

Problem 5.1.10 For the time-dependent force

$$F(t) = \begin{cases} 0 & t < 0 \\ mf_0 & 0 \leq t < T \\ 0 & t \geq T \end{cases},$$

driving a harmonic oscillator, find the position as a function of time, $x(t)$, from (5.26).

Problem 5.1.11 Using the expression in (5.26), find the position as a function of time for a driving force $F(t) = mf_0e^{i\sigma t}$, with $x(0) = x_0$, $\dot{x}(0) = v_0$, and compare with the result from Section 2.2.

Problem 5.1.12 Use the impulsive delta force, $F(t) = mf_0T\delta(t)$ in (5.28) to find the position as a function of time for a damped harmonic oscillator that gets "kicked" at $t = 0$.

Problem 5.1.13 Do the same for the force from Problem 5.1.10 to generate the first cycle in Figure 5.2 (i.e. solve for $t = 0 \rightarrow T$).

5.2 Two-Dimensional Oscillator

Suppose we have a mass m that can move in two dimensions, and is connected to a spring that is attached to the origin, but can pivot in any direction. We'll take the spring constant to be k, and set its equilibrium location to the origin. The equations of motion are just two copies of the usual one-dimensional case, one for each direction

$$m\ddot{x}(t) = -kx(t) \quad m\ddot{y}(t) = -ky(t) \tag{5.30}$$

and letting $\omega^2 \equiv k/m$, we can write the solutions

$$x(t) = A\cos(\omega t) + B\sin(\omega t) \quad y(t) = F\cos(\omega t) + G\sin(\omega t), \tag{5.31}$$

with constants A, B, F, and G waiting for initial or boundary values to pin them down. We can achieve uniform circular motion, of radius R, for the mass by taking $B = F = 0$ and setting $A = G = R$, so that

$$x(t) = R\cos(\omega t) \quad y(t) = R\sin(\omega t). \tag{5.32}$$

The vector pointing from the origin to the current (at time t) location of the mass is

$$\mathbf{r}(t) = x(t)\,\hat{\mathbf{x}} + y(t)\,\hat{\mathbf{y}} = R(\cos(\omega t)\,\hat{\mathbf{x}} + \sin(\omega t)\,\hat{\mathbf{y}}). \tag{5.33}$$

The initial conditions associated with this solution are $x(0) = R$, $\dot{x}(0) = 0$, $y(0) = 0$, and $\dot{y}(0) = \omega R$. We know everything about this solution, the perimeter of the motion is $2\pi R$, the period of the motion is $T = 2\pi/\omega$, all as usual.

What happens if we start the mass off at the same position, but give an initial velocity in the $\hat{\mathbf{y}}$ direction that is *not* carefully tuned to produce uniform circular motion? Suppose we take $\dot{y}(0) = \alpha \omega R$ for α some dimensionless constant – what type of motion do we have? What is the perimeter of the motion in this case? What is the period?

5.2.1 Elliptical Motion

If we take the more general solution, associated with $\dot{y}(0) = \alpha \omega R$,

$$x(t) = R\cos(\omega t) \quad y(t) = \alpha R \sin(\omega t), \tag{5.34}$$

the vector pointing from the origin to a point along the curve at time t is

$$\mathbf{r}(t) = R(\cos(\omega t)\,\hat{\mathbf{x}} + \alpha\,\sin(\omega t)\,\hat{\mathbf{y}}), \tag{5.35}$$

and along this one-dimensional curve, the relationship between the x and y coordinates is

$$\frac{x^2}{R^2} + \frac{y^2}{\alpha^2 R^2} = 1. \tag{5.36}$$

An ellipse is defined as the set of points satisfying

$$\frac{x^2}{a^2} + \frac{y^2}{b^2} = 1. \tag{5.37}$$

The constants a and b set the horizontal and vertical extent of the ellipse. The smaller of the two is called the "semi-minor" axis, and the larger is the "semi-major" axis. Comparing (5.36) with this general form, it is clear that we have an ellipse with semi-minor axis R and semi-major axis $\alpha R > R$ (for $\alpha > 1$). A sketch is shown in Figure 5.4.

5.2.2 Perimeter

For the elliptical $x(t)$ and $y(t)$ from (5.34), we have $\mathbf{r}(t) \equiv x(t)\,\hat{\mathbf{x}} + y(t)\,\hat{\mathbf{y}}$, and the tangent vector to the curve is $d\boldsymbol{\ell} = \dot{\mathbf{r}}(t)dt$ as shown in Figure 5.5. We want to find the length of the tangent to the curve as t goes from t to $t + dt$,

$$d\ell = \sqrt{\dot{\mathbf{r}}(t) \cdot \dot{\mathbf{r}}(t)}\,dt = \sqrt{(\omega R)^2 \sin^2(\omega t) + (\alpha \omega R)^2 \cos^2(\omega t)}\,dt. \tag{5.38}$$

Then the total distance traveled is obtained by integrating (over one full cycle, here)

$$L \equiv \oint d\ell = \omega R \int_0^T \sqrt{\sin^2(\omega t) + \alpha^2 \cos^2(\omega t)}\,dt. \tag{5.39}$$

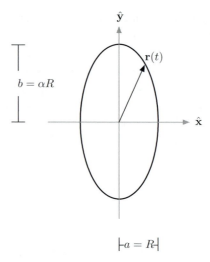

Fig. 5.4 An ellipse, centered at the origin, with semi-major axis b and semi-minor axis a. Points on the ellipse have x and y coordinates that satisfy (5.37), with the values of a and b coming from (5.36).

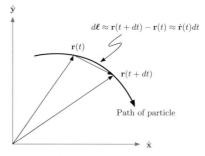

Fig. 5.5 A particle moves along a path in two dimensions. Between time t and $t + dt$, the particle moves from $\mathbf{r}(t)$ to $\mathbf{r}(t + dt)$ and the infinitesimal displacement over that time interval is $d\boldsymbol{\ell} = \dot{\mathbf{r}}(t)dt$.

We know the time it takes for the particle to return to its starting location/velocity, that's just the usual period $T = 2\pi/\omega$, so we are interested in the integral

$$L = \omega R \int_0^{\frac{2\pi}{\omega}} \sqrt{\sin^2(\omega t) + \alpha^2 \cos^2(\omega t)}\, dt. \qquad (5.40)$$

If $\alpha = 1$, we should recover uniform circular motion, and we can check the perimeter in this case:

$$L = \omega R \int_0^{\frac{2\pi}{\omega}} dt = 2\pi R \qquad (5.41)$$

as expected.

Going back, we want to evaluate the integral in (5.40) for arbitrary α. If we change variables to $p \equiv \omega t$, then the length becomes

$$L = \alpha R \int_0^{2\pi} \sqrt{1 - \beta \sin^2(p)}\, dp \qquad (5.42)$$

with $\beta \equiv 1 - 1/\alpha^2$. This equation for the total length along the curve does not make any reference to the angular frequency. That is sensible, since it's the distance travelled in one full cycle that we are probing, not how long it took the particle to traverse that distance. For a circular orbit, we have $\beta = 0$ and recover the usual circumference. The integral itself is called an "elliptic integral of the second kind," and is defined by

$$E(\phi, \beta) \equiv \int_0^\phi \sqrt{1 - \beta \sin^2(p)} \, dp, \tag{5.43}$$

so that the total length in (5.42) is

$$L = \oint d\ell = \alpha R E(2\pi, \beta) = \alpha R E\left(2\pi, 1 - \frac{1}{\alpha^2}\right). \tag{5.44}$$

Approximation

If we have $\beta \ll 1$, we can expand the integrand of $E(\phi, \beta)$,

$$\sqrt{1 - \beta \sin^2(p)} \approx 1 - \frac{1}{2} \sin^2(p)\beta - \frac{1}{8} \sin^4(p)\beta^2 - \frac{1}{16} \sin^6(p)\beta^3 - \frac{5}{128} \sin^8(p)\beta^4 - \cdots \tag{5.45}$$

and integrate each term by itself to get the approximation,

$$L = \oint d\ell \approx \frac{1}{\sqrt{1 - \beta}} R \left(2\pi - \frac{\pi}{2}\beta - \frac{3\pi}{32}\beta^2 - \frac{5\pi}{128}\beta^3 - \frac{175\pi}{8192}\beta^4 - \cdots\right), \tag{5.46}$$

and we could use $\beta \equiv 1 - 1/\alpha^2$ to write the approximation entirely in terms of α.

Problem 5.2.1   What does the parameter β do to our ellipse? What type of ellipse is described by $\beta = 0$, $\beta = 1$? Find the correction, for $\beta = 1/10$, to the circumference for a circle of radius R.

Problem 5.2.2   For the elliptical trajectories in this section, what type of motion occurs when $\alpha = 1$? Evaluate the length integral (5.44) in this case and make sure you get what you expect. What motion occurs when $\alpha = 0$? Again, evaluate the total distance travelled to make sure it makes sense (you will have to go back to (5.40) to do this).

Problem 5.2.3   Characterize the curve with $x(t) = R\cos(t)$, $y(t) = R\sin(t + \phi)$ where $\phi \in [0, \pi/2)$ is a constant and the parameter t goes from $0 \rightarrow 2\pi$.

5.3 Period of Motion

Conservation of energy is a powerful tool for characterizing the motion of a particle under the influence of some conservative force. Even if the position as a function of time cannot be obtained directly, we can use energy conservation to find the period of motion, provided we are willing to carry out some integration. We'll set up the period calculation in a spherically symmetric setting where only the distance to the origin, r, is relevant. But the same manipulations can be used to develop virtually identical expressions in a pure one-dimensional setting.[2]

Suppose we have a potential energy function in two or three dimensions that depends only on the distance to some origin. In that case the potential energy can be written $U(x,y,z) = U(r)$ with $r \equiv (x^2 + y^2 + z^2)^{1/2}$. We already have an example of this for our spring, $U(r) = \frac{1}{2}kr^2$ describes the potential energy of a spring with constant k and equilibrium at the origin. Quite generally, the equations of motion for a "spherically symmetric" potential energy of this form are

$$m\ddot{x}(t) = -U'(r)\frac{x}{r} \qquad m\ddot{y}(t) = -U'(r)\frac{y}{r} \qquad m\ddot{z}(t) = -U'(r)\frac{z}{r}. \qquad (5.47)$$

The statement of conservation of energy itself gives us a relation between the speed $v \equiv \sqrt{\dot{\mathbf{r}}(t) \cdot \dot{\mathbf{r}}(t)}$ and the potential. For a particle of mass m and total energy E, we have

$$\frac{1}{2}mv^2 + U(r) = E, \qquad (5.48)$$

and we can evaluate the speed in spherical coordinates, $\{r, \theta, \phi\}$, where (see Figure B.2)

$$x = r\sin\theta\cos\phi \qquad y = r\sin\theta\sin\phi \qquad z = r\cos\theta. \qquad (5.49)$$

Taking the time derivatives and forming $v^2 = \dot{x}^2 + \dot{y}^2 + \dot{z}^2$ in these new coordinates gives

$$\frac{1}{2}m\left(\dot{r}^2 + r^2\dot{\theta}^2 + r^2\sin^2\theta\dot{\phi}^2\right) + U(r) = E. \qquad (5.50)$$

From the rotational form of Newton's second law,

$$\frac{d\mathbf{L}}{dt} = \boldsymbol{\tau} \equiv \mathbf{r} \times \mathbf{F} \qquad (5.51)$$

with $\mathbf{F} \parallel \hat{\mathbf{r}}$ (for $\mathbf{F}$ coming from $U(r)$), we see that the torque is zero, so the angular momentum $\mathbf{L} \equiv \mathbf{r} \times \mathbf{p}$ is constant, $\dot{\mathbf{L}} = 0$. If we set the motion initially in the xy plane by taking $z = 0$ and $p_z = 0$, then $\mathbf{L}(t = 0) = L_z\hat{\mathbf{z}}$. From this initial condition, we see that the motion will *remain* in the xy plane since $\mathbf{L}(t) = \mathbf{L}(t = 0)$ for all times. We can, then, set $\theta = \pi/2$ and $\dot{\theta}(t) = 0$ (amounting to $z(t) = 0$, $\dot{z}(t) = 0$ here). In spherical coordinates, the z-component of angular momentum is

$$L_z = p_y x - p_x y = m\left(\dot{r}\sin\phi + r\cos\phi\dot{\phi}\right)r\cos\phi - m\left(\dot{r}\cos\phi - r\sin\phi\dot{\phi}\right)r\sin\phi$$

$$= mr^2\dot{\phi}$$

$$(5.52)$$

giving us an expression for $\dot{\phi}$ in terms of the constant L_z: $\dot{\phi} = L_z/(mr^2)$.

With these dynamical observations in place, conservation of energy reads

$$\frac{1}{2}m\dot{r}^2 + \frac{L_z^2}{mr^2} + U(r) = E. \qquad (5.53)$$

---

[2] Indeed, systems in which the angles of spherical coordinates do not appear, like spherically symmetric ones, are almost one-dimensional. The difference is that in one dimension, we have a variable $x: -\infty \to \infty$ whereas the radial coordinate is confined to the half-line, $r: 0 \to \infty$.

Suppose we start at r_0 with $\dot{r} = 0$. Then the constant value of E is just

$$E = \frac{L_z^2}{mr_0^2} + U(r_0), \tag{5.54}$$

and we can solve (5.53) for $\dot{r}$ up to sign:

$$\dot{r} = \pm\sqrt{\frac{2}{m}}\sqrt{\frac{L_z^2}{m}\left(\frac{1}{r_0^2} - \frac{1}{r^2}\right) + U(r_0) - U(r)}. \tag{5.55}$$

For periodic motion, we expect a return to the starting radius in some amount of time. That means the sign of $\dot{r}$ has to change (the mass goes in towards the center, then turns around and comes back, for example). We can find the time it takes to make it to that change of sign by integrating (5.55), assuming an initially negative value for $\dot{r}$,

$$\sqrt{\frac{m}{2}}\int_{r_0}^{R}\left[\frac{L_z^2}{m}\left(\frac{1}{r_0^2} - \frac{1}{r^2}\right) + U(r_0) - U(r)\right]^{-1/2} dr = -\int_0^{\bar{T}} dt = -\bar{T}, \tag{5.56}$$

where R is the value for which $\dot{r}$ is zero as determined by (5.55), and $\bar{T}$ is how long it takes to get to R. We are looking at half the total period, then. There are other values of R we could use, and those select different fractions of the period, but it's hard to know what fraction if we don't already have a good sense of the motion.

5.3.1 Harmonic Oscillator Period

Let's think about the harmonic oscillator again in this new language. For a one-dimensional oscillator, we would set $L_z = 0$, no angular momentum, and take $U(r) = kr^2/2$, with r playing the role of x here. The left-hand side of (5.56) is

$$\sqrt{\frac{m}{2}}\int_{r_0}^{R}\left[\frac{1}{2}k(r_0^2 - r^2)\right]^{-1/2} dr = \frac{\sqrt{m}}{\sqrt{kr_0}}\int_{r_0}^{R}\frac{1}{\sqrt{1 - \left(\frac{r}{r_0}\right)^2}} dr. \tag{5.57}$$

We can evaluate the integral by letting $x \equiv r/r_0$,

$$\frac{\sqrt{m}}{\sqrt{kr_0}}\int_{r_0}^{R}\frac{1}{\sqrt{1 - \left(\frac{r}{r_0}\right)^2}} dr = \sqrt{\frac{m}{k}}\int_{1}^{R/r_0}\frac{1}{\sqrt{1 - x^2}} dx. \tag{5.58}$$

Integrals that involve terms like $1 - x^2$ benefit from trigonometric substitution (since $\cos^2\theta + \sin^2\theta = 1$) as we shall see in Section 5.4. If we take $x = \sin\theta$, then

$$\int_{1}^{R/r_0}\frac{1}{\sqrt{1 - x^2}} dx = \int_{\sin^{-1}(1)}^{\sin^{-1}(R/r_0)} d\theta = \sin^{-1}\left(\frac{R}{r_0}\right) - \sin^{-1}(1) \tag{5.59}$$

and putting this back into (5.56), we have

$$\sqrt{\frac{m}{k}}\left(\sin^{-1}\left(\frac{R}{r_0}\right) - \frac{\pi}{2}\right) = -\bar{T}. \tag{5.60}$$

Finally, we need to know at what value of R does $\dot{r} = 0$? We can solve for R algebraically by setting $\dot{r} = 0$ in (5.55),

$$0 = \sqrt{\frac{2}{m}} \sqrt{\frac{1}{2} k (r_0^2 - R^2)} \tag{5.61}$$

so that $R = -r_0$ is the first solution that occurs, and then

$$\bar{T} = -\sqrt{\frac{m}{k}} \left(\sin^{-1}(-1) - \frac{\pi}{2} \right) = \pi \sqrt{\frac{m}{k}}. \tag{5.62}$$

This is, of course, only half of the full period, it takes just as long to get back to r_0 from $-r_0$, so the period is $T = 2\bar{T} = 2\pi \sqrt{m/k}$ as usual.

It is interesting to see that the period has emerged without reference to the oscillatory solution itself. We don't need to know the solution in order to find the period. We also note the lack of dependence on the initial extension, a famous property of (non-relativistic) simple harmonic oscillators.

5.3.2 Pendulum Period

Another physical system that reduces to simple harmonic motion in some appropriate limit is the "real" pendulum. We've already treated the pendulum in the small-angle approximation where the motion is approximately harmonic in Section 1.3. Referring to Figure 1.7, the energy, from (1.37) is

$$E = \frac{1}{2} m \left(L \dot{\theta} \right)^2 + mgL(1 - \cos \theta). \tag{5.63}$$

Initially, we'll let the pendulum bob go from rest at an angle of θ_0 so that $E = mgL(1 - \cos \theta_0)$ is the total energy. Then we can find $\dot{\theta}$ in terms of θ using (5.63),

$$\dot{\theta}^2 = \frac{2g}{L} (\cos \theta - \cos \theta_0). \tag{5.64}$$

We'll again consider just a portion of the motion, from $\theta = \theta_0 \rightarrow 0$ (corresponding to a quarter of the full period), where $\dot{\theta} \leq 0$ for the whole integration. Using (5.64) to generate the analogue of (5.56), we have

$$\sqrt{\frac{L}{2g}} \int_{\theta_0}^{0} \frac{d\theta}{\sqrt{\cos \theta - \cos \theta_0}} = -\bar{T}. \tag{5.65}$$

Now for the change of variables. If we could turn $\cos \theta$ into $-\sin^2 \phi$, we would be in more familiar territory. Let[3] $\theta = 2\phi$, then

$$\begin{aligned}
\bar{T} &= \sqrt{\frac{L}{2g}} \int_0^{\theta_0/2} \frac{2d\phi}{\sqrt{1 - 2\sin^2 \phi - \cos \theta_0}} \\
&= \sqrt{\frac{L}{2g}} \int_0^{\theta_0/2} \frac{2d\phi}{\sqrt{1 - \cos \theta_0} \sqrt{1 - \frac{2}{1 - \cos \theta_0} \sin^2 \phi}} \\
&= \sqrt{\frac{2L}{g(1 - \cos \theta_0)}} \int_0^{\theta_0/2} \frac{d\phi}{\sqrt{1 - \csc^2(\theta_0/2) \sin^2 \phi}}
\end{aligned} \tag{5.66}$$

---

[3] So that we can use the identity $\cos(2\phi) = 1 - 2\sin^2 \phi$.

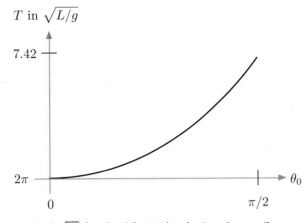

T in $\sqrt{L/g}$

Fig. 5.6   The pendulum period (in units of $\sqrt{L/g}$) from (5.68) for initial angles $\theta_0 = 0 \rightarrow \pi/2$.

The integral in (5.66) is again a named one, it is called the "elliptic integral of the first kind," and defined by

$$F(\phi, \beta) = \int_0^\phi \frac{1}{\sqrt{1 - \beta \sin^2(p)}} \, dp, \tag{5.67}$$

where the integrand is just the inverse of the integrand in (5.43). In terms of the elliptic integral, the period of the pendulum, $T = 4\bar{T}$ is

$$T = 4\sqrt{\frac{2L}{g(1 - \cos\theta_0)}} F\left(\frac{\theta_0}{2}, \csc^2\left(\frac{\theta_0}{2}\right)\right). \tag{5.68}$$

We can plot the period as a function of initial angle θ_0, that is shown in Figure 5.6, where T is in units of $\sqrt{L/g}$. The period starts at 2π (associated with the period of the simple pendulum at all starting angles) in those units, and rises to ≈ 7.42 at $\theta_0 = \pi/2$.

Problem 5.3.1   What is the period of a pendulum of length 1 m that starts from rest at an initial angle $\theta_0 = \pi/4$? What is the "simple" pendulum period in this case? What percent error does one make in using the simple pendulum period?

5.4 Techniques of Integration

In the last section, we encountered two different types of integration. In the case of the harmonic oscillator period, the integral was familiar once a trigonometric substitution was in place, and we could evaluate the integral in terms of simple functions that we recognized. In the case of the period of a pendulum, the integral itself was defined to be an "elliptic integral," and could not be simplified into familiar functions. In this section, we'll review some standard integral substitutions that can help in the former case.

Given an integral to evaluate,

$$I = \int_a^b f(x)\, dx, \tag{5.69}$$

where the function $f(x)$ and the limits a and b are given, what tools are available to us? Most one-dimensional simplifications start (and many end) with a "change of variables": Take some function of x (inspired by the integrand) $g(x)$ and let $y = g(x)$ relate x to the new variable y. Then

$$dy = \frac{dg(x)}{dx} dx \longrightarrow dx = \frac{dy}{\frac{dg(x)}{dx}}. \tag{5.70}$$

If we define the values $y_a \equiv g(a)$ and $y_b \equiv g(b)$, then we can write the integral as

$$I = \int_{y_a}^{y_b} f(x) \frac{dy}{\frac{dg(x)}{dx}} \tag{5.71}$$

with the understanding that we must replace any x appearing in the integrand with its expression in terms of y, obtained by inverting the defining relationship $y = g(x)$. The goal of this type of approach is to find a function $g(x)$ that makes the integral easy to carry out in the new variable y. This procedure is what took us from the integral in (5.58) to the significantly simpler one in (5.59).

As an example, take $f(x) = 1/(x + c)^\alpha$ for some constant c and $\alpha \neq 1$. We'll let $g(x) = x+c$, simplifying the denominator of $f(x)$. Then $dy = dx$ and we need the "inverse," $y = x+c \rightarrow x = y-c$, easy in this case. The limit points become $y_a = a+c$ and $y_b = b+c$, and using these in (5.71) gives

$$I = \int_{a+c}^{b+c} y^{-\alpha}\, dy = \frac{1}{-\alpha + 1} y^{-\alpha+1} \Big|_{y=a+c}^{y=b+c} = \frac{1}{1-\alpha}\left(\frac{1}{(b+c)^{\alpha-1}} - \frac{1}{(a+c)^{\alpha-1}}\right). \tag{5.72}$$

For a slightly more involved example, let $f(x) = x/(x^2 + c)^\alpha$ (again with $\alpha \neq 1$), so we want to evaluate

$$I = \int_a^b \frac{x}{(x^2 + c)^\alpha}\, dx. \tag{5.73}$$

This time, take $y = g(x) \equiv x^2 + c$. The choice is motivated by the fact that $dy = 2x\,dx$ and $x\,dx$ appears in the numerator of (5.73). Then working through the same steps gives

$$I = \frac{1}{2} \int_{a^2+c}^{b^2+c} y^{-\alpha}\, dy = \frac{1}{2(1-\alpha)} y^{-\alpha+1} \Big|_{y=a^2+c}^{y=b^2+c} = \frac{1}{2(1-\alpha)}\left(\frac{1}{(b^2+c)^{\alpha-1}} - \frac{1}{(a^2+c)^{\alpha-1}}\right). \tag{5.74}$$

5.4.1 Trigonometric Substitution

There is a special class of integrals where the most useful substitution involves trigonometric functions. In general, if an integrand depends on the special combination $1 - x^2$, then substitutions of the form $y \equiv g(x) = \sin^{-1}(x)$ can be useful. We've already seen an example of this type of substitution in Section 5.3.1 with the integrand $1/\sqrt{1-x^2}$, where taking $x = \sin(y)$ gives

$$I = \int_a^b \frac{1}{\sqrt{1-x^2}} \, dx = \int_{\sin^{-1}(a)}^{\sin^{-1}(b)} \frac{1}{\cos(y)} \cos(y) \, dy = \sin^{-1}(b) - \sin^{-1}(a) \qquad (5.75)$$

as in (5.59). Of course, we have the potential problem that if a or b is larger than 1, the arcsine will return a complex value.[4]

As another example, take $f(x) = 1/(1-x^2)^{3/2}$, just one additional power in the denominator. This time, if we take $x = \sin(y)$, we get

$$I = \int_a^b \frac{1}{(1-x^2)^{3/2}} \, dx = \int_{\sin^{-1}(a)}^{\sin^{-1}(b)} \frac{1}{\cos^3(y)} \cos(y) \, dy = \int_{\sin^{-1}(a)}^{\sin^{-1}(b)} \frac{1}{\cos^2(y)} \, dy. \qquad (5.76)$$

The integral of $1/\cos^2(y)$ appearing on the far right can be simplified by noting that

$$\frac{d\tan(y)}{dy} = \frac{d}{dy}\frac{\sin(y)}{\cos(y)} = 1 + \frac{\sin^2(y)}{\cos^2(y)} = \frac{1}{\cos^2(y)} \qquad (5.77)$$

so

$$I = \int_{\sin^{-1}(a)}^{\sin^{-1}(b)} \frac{1}{\cos^2(y)} \, dy = \tan(y)\Big|_{y=\sin^{-1}(a)}^{\sin^{-1}(b)}. \qquad (5.78)$$

Before evaluating the limits, we can write $\tan(y)$ in terms of $x = \sin(y)$: $\tan(y) = x/\sqrt{1-x^2}$, and

$$I = \frac{x}{\sqrt{1-x^2}}\Big|_{x=a}^{b} = \frac{b}{\sqrt{1-b^2}} - \frac{a}{\sqrt{1-a^2}}. \qquad (5.79)$$

There are some integrals that do not clearly depend on $1-x^2$, but have integrands that can be transformed to depend on $1-x^2$. Take

$$I = \int_a^b \frac{1}{\sqrt{A+Bx^2}} \, dx \qquad (5.80)$$

for constants A and B. If we factor out the A and let $y = i\sqrt{B/A}x$, then using Problem 1.6.4 to evaluate sine with complex argument,

$$I = \int_{i\sqrt{B/A}a}^{i\sqrt{B/A}b} \frac{1}{\sqrt{A}}\frac{1}{\sqrt{1-y^2}}\frac{dy}{i\sqrt{\frac{B}{A}}} = \frac{1}{i\sqrt{B}}\left(\sin^{-1}\left(i\sqrt{\frac{B}{A}}b\right) - \sin^{-1}\left(i\sqrt{\frac{B}{A}}a\right)\right).$$

$$\qquad (5.81)$$

$$= \frac{1}{\sqrt{B}}\left(\sinh^{-1}\left(\sqrt{\frac{B}{A}}b\right) - \sinh^{-1}\left(\sqrt{\frac{B}{A}}a\right)\right).$$

The inverse trigonometric functions like arcsine and arccosine are related to logarithms, which is a sort of generalized inverse for both trigonometric and hyperbolic trigonometric

---

[4] Of course, if $|a|$ or $|b|$ is greater than one, the integrand itself will be complex for some values of x, a warning sign.

functions. It is sometimes useful to write integrals involving, for example, arcsine in terms of logarithms. To do that, note that $\log(e^{i\theta}) = i\theta$, so

$$\sin^{-1}(\sin\theta) = -i\log(e^{i\theta}) = -i\log(\cos\theta + i\sin\theta) \tag{5.82}$$

and if we let $x \equiv \sin\theta$, then

$$\sin^{-1}x = -i\log\left(\sqrt{1-x^2} + ix\right). \tag{5.83}$$

To get the hyperbolic form, you could let $\theta \to -i\eta$ in (5.82) or equivalently, start with

$$\sinh^{-1}(\sinh(\eta)) = \log(e^{\eta}) = \log(\cosh\eta + \sinh\eta), \tag{5.84}$$

let $x = \sinh(\eta)$ and use the hyperbolic relation $\cosh^2\eta - \sinh^2\eta = 1$ to get

$$\sinh^{-1}x = \log\left(\sqrt{1+x^2} + x\right). \tag{5.85}$$

Going back to our integrand of interest from (5.81), this time in indefinite integral form,

$$\int \frac{1}{\sqrt{A+Bx^2}}\,dx = \frac{1}{\sqrt{B}}\sinh^{-1}\left(\sqrt{\frac{B}{A}}x\right) + C$$
$$= \frac{1}{\sqrt{B}}\log\left[F\left(\sqrt{\frac{B}{A}}x + \sqrt{1+\frac{B}{A}x^2}\right)\right] \tag{5.86}$$

where C was a constant of integration that we then took to be $C = \log(F)$ giving a multiplicative F inside the argument of log.

5.4.2 Integration by Parts

Another important tool is integration by parts. Given two functions $f(x)$ and $g(x)$, integration by parts reads

$$\int_a^b f'(x)g(x)\,dx = f(x)g(x)\Big|_{x=a}^b - \int_a^b f(x)g'(x)\,dx \tag{5.87}$$

where primes refer to x-derivatives. The proof of the formula comes directly from the fundamental theorem of calculus and the product rule. The fundamental theorem of calculus tells us that

$$\int_a^b \frac{d}{dx}(f(x)g(x))\,dx = f(x)g(x)\Big|_{x=a}^b \tag{5.88}$$

and at the same time, using the product rule, we have

$$\int_a^b \frac{d}{dx}(f(x)g(x))\,dx = \int_a^b f'(x)g(x)\,dx + \int_a^b f(x)g'(x)\,dx. \tag{5.89}$$

Since the left-hand sides of (5.88) and (5.89) are the same, the right-hand sides must also be equal, giving

$$f(x)g(x)\Big|_{x=a}^b = \int_a^b f'(x)g(x)\,dx + \int_a^b f(x)g'(x)\,dx \tag{5.90}$$

whence (5.87) follows.

We'll apply integration by parts to an example integral,

$$I = \int_0^{2\pi} x \sin x \, dx \tag{5.91}$$

where we take $f'(x) = \sin x$, $g(x) = x$, then using (5.87)

$$I = -x \cos x \big|_{x=0}^{2\pi} - \int_0^{2\pi} (-\cos x) \, dx = -2\pi + \int_0^{2\pi} \cos x \, dx = -2\pi. \tag{5.92}$$

In the next example, we'll use integration by parts to evaluate the integral

$$J \equiv \int_a^b \frac{x^2}{(1 - x^2)^{3/2}} \, dx. \tag{5.93}$$

The sneaky trick is to start by considering an integral that we already know how to evaluate,

$$I = \int_a^b \frac{1}{\sqrt{1 - x^2}} \, dx = \sin^{-1}(a) - \sin^{-1}(b). \tag{5.94}$$

Now reimagine the integrand on the left in the context of integration by parts, take $f(x) = x$ and $g(x) = 1/\sqrt{1 - x^2}$, then $f'(x)g(x)$ is the integrand in I. Applying integration by parts,

$$I = \frac{x}{\sqrt{1 - x^2}} \Big|_{x=a}^b - \int_a^b \frac{x^2}{(1 - x^2)^{3/2}} \, dx \tag{5.95}$$

and the second term is precisely J. Using our expression for I from (5.94), we have

$$\int_a^b \frac{x^2}{(1 - x^2)^{3/2}} \, dx = \frac{b}{\sqrt{1 - b^2}} - \frac{a}{\sqrt{1 - a^2}} - \sin^{-1}(a) + \sin^{-1}(b). \tag{5.96}$$

Problem 5.4.1 For the example $f(x) = 1/(x + c)^\alpha$, why doesn't the expression in (5.72) apply when $\alpha = 1$? Evaluate the integral in that case.

Problem 5.4.2 It is often the case that we need to evaluate derivatives of inverse trigonometric functions like $\sin^{-1}(x)$. A nice way to find the expressions for these derivatives is to let $f(x) \equiv \sin^{-1}(x)$, then take the derivative of both sides of the equation $\sin(f(x)) = x$ with respect to x and use the chain rule to isolate the derivative of $f(x)$ with respect to x. Carry out this program to find the derivative of $\sin^{-1}(x)$, $\cos^{-1}(x)$, and $\tan^{-1}(x)$.

Problem 5.4.3 Sometimes an integrand's dependence on $1 - x^2$ can be obscured by decorative constants. Evaluate the integral

$$I = \int_a^b \frac{1}{\sqrt{Au^2 + B}} \, du$$

for constants A and B by first changing variables to get an integrand that clearly depends on the combination $1 - x^2$ (don't be afraid to use complex numbers in your change of variables).

Problem 5.4.4 For integrands that depend on the combination $1 - x^2$, we took $\sin(y) = x$ since $1 - x^2$ was easy to evaluate in that case. Just as easy is to take $\cos(y) = x$, what happens to the integral

$$I = \int_a^b \frac{1}{\sqrt{1 - x^2}}\, dx$$

if you use this substitution instead?

Problem 5.4.5 Write $\cos^{-1}(x)$ and $\cosh^{-1}(y)$ in terms of the logarithm function.

Problem 5.4.6 Evaluate the indefinite integrals (don't forget the constant of integration), write your results in terms of logarithms when inverse trigonometric functions appear,

$$\int \frac{1}{\sqrt{1 - x^2}}\, dx \quad \int \frac{1}{(1 - x^2)^{3/2}}\, dx \quad \int \frac{1}{\sqrt{1 + x^2}}\, dx$$

$$\int \frac{1}{(1 + x^2)^{3/2}}\, dx \quad \int \frac{x}{\sqrt{1 - x^2}}\, dx \quad \int \frac{x}{(1 - x^2)^{3/2}}\, dx$$

check your results by differentiating each and verifying that you recover the integrand.

Problem 5.4.7 Evaluate the integral

$$I = \int_a^b \frac{x^2}{\sqrt{1 - x^2}}\, dx$$

using integration by parts.

Problem 5.4.8 Evaluate the definite integral:

$$I = \int_0^1 \sqrt{1 - x^2}\, dx.$$

Problem 5.4.9 We can use integrals similar to the one-dimensional ones in this section to find areas for axially symmetric "surfaces of revolution." Given a surface extending from $z = 0$ to $z = \ell$, and with a provided function $s(z)$ that tells us the radius of the surface at a particular height z, the infinitesimal surface area element can be written, referring to Figure 5.7, as

$$da = \sqrt{dz^2 + (s'(z)dz)^2}\, 2\pi s(z) = 2\pi \sqrt{1 + s'(z)^2}\, s(z) dz$$

and the area integral for the entire surface is

$$A = 2\pi \int_0^\ell \sqrt{1 + s'(z)^2}\, s(z)\, dz.$$

Work out the function $s(z)$ for a cylinder of radius R and a sphere of radius R, and check that you get the correct surface areas in both those cases.

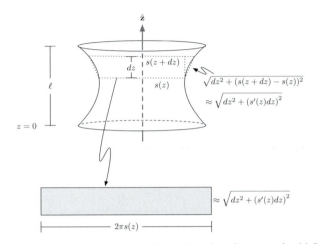

Fig. 5.7 A surface obtained by rotating a curve around the z axis. At a height z, the radius is given by $s(z)$. The strip shown on the upper figure, and blown up in the lower (by cutting it and unrolling) has infinitesimal area given by $da = 2\pi\sqrt{1 + s'(z)^2}s(z)dz$.

5.5 Relativistic Oscillator

Returning now to oscillator physics – as another modification that reduces to the simple harmonic oscillator in some limit, we can think about a mass moving at relativistic speeds (see [6] for a broader discussion). One immediate problem with simple harmonic oscillator motion is its maximum speed. Working in one dimension, conservation of energy gives (for a mass m that starts at x_0 from rest)

$$\frac{1}{2}mv^2 + \frac{1}{2}kx^2 = \frac{1}{2}kx_0^2 \longrightarrow v^2 = \frac{k}{m}\left(x_0^2 - x^2\right) \tag{5.97}$$

which is maximized at $x = 0$. The maximum speed is $\sqrt{k/m}x_0$ which could be greater than the speed of light. That is forbidden by special relativity, and the fundamental shift that avoids this problem is a redefinition of the total energy. In nonrelativistic mechanics, the total energy has a kinetic, $mv^2/2$ piece, and a potential energy $U(x)$ that is given. There, $mv^2/2 + U(x) = E$, while in special relativity, energy conservation for the same potential energy function is:

$$\frac{mc^2}{\sqrt{1 - \frac{v^2}{c^2}}} + U(x) = E. \tag{5.98}$$

This form has two important properties: (1) for $v \ll c$, it reduces to the nonrelativistic expression,

$$\frac{mc^2}{\sqrt{1 - \frac{v^2}{c^2}}} + U(x) \approx mc^2\left(1 + \frac{1}{2}\frac{v^2}{c^2}\right) + U(x) = mc^2 + \frac{1}{2}mv^2 + U(x) \tag{5.99}$$

with an undetectable constant offset (the mc^2 out front on the right) and (2) it prevents speeds greater than c. We can see this by solving (5.98) for v^2

$$v^2 = c^2 \left[1 - \left(\frac{mc^2}{E - U(x)} \right)^2 \right] \tag{5.100}$$

with $E - U(x) \geq mc^2$ so that the term we subtract from 1 is itself less than or equal to one.

Let's find the period of an oscillating mass in this relativistic setting. We can use (5.100) directly as we did in the nonrelativistic case. Suppose the mass starts from rest at an initial extension of x_0, then $E = mc^2 + (k/2)x_0^2$, and initially (just after $t = 0$) the mass has $v < 0$, so that

$$\int_{x_0}^0 \frac{dx}{c\sqrt{1 - \left(\frac{mc^2}{E - U(x)} \right)^2}} = -\int_0^{\bar{T}} dt = -\bar{T} \tag{5.101}$$

where we know that the time it takes for the mass to go from $x_0 \to 0$ is a quarter of the full period. Now for some substitutions, let $\alpha \equiv E/(mc^2)$, then

$$\bar{T} = \frac{1}{c} \int_0^{x_0} \frac{dx}{\sqrt{1 - \left(\frac{1}{\alpha - kx^2/(2mc^2)} \right)^2}}, \tag{5.102}$$

and take $q \equiv \sqrt{k/(2m)}x/c$ to get

$$\bar{T} = \sqrt{\frac{2m}{k}} \int_0^{q_0 \equiv \sqrt{\frac{k}{2m}} \frac{x_0}{c}} \frac{\alpha - q^2}{\sqrt{(\alpha - q^2)^2 - 1}} \, dq. \tag{5.103}$$

The integral is ready for trigonometric substitution, define θ by $q = \sqrt{1 + \alpha} \sin \theta$, then

$$\bar{T} = \sqrt{\frac{2m}{k}} \int_0^{\theta_0} \frac{\alpha \cos^2 \theta - \sin^2 \theta}{\sqrt{(\alpha^2 + 1) \cos^2 \theta - 2(\alpha \sin^2 \theta + 1)}} \sqrt{1 + \alpha} \, d\theta$$

$$= \sqrt{\frac{2m}{k}} \int_0^{\theta_0} \frac{\alpha - (1 + \alpha) \sin^2 \theta}{\sqrt{(\alpha - 1)(\alpha + 1)} \sqrt{1 - \frac{\alpha+1}{\alpha-1} \sin^2 \theta}} \sqrt{1 + \alpha} \, d\theta \tag{5.104}$$

with $\theta_0 \equiv \sin^{-1}(q_0/\sqrt{1 + \alpha})$. The integrand here is made up of the two pieces in the numerator. The first term is again an elliptic integral familiar from (5.67). We have another term that looks like

$$\int_0^\phi \frac{\sin^2(p)}{\sqrt{1 - \beta \sin^2(p)}} \, dp, \tag{5.105}$$

and writing the integrand's numerator as $\sin^2(p) = 1/\beta(1 - \beta \sin^2(p)) - 1/\beta$, we can write the integrand in terms of a sum

$$\int_0^\phi \frac{\sin^2(p)}{\sqrt{1 - \beta \sin^2(p)}} \, dp = \frac{1}{\beta} \int_0^\phi \sqrt{1 - \beta \sin^2(p)} \, dp - \frac{1}{\beta} \int_0^\phi \frac{1}{\sqrt{1 - \beta \sin^2(p)}} \, dp$$

$$= \frac{1}{\beta} (E(\phi, \beta) - F(\phi, \beta)). \tag{5.106}$$

Returning to (5.104) with these in place (in this setting, the generic β takes on the value $\beta = (\alpha + 1)/(\alpha - 1)$),

$$\bar{T} = \sqrt{\frac{2m}{k}} \frac{1}{\sqrt{\alpha - 1}} \left(\alpha F\left(\theta_0, \frac{\alpha + 1}{\alpha - 1}\right) + (\alpha - 1)\left(E\left(\theta_0, \frac{\alpha + 1}{\alpha - 1}\right) - F\left(\theta_0, \frac{\alpha + 1}{\alpha - 1}\right)\right)\right)$$

$$= \sqrt{\frac{2m}{k}} \frac{1}{\sqrt{\alpha - 1}} \left(F\left(\theta_0, \frac{\alpha + 1}{\alpha - 1}\right) + (\alpha - 1)\left(E\left(\theta_0, \frac{\alpha + 1}{\alpha - 1}\right)\right)\right).$$

$$(5.107)$$

Noting that $E = mc^2 + kx_0^2/2$ by the initial conditions, we have $\alpha = 1 + q_0^2$ with $q_0 \equiv \sqrt{k/(2m)}x_0/c$, then the period of the motion is

$$T = 4\bar{T} = 4\sqrt{\frac{2m}{k}} \frac{1}{q_0} \left[F\left(\sin^{-1}\left(\frac{q_0}{\sqrt{2 + q_0^2}}\right), 1 + \frac{2}{q_0^2}\right) \right.$$

$$\left. + q_0^2 \left(E\left(\sin^{-1}\left(\frac{q_0}{\sqrt{2 + q_0^2}}\right), 1 + \frac{2}{q_0^2}\right)\right)\right].$$

$$(5.108)$$

If we take the $q_0 \to 0$ limit, we recover the familiar $T = 2\pi\sqrt{m/k}$. In the other limit, as $q_0 \to \infty$, we can gain insight by thinking about the physics of the motion. If you pull the mass way way back and release it, there is a huge force initially, and a large initial acceleration. So the mass speeds up to its maximum (at c) quickly. Once the mass has gone to $-x_0$, it slows, stops, and turns around. In the ultra-relativistic limit, then, the mass moves back and forth between x_0 and $-x_0$ at the speed of light. Its period, in that setting, is just $T = 4x_0/c$, independent of k and m. In the nonrelativistic limit, the period is independent of x_0 and in the ultra-relativistic limit, it depends *only* on x_0. In between, the dependence is on x_0, k and m together in the complicated fashion expressed in (5.108), and displayed in Figure 5.8, where we plot $T(q_0)$ in units of $\sqrt{m/k}$. Beyond the information the period provides, the particle's trajectory itself can be found numerically, and you will explore that process in Section 8.2.

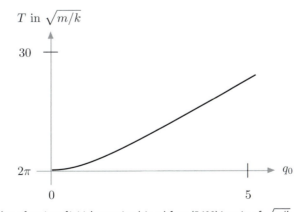

Fig. 5.8   The relativistic period as a function of initial extension (via q_0) from (5.108) in units of $\sqrt{m/k}$.

Problem 5.5.1   Show that $E - U(x) \geq mc^2$ for all x in special relativity. This property ensures that the expression on the right in (5.100) is never negative (at which point the particle would have to travel with imaginary speed).

5.6 Relativistic Lengths

In special relativity, there is a different notion of length, one that extends the usual Pythagorean sum-of-squares to include time, and enforces the constancy of the speed of light (in inertial frames moving with constant relative velocity). Suppose we have a curve with parameter λ in one spatial dimension and time, so that we are given $t(\lambda)$ and $x(\lambda)$. Then the infinitesimal "Minkowski" length, squared, is given by

$$ds^2 = -c^2 dt^2 + dx^2$$
$$= \left[-c^2 \left(\frac{dt(\lambda)}{d\lambda} \right)^2 + \left(\frac{dx(\lambda)}{d\lambda} \right)^2 \right] d\lambda^2. \tag{5.109}$$

The relative sign of the temporal and spatial-piece is important, and is a defining property of Minkowski space–time, but we could put minus signs on the spatial piece, and use a "+" for the temporal piece. The choice makes a difference in certain expressions, but not, of course, for any physical predictions.

As an example, suppose a particle travels at constant speed from the origin to location a along the x axis at time T. The motion can be written in temporal-parametrization as $x(t) = at/T$. The Pythagorean distance travelled is, of course, a. What is the relativistic length of this curve? Since we are using time to parametrize the motion, we have $\lambda = t$ and $\frac{dt(\lambda)}{d\lambda} = 1$. Then the infinitesimal length is

$$ds = \sqrt{ -c^2 + \left(\frac{dx(t)}{dt} \right)^2 } \, dt = ic\sqrt{ 1 - \frac{a^2}{c^2 T^2} } \, dt. \tag{5.110}$$

The imaginary number i shows up out front, but we could eliminate it by taking the other sign convention in (5.109). To find the total Minkowski length of the path, we integrate

$$S = ic \int_0^T \sqrt{ 1 - \frac{a^2}{c^2 T^2} } \, dt = \sqrt{a^2 - c^2 T^2} \tag{5.111}$$

and the temporal motion makes a (big) difference here.

It is, of course, possible to move through time by itself, with no spatial motion at all (that happens when, for example, you stand still). If $x(t) = 0$ for $t = 0 \rightarrow T$, we would say that the Pythagorean distance travelled is zero (remaining at the origin for the whole time of interest), but the Minkowski distance is

$$S = ic \int_0^T dt = icT, \tag{5.112}$$

which is related to the distance travelled by light in a time T. This special type of motion, through time and not space, is always achievable for massive particles, and the corresponding temporal evolution is referred to as "proper time," usually denoted τ. In special relativity, then, there are two natural types of time: (1) the coordinate time that is one of the elements of the new "spacetime" coordinate system (x, y, z, and ct) and (2) proper time, the time in the rest frame of a moving particle. The rest frame of a moving particle is an interesting concept. Think of wearing a watch on your wrist. As you move through spacetime, along whatever trajectory you like, the watch moves with you – an external observer would see it moving along with you, but relative to you, the watch is at rest (as is evidenced by the fact that it remains at the same spot on your wrist). You are the "rest frame" of the watch. This doesn't mean that you have to be at rest relative to some other observer, just that the watch is at rest relative to you. From the watch's point of view, staring at the same patch of your arm, it is moving only through time.

The relation that defines proper time mathematically is:

$$-c^2 d\tau^2 = -c^2 dt^2 + dx^2, \tag{5.113}$$

enforcing the notion that the total Minkowski distance travelled is the same in the watch's rest frame (on the left) or any other (the right-hand side, an external observer would say that the watch is moving through time and space). The presence of two different times leads to all sorts of strange behavior. One of the most famous is the "twin paradox."

The Twin Paradox

You have two people who are the same age standing on the surface of the earth. One of them blasts off, heads away from the earth, then turns around and comes back while the other stays on the earth at rest. Question: How much time has elapsed for each twin? To be concrete, suppose the twin that leaves moves according to

$$x(t) = R + a(1 - \cos(\omega t)), \tag{5.114}$$

starting at the surface of the earth, R, at $t = 0$, going out to $R + 2a$, and returning to R at $t^* = 2\pi/\omega$. These times all refer to the coordinate time, and we can find the Minkowski distance travelled by the moving twin in either the rest frame of the moving twin, which has $ds^2 = -c^2 d\tau^2$ or in the frame of the twin on earth, in which case we need to use $ds^2 = -c^2 dt^2 + dx^2$. The relation (5.113) tells us the two distances must be the same. If we use the proper time, then the integral is easy to set up:

$$S = ic \int_0^{\tau^*} d\tau = ic\tau^* \tag{5.115}$$

where τ^* is the proper time associated with the coordinate time t^*. But we don't *know* τ^*, that's not given in the problem, and the relation between t and τ is, in a sense, what we are trying to identify. So we must instead compute the total distance travelled using the known t and $x(t)$ in the earth's frame,

$$S = ic \int_0^{t^* = 2\pi/\omega} \sqrt{1 - \frac{1}{c^2}\left(\frac{dx(t)}{dt}\right)^2}\, dt \tag{5.116}$$

with $\frac{dx(t)}{dt} = a\omega \sin(\omega t)$, so that

$$S = ic \int_0^{t^* = 2\pi/\omega} \sqrt{1 - \frac{a^2\omega^2}{c^2} \sin^2(\omega t)}\, dt. \tag{5.117}$$

In this form, it is clear that an elliptic integral will play a role, and we perform the usual substitutions to get the integral to look like the right-hand side of (5.43), let $\omega t \equiv \phi$, then

$$S = \frac{ic}{\omega} \int_0^{2\pi} \sqrt{1 - \frac{a^2\omega^2}{c^2} \sin^2 \phi}\, d\phi. \tag{5.118}$$

The maximum speed of the moving twin is $a\omega$, so let $\beta_{\max} \equiv a\omega/c$,

$$S = \frac{ic}{\omega} E(2\pi, \beta_{\max}^2). \tag{5.119}$$

Now that we have the total (Minkowski) distance travelled, calculated from the earth's point of view, we can use the equality from (5.113) to find τ^*, since $S = ic\tau^*$,

$$\tau^* = \frac{1}{\omega} E(2\pi, \beta_{\max}^2) = \frac{t^*}{2\pi} E(2\pi, \beta_{\max}^2). \tag{5.120}$$

A plot of τ^* in units of t^*, as a function of the maximum speed (ratio with c) $\beta_{\max}$ is shown in Figure 5.9. For $\beta_{\max}$ small, $\tau^* \approx t^*$ as should be expected from our low-speed experience where the proper time and coordinate time are indistinguishable. But as the maximum speed of the moving twin goes up, the amount of time elapsed in its rest frame, τ^*, *goes down*. The moving twin returns to earth younger than the one that remained on the earth.

Problem 5.6.1   In space–time, the manner in which we travel through time matters. Suppose we have motion in one spatial dimension: $x(t) = at^2/T^2$ for a length a and time T. What is the Pythagorean distance travelled by a particle moving according to this $x(t)$ as t goes from $0 \to T$? What is the Minkowski distance travelled in the same time interval?

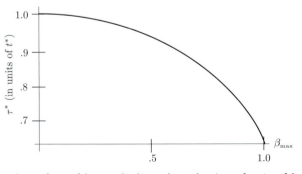

Fig. 5.9   The total time elapsed in the rest frame of the twin that leaves the earth, τ^*, as a function of the maximum speed of the moving twin ($\beta_{\max} \equiv v_{\max}/c$).

Problem 5.6.2 For uniform circular motion of radius R, $x(t) = R\cos(\omega t)$ and $y(t) = R\sin(\omega t)$. In this two-dimensional case, the infinitesimal Minkowski length is

$$ds^2 = -c^2 dt^2 + dx^2 + dy^2.$$

Using this expression find the Minkowski distance travelled in one full cycle. Compare with the "usual" distance travelled by a particle undergoing circular motion.

Problem 5.6.3 Take the time derivative of (5.98) and show that you can write the result as

$$m\ddot{x}(t) = F(x(t))\left(1 - \frac{\dot{x}(t)^2}{c^2}\right)^{3/2} \qquad (5.121)$$

given a conservative force $F(x) = -\frac{dU(x)}{dx}$. The right-hand side defines a relativistic "effective" force, what happens to it when the particle speed approaches c?

Problem 5.6.4 What is the force $F(x)$ that causes the motion (5.114) (for this relativistic problem, you must use the relativistic form of Newton's second law (5.121))?

Problem 5.6.5 Solve (5.121) for a constant force $F(x) = F_0$ given the initial conditions: $x(0) = b$ and $\dot{x}(0) = 0$.

Waves in Three Dimensions

Most of the work we have done thus far has been in one spatial dimension. It is easy to imagine a network of masses connected by springs such that oscillation can occur in multiple directions at once. In this chapter, we will move the discussion of both oscillations and waves to three dimensions. There are a number of changes that happen in this expanded setting. First, the equations of motion for springs connecting masses become nonlinear and thus much harder to solve (we cannot use the linear algebraic solutions from Chapter 3). Second, and the focus of this chapter, the wave equation has solutions that are more interesting in three dimensions, even in the static case, than in one dimension. Those higher-dimensional solutions require vector calculus, and we will develop some of the fundamental results from vector calculus slowly, taking time to make contact with problems of interest that come up as we go.

6.1 Vectors in Three Dimensions

The vector calculus we need exists in three dimensions, and to start, we'll briefly review the discussion from Section 3.1 in the specific $\mathbb{R}^3$ setting. In three dimensions, we have three orthogonal, independent directions. To each, we assign a basis vector that points in the increasing-coordinate direction and has unit length. These basis vectors are denoted $\mathbf{e}^1 \equiv \hat{\mathbf{x}}$, $\mathbf{e}^2 \equiv \hat{\mathbf{y}}$ and $\mathbf{e}^3 \equiv \hat{\mathbf{z}}$. Any three-dimensional vector can then be expressed as a linear combination of these three. As an example, $\mathbf{v} = 3\,\hat{\mathbf{x}} + 2\,\hat{\mathbf{y}} - \hat{\mathbf{z}}$ represents a vector whose tail is at the origin, and whose tip is at $x = 3$, $y = 2$ and $z = -1$.

In general, we will use variables like v_1, v_2, and v_3 (as in Section 3.1[1]) to indicate the components of $\mathbf{v} = v_1\,\hat{\mathbf{x}} + v_2\,\hat{\mathbf{y}} + v_3\,\hat{\mathbf{z}}$ (equivalently, we sometimes indicate the direction in the component name, so you will also see v_x, v_y, and v_z in $\mathbf{v} = v_x\,\hat{\mathbf{x}} + v_y\,\hat{\mathbf{y}} + v_z\,\hat{\mathbf{z}}$). All of the operations we defined in general hold in three dimensions. Given two vectors $\mathbf{v} = v_1\,\hat{\mathbf{x}} + v_2\,\hat{\mathbf{y}} + v_3\,\hat{\mathbf{z}}$ and $\mathbf{w} = w_1\,\hat{\mathbf{x}} + w_2\,\hat{\mathbf{y}} + w_3\,\hat{\mathbf{z}}$, the dot product is

$$\mathbf{v} \cdot \mathbf{w} \equiv \sum_{j=1}^{3} v_j w_j. \tag{6.1}$$

---

[1] Aside from consistency with our previous vector discussions, keeping the indices lowered allows us to unambiguously raise the components of a vector to powers. There is mathematical significance to the up/down placement of component indices, and technically, all component indices should be up, but we will have no reason to distinguish between the up/down placement in any technical sense.

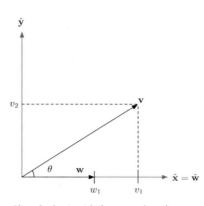

Fig. 6.1 Two vectors **v** and **w** define a plane. Align the $\hat{\mathbf{x}}$ axis with the vector $\hat{\mathbf{w}}$ as shown.

The length of a vector **v** is given by its dot product with itself,

$$v \equiv \sqrt{\mathbf{v} \cdot \mathbf{v}} = \sqrt{\sum_{j=1}^{3} (v_j)^2}. \tag{6.2}$$

The dot product can be understood as a projection of one vector onto another. Given two vectors **v** and **w**, define the unit vectors $\hat{\mathbf{v}} \equiv \mathbf{v}/v$, $\hat{\mathbf{w}} \equiv \mathbf{w}/w$. Any two nonparallel vectors define a plane, and take the plane spanned by **v** and **w** to be the xy plane. In addition, align the $\hat{\mathbf{x}}$ axis with $\hat{\mathbf{w}}$, so that we have the setup shown in Figure 6.1.

With this orientation and alignment, we have $\mathbf{w} = w_1 \hat{\mathbf{x}}$ and $\mathbf{v} = v_1 \hat{\mathbf{x}} + v_2 \hat{\mathbf{y}}$. The dot product is then

$$\mathbf{v} \cdot \mathbf{w} = v_1 w_1. \tag{6.3}$$

Referring to Figure 6.1 again, we can write $v_1 = v \cos \theta$ where θ is the angle between **v** and **w** and v is the length of **v**, as always. Then since the length of **w** is just w_1, we have

$$\mathbf{v} \cdot \mathbf{w} = vw \cos \theta \tag{6.4}$$

with $v \cos \theta$ telling us the amount of **v** that points in the $\hat{\mathbf{w}}$ direction.

In addition to the dot product, there is a special vector product found in three dimensions, the "cross product." For **v** and **w** in the usual Cartesian coordinates, the cross product is defined to be

$$\mathbf{v} \times \mathbf{w} \equiv \det \begin{pmatrix} \hat{\mathbf{x}} & \hat{\mathbf{y}} & \hat{\mathbf{z}} \\ v_1 & v_2 & v_3 \\ w_1 & w_2 & w_3 \end{pmatrix} = (v_2 w_3 - w_2 v_3) \, \hat{\mathbf{x}} - (v_1 w_3 - w_1 v_3) \, \hat{\mathbf{y}} + (v_1 w_2 - w_1 v_2) \, \hat{\mathbf{z}}. \tag{6.5}$$

That's the general definition, if we go back to the setup in Figure 6.1, we have

$$\mathbf{v} \times \mathbf{w} = -w_1 v_2 \, \hat{\mathbf{z}}. \tag{6.6}$$

This time, $v_2 = v \sin \theta$, so we can write the cross product geometrically

$$\mathbf{v} \times \mathbf{w} = -vw \sin \theta \, \hat{\mathbf{z}}. \tag{6.7}$$

The magnitude of the cross product gives the projection of $\mathbf{v}$ onto the $\hat{\mathbf{y}}$ axis, with direction given by $-\hat{\mathbf{z}}$, into the page in Figure 6.1. In general, the direction of the cross product is obtained using the right-hand rule: Point your fingers in the direction of $\mathbf{v}$, bend them in the direction of $\mathbf{w}$, then your thumb points in the direction of $\mathbf{v} \times \mathbf{w}$. From the right-hand rule it should be clear that $\mathbf{w} \times \mathbf{v} = -\mathbf{v} \times \mathbf{w}$.

Any two vectors that aren't parallel (or anti-parallel) can be used to define a set of three directions that spans the space $\mathbb{R}^3$. Given $\mathbf{v}$ and $\mathbf{w}$, which span a two-dimensional plane, we augment with $\mathbf{v} \times \mathbf{w}$, which is perpendicular to both $\mathbf{v}$ and $\mathbf{w}$, to get an orthogonal third dimension. The dot and cross product magnitudes tell us the projection of the vectors onto each other.

Problem 6.1.1 Given $\mathbf{v} = \hat{\mathbf{x}} - \hat{\mathbf{y}}$ and $\mathbf{w} = 2\,\hat{\mathbf{x}} + 3\,\hat{\mathbf{y}}$, what is the angle between $\mathbf{v}$ and $\mathbf{w}$?

Problem 6.1.2 Show that $\mathbf{A} \times (\mathbf{B} \times \mathbf{C}) = \mathbf{B}(\mathbf{A} \cdot \mathbf{C}) - \mathbf{C}(\mathbf{A} \cdot \mathbf{B})$.

Problem 6.1.3 Given any two nonparallel unit (length one) vectors, $\hat{\mathbf{a}}$ and $\hat{\mathbf{v}}$, what is the angle between $\hat{\mathbf{a}}$ and $\mathbf{b} \equiv \hat{\mathbf{v}} - (\hat{\mathbf{a}} \cdot \hat{\mathbf{v}})\hat{\mathbf{a}}$?

Problem 6.1.4 Suppose that you made a unit vector out of $\mathbf{b}$ from the previous problem, $\hat{\mathbf{b}} = \mathbf{b}/b$. Show that taking $\hat{\mathbf{c}} \equiv \hat{\mathbf{a}} \times \hat{\mathbf{b}}$, you have a three-dimensional basis (orthogonal unit vectors). What are $\hat{\mathbf{a}} \times \hat{\mathbf{c}}$ and $\hat{\mathbf{b}} \times \hat{\mathbf{c}}$?

Problem 6.1.5 Given two (nonparallel) vectors $\mathbf{v}$ and $\mathbf{w}$, show that the magnitude of the cross product, $\mathbf{v} \times \mathbf{w}$ is equal to the area of the parallelogram spanned by the two vectors (shown here).

6.2 Derivatives

For one-dimensional functions like $f(x)$, there is only one derivative to consider, $\frac{df(x)}{dx}$ defined in the usual way. In two or more dimensions, functions can depend on multiple independent arguments. Then we must be able to probe the derivative in each direction independently. For a function $f(x, y, z)$ there are three *partial* derivatives (as defined in Section 4.1). These can be usefully made into a vector of sorts, the "gradient," defined by

$$\nabla f(x, y, z) \equiv \frac{\partial f(x, y, z)}{\partial x}\,\hat{\mathbf{x}} + \frac{\partial f(x, y, z)}{\partial y}\,\hat{\mathbf{y}} + \frac{\partial f(x, y, z)}{\partial z}\,\hat{\mathbf{z}}. \tag{6.8}$$

6.2.1 Gradient

The gradient has a natural interpretation. Suppose we want to compare the value of the function f at x, y, and z to its value at a nearby point with coordinates $x + \Delta x$, $y + \Delta y$,

Fig. 6.2 A parallelogram spanned by the vectors $\mathbf{v}$ and $\mathbf{w}$. In this problem, you will show that the area of the parallelogram is the magnitude of $\mathbf{v} \times \mathbf{w}$.

$z + \Delta z$, for "small" $\{\Delta x, \Delta y, \Delta z\}$. We can Taylor expand $f(x + \Delta x, y + \Delta y, z + \Delta z)$ to first order in the small variables,

$$f(x + \Delta x, y + \Delta y, z + \Delta z) \approx f(x, y, z) + \frac{\partial f(x, y, z)}{\partial x} \Delta x + \frac{\partial f(x, y, z)}{\partial y} \Delta y + \frac{\partial f(x, y, z)}{\partial z} \Delta z.$$

(6.9)

Let $\mathbf{\Delta} \equiv \Delta x \,\hat{\mathbf{x}} + \Delta y \,\hat{\mathbf{y}} + \Delta z \,\hat{\mathbf{z}}$, then we can write the expansion as

$$f(x + \Delta x, y + \Delta y, z + \Delta z) \approx f(x, y, z) + \nabla f(x, y, z) \cdot \mathbf{\Delta}.$$

(6.10)

Now let's say you wanted to *choose* the vector $\mathbf{\Delta}$ so as to maximize the difference $f(x + \Delta x, y + \Delta y, z + \Delta z) - f(x, y, z)$. Given the dot product's geometric interpretation, we have $\nabla f \cdot \mathbf{\Delta} = |\nabla f| \Delta \cos \theta$ where θ is the angle between the gradient of f and the vector $\mathbf{\Delta}$. To maximize the difference, we take $\theta = 0$ which makes $\mathbf{\Delta} \parallel \nabla f$. The gradient, then, points in the direction of maximum increase for the function $f(x, y, z)$.

If we take the vector operator

$$\nabla \equiv \hat{\mathbf{x}} \frac{\partial}{\partial x} + \hat{\mathbf{y}} \frac{\partial}{\partial y} + \hat{\mathbf{z}} \frac{\partial}{\partial z},$$

(6.11)

which is waiting to act on functions like $f(x, y, z)$ (hence the placement of the derivatives to the right of the basis vectors), and treat it as if it were itself a vector, then we can introduce dot and cross products involving ∇.

Springs in Three Dimensions

The gradient can be used to find the spring force in three dimensions. In one dimension, the potential energy associated with a spring is $U = 1/2k(x - a)^2$ where k is the spring constant, and a is the equilibrium length. In three dimensions, the potential energy is $U = 1/2k(\sqrt{\mathbf{r} \cdot \mathbf{r}} - a)^2$ where $\mathbf{r} \equiv x\,\hat{\mathbf{x}} + y\,\hat{\mathbf{y}} + z\,\hat{\mathbf{z}}$. Then the associated force is given by the negative gradient of the potential energy,

$$\mathbf{F} = -\nabla U = -k(r - a)\,\hat{\mathbf{r}} = -k(\mathbf{r} - a\hat{\mathbf{r}}).$$

(6.12)

Notice that the force here is no longer linear in $\mathbf{r}$, the position of the particle attached to the spring, with the nonlinear $\hat{\mathbf{r}}$ as the culprit.

For some initial conditions, the behavior of the mass attached to the spring is the familiar one-dimensional, oscillatory motion. That happens when the initial velocity is zero, or parallel to the initial position vector. In that case, you may as well define that direction to be $\hat{\mathbf{x}}$ and rewrite the problem as a one-dimensional one. If the initial velocity vector is not parallel to the initial position, we can get other behaviors like the elliptical orbits of Section 5.2.1.

Another place we are forced to take a higher-dimensional view of the particle motion is when a mass is attached to a spring that is not attached to the origin. Suppose we start with a spring that has one end fixed at $\mathbf{b}$, the other end attached to a mass m at $\mathbf{r}$. The potential energy function must now depend on the difference between the magnitude of $\mathbf{r} - \mathbf{b}$ and the equilibrium length a. The energy takes the form

$$U = \frac{1}{2}k(|\mathbf{r} - \mathbf{b}| - a)^2$$

(6.13)

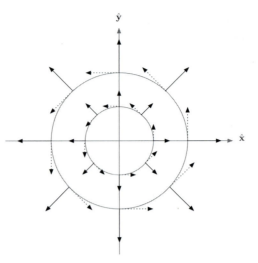

Some representative points with the vector directions of $\mathbf{r} = x\,\hat{\mathbf{x}} + y\,\hat{\mathbf{y}} + z\,\hat{\mathbf{z}}$ (solid) and $\boldsymbol{\phi} = -y\,\hat{\mathbf{x}} + x\,\hat{\mathbf{y}}$ (dotted) shown. In evaluating $\mathbf{r}$, we have set $z = 0$ to display the vectors in the xy plane.

with force

$$\mathbf{F} = -\nabla U = -k\left(\mathbf{r} - \mathbf{b} - a\frac{\mathbf{r} - \mathbf{b}}{|\mathbf{r} - \mathbf{b}|}\right), \tag{6.14}$$

which is again nonlinear, and has the direction $\mathbf{b}$ built in. The nonlinearity in $\mathbf{r}$ means that we cannot just write the equations of motion in matrix-vector form, with integral solution given by expressions analogous to those developed in Section 5.1, nor are their natural normal modes we can use to build solutions as in Section 3.4.

6.2.2 Divergence and Curl

Given a vector valued function $\mathbf{f}(x, y, z) \equiv f_x(x, y, z)\,\hat{\mathbf{x}} + f_y(x, y, z)\,\hat{\mathbf{y}} + f_z(x, y, z)\,\hat{\mathbf{z}}$ (using coordinate labels for the components of $\mathbf{f}$), we can take the dot product of ∇ with $\mathbf{f}$:

$$\nabla \cdot \mathbf{f} = \frac{\partial f_x}{\partial x} + \frac{\partial f_y}{\partial y} + \frac{\partial f_z}{\partial z}. \tag{6.15}$$

This derivative, itself a single function of position, is called the "divergence." To understand the name, and gain some intuition for its operation, consider two vector functions: $\mathbf{r} \equiv x\,\hat{\mathbf{x}} + y\,\hat{\mathbf{y}} + z\,\hat{\mathbf{z}}$ and $\boldsymbol{\phi} \equiv -y\,\hat{\mathbf{x}} + x\,\hat{\mathbf{y}}$. The vector $\mathbf{r}$ points from the origin to any point with coordinates $\{x, y, z\}$ as shown in Figure 6.3. It "diverges" from the origin with magnitude that gets larger as the points get further away from the origin. The divergence of $\mathbf{r}$ is

$$\nabla \cdot \mathbf{r} = \frac{\partial x}{\partial x} + \frac{\partial y}{\partial y} + \frac{\partial z}{\partial z} = 3 \tag{6.16}$$

a constant.[2]

[2] Notice its dimensional significance. In four dimensions with coordinates x, y, z, and w, you'd have $\mathbf{r} = x\,\hat{\mathbf{x}} + y\,\hat{\mathbf{y}} + z\,\hat{\mathbf{z}} + w\,\hat{\mathbf{w}}$ with $\nabla \cdot \mathbf{r} = 4$.

The other vector function, ϕ, also shown in Figure 6.3, points counter-clockwise along a circle of radius $s \equiv \sqrt{x^2 + y^2}$ and increases in magnitude with distance from the origin. The divergence of this function is

$$\nabla \cdot \phi = -\frac{\partial y}{\partial x} + \frac{\partial x}{\partial y} = 0. \tag{6.17}$$

The divergence operation returns zero for vectors like ϕ that "curl around" a point, and nonzero for vectors like $\mathbf{r}$ that "diverge" from a point.

If we act on our generic $\mathbf{f}(x, y, z)$ with ∇ using the cross product, then we have the "curl" of $\mathbf{f}$:

$$\nabla \times \mathbf{f} \equiv \left(\frac{\partial f_z}{\partial y} - \frac{\partial f_y}{\partial z}\right) \hat{\mathbf{x}} - \left(\frac{\partial f_z}{\partial x} - \frac{\partial f_x}{\partial z}\right) \hat{\mathbf{y}} + \left(\frac{\partial f_y}{\partial x} - \frac{\partial f_x}{\partial y}\right) \hat{\mathbf{z}}. \tag{6.18}$$

The curl of $\mathbf{r}$ is zero, $\nabla \times \mathbf{r} = 0$. The curl is insensitive to functions that "diverge" from a point. Meanwhile,

$$\nabla \times \phi = 2\,\hat{\mathbf{z}} \tag{6.19}$$

a constant.[3] The direction of the curl of ϕ is given by the right-hand rule: Curl the fingers of your right hand in the direction of the vector function, your thumb points in the direction of the curl of the vector. Since ϕ points counter-clockwise in the xy plane, the right-hand rule predicts a direction out of the page, the $\hat{\mathbf{z}}$ direction. The curl tells us the extent to which a vector function "curls around" a point (more appropriately, an infinite line going through the point perpendicular to the plane of the "curling" vector function).

Suppose we had a vector function whose magnitude depends only on the distance from the origin, and whose direction is radially outward, $\mathbf{f} = f(\sqrt{x^2 + y^2 + z^2})\,\hat{\mathbf{r}}$. We could write this as $\mathbf{f} = f(r)\,\hat{\mathbf{r}}$ since the magnitude of the vector $\mathbf{r}$ is precisely $r = \sqrt{x^2 + y^2 + z^2}$. What is the curl of this $\mathbf{f}$? Because $\mathbf{f}$ points radially away from the origin, we expect its curl to be zero, but let's check,

$$\nabla \times \mathbf{f} = \nabla \times \left(\frac{f(r)}{r}\mathbf{r}\right) = \nabla \times \left(\frac{f(r)}{r}(x\,\hat{\mathbf{x}} + y\,\hat{\mathbf{y}} + z\,\hat{\mathbf{z}})\right) \tag{6.20}$$

where we have used the fact that $\mathbf{r} = r\hat{\mathbf{r}}$ to isolate the components of $\mathbf{f}$. The x-derivative of r is

$$\frac{\partial r}{\partial x} = \frac{\partial\sqrt{x^2 + y^2 + z^2}}{\partial x} = \frac{\frac{1}{2}2x}{\sqrt{x^2 + y^2 + z^2}} = \frac{x}{r} \tag{6.21}$$

and similarly for the y and z derivatives. Using these to evaluate the right-hand side of (6.20) gives zero, as expected.

For the divergence of this specialized $\mathbf{f}$, we have

$$\nabla \cdot \mathbf{f} = \nabla \cdot \left(\frac{f(r)}{r}(x\,\hat{\mathbf{x}} + y\,\hat{\mathbf{y}} + z\,\hat{\mathbf{z}})\right) = \frac{2r^2 f(r) + r^3 f'(r)}{r^3}. \tag{6.22}$$

Suppose we put in the magnitude $f(r) = A/r^2$, for constant A, so that $\mathbf{f} = A/r^2\,\hat{\mathbf{r}}$. This example comes up a lot, and represents a highly divergent vector function, with magnitude

---

[3] Given that ϕ is defined in a plane, the 2 is ... interesting.

decreasing as the function is evaluated at points further and further from the origin. Running this through the divergence in the form (6.22) gives $\nabla \cdot \mathbf{f} = 0$. This is problematic, since the function clearly diverges, and the resolution will come in the next section.

We can also study the curl of $\mathbf{g} = g(s)\,\hat{\boldsymbol{\phi}}$ where $s \equiv \sqrt{x^2 + y^2}$. This function has no divergence, befitting a "curly" function, and the curl can be written in terms of the magnitude $g(s)$. Noting that $\hat{\boldsymbol{\phi}} = (-y\,\hat{\mathbf{x}} + x\,\hat{\mathbf{y}})/s$, we have

$$\nabla \times \mathbf{g} = \nabla \times \left(\frac{g(s)(-y\,\hat{\mathbf{x}} + x\,\hat{\mathbf{y}})}{s} \right) = \frac{s^2 g(s) + s^3 g'(s)}{s^3}\,\hat{\mathbf{z}}. \tag{6.23}$$

Now again take a special case, $g(s) = A/s$ for constant A. Using this form for $g(s)$ in (6.23) gives a curl of zero, and yet we expect the curl to be nonzero since $\mathbf{g}$ is quintessentially curly. This case complements the special case for the divergence of $\hat{\mathbf{r}}/r^2$. The major problem here is one of definition and existence, since in neither of these cases is the function, or its derivative, well-defined at the origin.

Problem 6.2.1 What is ∇r for $r \equiv \sqrt{x^2 + y^2 + z^2}$? What is the gradient of r^p?

Problem 6.2.2 Given the function $f(x, y, z) = \sqrt{x^2 + y^2}$, what is the direction of greatest increase at location $x = 1$, $y = 1$ computed "visually"? Does this agree with the direction you get from the gradient of f?

Problem 6.2.3 Given three masses, m_1, m_2, and m_3 with associated position vectors $\mathbf{x}^1(t)$, $\mathbf{x}^2(t)$, and $\mathbf{x}^3(t)$, write the equations of motion if the masses m_1 and m_2, m_2 and m_3, and m_1 and m_3 are attached by identical springs with constant k and equilibrium a.

Problem 6.2.4 Think up a vector $\mathbf{f}$ that has zero curl, nonzero divergence, and vice-versa.

Problem 6.2.5 Work out the details of the curl in (6.20) to establish that the curl really is zero.

Problem 6.2.6 Verify that for $\mathbf{g} = g(s)\,\hat{\boldsymbol{\phi}}$, $\nabla \cdot \mathbf{g} = 0$, the vector function is divergenceless.

Problem 6.2.7 What are the divergence and curl of the vector $\mathbf{v} = A(x, y, z)\mathbf{r} + B(x, y, z)\boldsymbol{\phi}$ where $A(x, y, z)$ and $B(x, y, z)$ are arbitrary functions of position, $\mathbf{r} \equiv x\,\hat{\mathbf{x}} + y\,\hat{\mathbf{y}} + z\,\hat{\mathbf{z}}$, and $\boldsymbol{\phi} \equiv -y\,\hat{\mathbf{x}} + x\,\hat{\mathbf{y}}$?

Problem 6.2.8 What is the curl of the gradient of a function $f(x, y, z)$: $\nabla \times \nabla f = ?$ Try to describe the result geometrically by thinking about the geometric role of the gradient (pointing in the direction of greatest increase) and the curl (measuring "curliness").

Problem 6.2.9 Find the divergence of the curl of a generic vector function $\mathbf{h}(x, y, z)$, $\nabla \cdot (\nabla \times \mathbf{h}) = ?$

6.3 Fundamental Theorem of Calculus for Vectors

The integral is the operation that "undoes" derivatives. The fundamental theorem of calculus provides the formal description. In one dimension, the fundamental theorem tells us that for a function $f(x)$, whose derivative is integrated between $x = a$ to b,

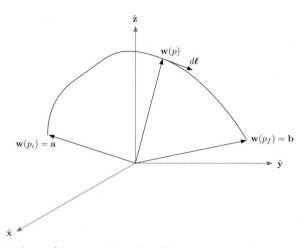

An example path given by the tip of the vector $\mathbf{w}(p)$ evaluated from $p = p_i \to p_f$. The tangent to the path, $d\boldsymbol{\ell}$, is shown at a point.

$$\int_a^b \frac{df(x)}{dx}\, dx = f(b) - f(a). \tag{6.24}$$

The integral of the derivative of $f(x)$ is related to the evaluation of $f(x)$ at the integration limits. The fundamental theorem turns an integral over a domain into an evaluation at the domain's boundary. Given the proliferation of derivatives found in vector calculus, there should be a variety of similar statements, for the integration of gradients, divergences, and curls.

6.3.1 Line Integrals

The work done by a force along a path is an example of a "line integral." You provide the force, $\mathbf{F}$, a vector function of position, and a path, described by a vector $\mathbf{w}(p)$, parametrized by[4] p, that points from the origin to locations along the path. Then the integral of interest involves the projection of the component of force along the path, accomplished by taking the dot-product of $\mathbf{F}$ with an infinitesimal tangent (to the path) vector $d\boldsymbol{\ell}$, the "line element." The work done by the force is given by

$$W = \int_\mathbf{a}^\mathbf{b} \mathbf{F} \cdot d\boldsymbol{\ell}, \tag{6.25}$$

where the starting point $\mathbf{a} = \mathbf{w}(p_i)$ and the ending point $\mathbf{b} = \mathbf{w}(p_f)$ for some initial p_i and final p_f. An example of a path, with defining $\mathbf{w}(p)$ vector and $d\boldsymbol{\ell}$ is shown in Figure 6.4.

The infinitesimal tangent vector can be written in terms of the derivative of $\mathbf{w}(p)$, since

$$d\boldsymbol{\ell} \propto \mathbf{w}(p + dp) - \mathbf{w}(p), \tag{6.26}$$

---

[4] The parameter is typically taken to be time, but that is not necessary.

and Taylor expanding on the right for small dp gives $d\ell = \frac{d\mathbf{w}(p)}{dp}dp$. Then we can write the work integral as

$$W = \int_{p_i}^{p_f} \mathbf{F}(\mathbf{w}(p)) \cdot \frac{d\mathbf{w}(p)}{dp} \, dp \qquad (6.27)$$

where $\mathbf{F}(\mathbf{w}(p))$ means evaluate the force vector at the three-dimensional location given by $\mathbf{w}(p)$.

Example

Take the force vector to be $\mathbf{F} = A\mathbf{r}$ for constant $A > 0$. This is a force that points away from the origin at all locations. For the path, we'll take the line from the origin to 1 on the x axis. The vector describing this path is $\mathbf{w}(p) = p\,\hat{\mathbf{x}}$ for $p = 0 \rightarrow 1$. The work done by the force is then

$$W = \int_0^1 \mathbf{F} \cdot \frac{d\mathbf{w}(p)}{dp} \, dp = \int_0^1 A \, dp = A. \qquad (6.28)$$

As another example, take the same force, but the path described by $\mathbf{w}(p) = A(\cos(p)\,\hat{\mathbf{x}} + \sin(p)\,\hat{\mathbf{y}} + p\,\hat{\mathbf{z}})$, with $p = 0 \rightarrow 2\pi$. This time, the work is

$$W = \int_0^{2\pi} A(-\sin(p) + \cos(p) + 1) \, dp = 2\pi A. \qquad (6.29)$$

Notice that the first two terms integrate to zero, those represent circular motion in a plane, while the $\hat{\mathbf{z}}$ component of $\mathbf{w}(p)$ goes straight up the z axis.

Conservative Forces

Given a potential energy function $U(x, y, z)$, a conservative force comes from its gradient: $\mathbf{F} = -\nabla U$. For conservative forces, the work integral simplifies considerably. Along the path, the values of x, y, and z are themselves functions of the parameter p through $\mathbf{w}(p)$, so that the total p-derivative of the potential energy function is

$$\frac{dU(\mathbf{w}(p))}{dp} = \frac{\partial U}{\partial x}\frac{dw_x(p)}{dp} + \frac{\partial U}{\partial y}\frac{dw_y(p)}{dp} + \frac{\partial U}{\partial z}\frac{dw_z(p)}{dp} = \nabla U \cdot \frac{d\mathbf{w}(p)}{dp}. \qquad (6.30)$$

The right-hand side here can be written as $-\mathbf{F} \cdot \frac{d\mathbf{w}(p)}{dp}$, just the negative of the integrand associated with the work. For conservative forces, then, the one-dimensional fundamental theorem of calculus applies,

$$W = \int_{\mathbf{a}}^{\mathbf{b}} \mathbf{F} \cdot d\ell = -\int_{p_i}^{p_f} \frac{dU}{dp} \, dp = -(U(\mathbf{w}(p_f)) - U(\mathbf{w}(p_i))) = U(\mathbf{a}) - U(\mathbf{b}), \quad (6.31)$$

and we can see that the work done by a conservative force is path-independent: only the locations of the end-points matter. As a corollary of that observation, it is clear that if the path is "closed," meaning that the starting and ending points are the same, $\mathbf{a} = \mathbf{b}$, then the work done by the force is zero.

The spring force in three dimensions from (6.12) is an example of a conservative force, coming from the potential energy function $U = 1/2k(\sqrt{\mathbf{r} \cdot \mathbf{r}} - a)^2$. The work done by this force as a particle moves from $\mathbf{f}$ to $\mathbf{g}$ (changing the labels of the initial and final points to avoid clashing with the equilibrium length a) is

$$W = \frac{1}{2}k\left[(f-a)^2 - (g-a)^2\right],$$

and, of course, if $\mathbf{f} = \mathbf{g}$, parametrizing a closed curve, $W = 0$.

6.3.2 Area and Volume Integrals

Conservative forces that come from the gradient of a potential energy function represent one type of vector function, ∇U. How about the integrals of the divergence and curl applied to vector functions? The relevant integrands and their "fundamental theorem" variants are most useful in higher-dimensional integration.

A single function of three variables, $f(x, y, z)$, can be integrated over a three-dimensional domain, call it Ω. The boundary of the domain is denoted $\partial\Omega$, and this boundary is a "closed" surface, one with a clear interior and exterior. There is a vector $\hat{\mathbf{n}}$ that points normal to the surface (and is called the unit normal), and away from the interior. For surfaces that do not have an interior and exterior, the vector $\hat{\mathbf{n}}$ points perpendicular to the surface, but there are two options (roughly, "upward" and "downward").

A generic volume integral is denoted

$$I = \int_\Omega f(x, y, z) \underbrace{dxdydz}_{\equiv d\tau}, \tag{6.32}$$

where we must be given the integrand, $f(x, y, z)$, and the domain of integration, Ω. The "volume element" $d\tau$ represents the volume of an infinitesimal box, so that we are adding up tiny box volumes weighted by the function $f(x, y, z)$. As an example, take $f(x, y, z) = x^2y^2z^2$, and for Ω, use a box of side length ℓ, centered at the origin. Then,

$$I = \int_{-\ell}^{\ell}\int_{-\ell}^{\ell}\int_{-\ell}^{\ell} x^2y^2z^2 \, dxdydz = \frac{8\ell^9}{27}, \tag{6.33}$$

performing the integrals one at a time.

Area integrals involve an infinitesimal area *vector* $d\mathbf{a} = da\,\hat{\mathbf{n}}$, the "area element." The magnitude of the area element is the area of an infinitesimal patch on the surface, but unlike the volume element, the area element has a direction, pointing perpendicular to the surface. As an example, if you have a flat surface lying in the xy plane, then the direction perpendicular to the surface is $\hat{\mathbf{n}} = \pm\hat{\mathbf{z}}$ (a flat surface has no clear interior/exterior, so there is a sign ambiguity in the unit normal). The area element is $d\mathbf{a} = dxdy\,\hat{\mathbf{z}}$, picking the $+$ sign.

There are many different types of area integral. You can integrate a function like $f(x, y, z)$ over some surface, and that integral will naturally return a vector. This can lead to some confusing results. For example, take a spherical surface, this has a unit normal vector $\hat{\mathbf{n}} = \hat{\mathbf{r}}$ which we take to point radially outward (a sphere has a clear inside and outside). Since this is a closed surface, we put a circle around the integral to indicate that we are adding up the area elements over the entire surface, and we'll refer to the surface itself as S. Consider the integral

$$\mathbf{I} = \oint_S d\mathbf{a} \tag{6.34}$$

for the surface of the sphere. Every little patch of area has an antipodal patch with area vector pointing in the opposite direction. If you add up all the area elements, you get $\mathbf{I} = 0$, they cancel in pairs.[5] We know the area of a sphere of radius R is $4\pi R^2$, not zero. But the *vector* area of a sphere is zero, because of the vector addition of infinitesimal area elements. If you take the magnitude of $d\mathbf{a}$ everywhere, you will get the expected result

$$\oint_S da = 4\pi R^2. \tag{6.35}$$

Just be careful what you are asking for.

Of the many types of area integral, one that is very useful is the integral of a vector dotted into the area element, a generalization of the line integral. Given a surface S with area element $d\mathbf{a}$ and a vector function $\mathbf{V}$, we can construct

$$\int_S \mathbf{V} \cdot d\mathbf{a} \text{ or } \oint_S \mathbf{V} \cdot da. \tag{6.36}$$

The first expression applies to open surfaces, the second to closed ones. In this type of integral, we project the vector function $\mathbf{V}$ onto the unit normal and add up over the entire surface, evaluating $\mathbf{V}$ at the surface.

To see how the evaluation works, let's take our surface to be a square of side length 2ℓ centered at the origin and sitting in the yz plane. The area element is $d\mathbf{a} = dy\,dz\,\hat{\mathbf{x}}$. For our vector function, take $\mathbf{V} = \mathbf{r} \equiv x\,\hat{\mathbf{x}} + y\,\hat{\mathbf{y}} + z\,\hat{\mathbf{z}}$. Then the integral over the surface is

$$\int_S \mathbf{r} \cdot d\mathbf{a} = \int_{-\ell}^{\ell} \int_{-\ell}^{\ell} 0\,dy\,dz = 0 \tag{6.37}$$

since the value of x at the surface is zero. If you instead took $\mathbf{V} = \boldsymbol{\phi} = -y\,\hat{\mathbf{x}} + x\,\hat{\mathbf{y}}$ and used the same surface, you'd get

$$\int_S \boldsymbol{\phi} \cdot d\mathbf{a} = \int_{-\ell}^{\ell} \int_{-\ell}^{\ell} (-y)\,dy\,dz = -2\ell \int_{-\ell}^{\ell} y\,dy = 0 \tag{6.38}$$

but this time, the integral is nontrivial even though it ultimately evaluates to zero.

As an example that does not vanish, take $\mathbf{V} = -y^2\,\hat{\mathbf{x}} + x^2\,\hat{\mathbf{y}} + z^2\,\hat{\mathbf{z}}$, for the same surface,

$$\int_S \mathbf{V} \cdot d\mathbf{a} = \int_{-\ell}^{\ell} (-y^2)\,dy\,dz = -2\ell \int_{-\ell}^{\ell} y^2\,dy = -\frac{4\ell^4}{3}. \tag{6.39}$$

6.3.3 The Divergence Theorem

Remember the structure of the fundamental theorem of calculus: The integral of a derivative of a function is equal to the function evaluated on the boundary. For the volume

---

[5] There is a "symmetry" argument we can make to establish that the vector area of a sphere is zero: Since a sphere is the same in all directions, where could such an area vector possibly point?

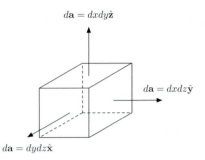

$$da = dxdy\hat{\mathbf{z}}$$

$$da = dxdz\hat{\mathbf{y}}$$

$$da = dydz\hat{\mathbf{x}}$$

Fig. 6.5   The domain Ω is a cube of side length 2ℓ centered at the origin. The boundary of the domain, $\partial\Omega$, is the surface of the cube, each face with its own outward normal and infinitesimal area element $d\mathbf{a}$. The left, back and bottom sides mirror the right, front, and top (shown).

integral in (6.32), a natural "derivative" for use as the integrand is the divergence of a vector function $\mathbf{V}$. If we integrate that over a volume Ω with closed boundary $\partial\Omega$, a reasonable guess for the form of the fundamental theorem here would be

$$\int_{\Omega} \nabla \cdot \mathbf{V} \, d\tau = \oint_{\partial\Omega} \mathbf{V} \cdot d\mathbf{a}. \qquad (6.40)$$

This expression has the correct structure: on the left, the integrand is a derivative of sorts, and on the right, we have the function $\mathbf{V}$ evaluated over the boundary of the domain.

As a plausibility argument and example, take Ω to be the interior of a cube of side length 2ℓ centered at the origin. The surface of the cube is $\partial\Omega$, and consists of the six faces, each with its own unit normal, as shown in Figure 6.5. Take $\mathbf{V} = z\hat{\mathbf{z}}$ so that $\nabla \cdot \mathbf{V} = 1$, then the volume integral on the left of (6.40) is just $(2\ell)^3$. For the surface integral on the right, we have

$$\oint_{\partial\Omega} \mathbf{V} \cdot d\mathbf{a} = \int_{\text{top}} \ell \, dxdy + \int_{\text{bottom}} (-\ell)(-1) \, dxdy = \ell(2\ell)^2 + \ell(2\ell)^2 = (2\ell)^3, \quad (6.41)$$

(the other four sides do not contribute because their surface normals are perpendicular to $\hat{\mathbf{z}}$) matching the result from the volume integral.

We won't prove the theorem rigorously, but an argument can be sketched quickly. Any vector function can be Taylor-expanded in the vicinity of a point $\mathbf{r}_0$, using (6.9) applied to each component. Take $\mathbf{V}(\mathbf{r}_0 + \mathbf{r})$ where as usual $\mathbf{r} \equiv x\hat{\mathbf{x}} + y\hat{\mathbf{y}} + z\hat{\mathbf{z}}$, and represents our probe of the vicinity of the constant $\mathbf{r}_0$, then

$$\mathbf{V}(\mathbf{r}_0 + \mathbf{r}) \approx \mathbf{V}(\mathbf{r}_0) + (\mathbf{r} \cdot \nabla V_x)\,\hat{\mathbf{x}} + (\mathbf{r} \cdot \nabla V_y)\,\hat{\mathbf{y}} + (\mathbf{r} \cdot \nabla V_z)\,\hat{\mathbf{z}}. \qquad (6.42)$$

All of the derivatives in (6.42) are just constants, derivatives of $\mathbf{V}$ evaluated at the point $\mathbf{r}_0$. We can name them to make it clear that they do not change under the integration. The coordinate dependence that we need to integrate is all carried by $\mathbf{r}$. Let

$$\mathbf{a} \equiv \nabla V_x(\mathbf{r}_0) \quad \mathbf{b} \equiv \nabla V_y(\mathbf{r}_0) \quad \mathbf{c} \equiv \nabla V_z(\mathbf{r}_0), \qquad (6.43)$$

then we can write

$$\mathbf{V}(\mathbf{r}_0 + \mathbf{r}) \approx \mathbf{V}(\mathbf{r}_0) + (\mathbf{r} \cdot \mathbf{a})\,\hat{\mathbf{x}} + (\mathbf{r} \cdot \mathbf{b})\,\hat{\mathbf{y}} + (\mathbf{r} \cdot \mathbf{c})\,\hat{\mathbf{z}}. \qquad (6.44)$$

The divergence of $\mathbf{V}(\mathbf{r}_0 + \mathbf{r})$, taking derivatives with respect to the coordinates in $\mathbf{r}$ is just

$$\nabla \cdot \mathbf{V}(\mathbf{r}_0 + \mathbf{r}) \approx a_x + b_y + c_z \tag{6.45}$$

(referring to the components of $\mathbf{a}$, $\mathbf{b}$, and $\mathbf{c}$).

Imagine making a cube of side length 2ℓ as before, but small enough so that the approximation in (6.44) holds. For simplicity, take $\mathbf{r}_0 = 0$ and assume $\mathbf{V}(0) = 0$. Then we can perform the volume and surface integrals,

$$\int_\Omega \nabla \cdot \mathbf{V}\, d\tau = (a_x + b_y + c_z)(2\ell)^3 \tag{6.46}$$

and (performing the top, right, and front integrals, with each doubled to account for the other three sides)

$$\oint_{\partial\Omega} \mathbf{V} \cdot d\mathbf{a} = 2 \int_{-\ell}^{\ell} \int_{-\ell}^{\ell} \mathbf{r}|_{z=\ell} \cdot \mathbf{c}\, dxdy + 2 \int_{-\ell}^{\ell} \int_{-\ell}^{\ell} \mathbf{r}|_{y=\ell} \cdot \mathbf{b}\, dxdz + 2 \int_{-\ell}^{\ell} \int_{-\ell}^{\ell} \mathbf{r}|_{x=\ell} \cdot \mathbf{a}\, dydz$$

$$= 8a_x\ell^3 + 8b_y\ell^3 + 8c_z\ell^3$$

$$= (a_x + b_y + c_z)(2\ell)^3. \tag{6.47}$$

The two integrals are equal, establishing the theorem for a tiny domain over which the vector function is approximated by (6.42) (a physicist's assumption is that such a tiny domain exists). To build up larger domains, we just add together the small cubical ones. Take two cubical domains for which we have established the theorem, call them Ω_1 and Ω_2, and let them touch at a face as shown in Figure 6.6. We have a new domain, Ω, that is the union of Ω_1 and Ω_2. Now, given $\mathbf{V}$, we know that the theorem holds for Ω_1 and Ω_2, so that

$$\int_{\Omega_1} \nabla \cdot \mathbf{V}\, d\tau_1 = \oint_{\partial\Omega_1} \mathbf{V} \cdot d\mathbf{a}_1, \qquad \int_{\Omega_2} \nabla \cdot \mathbf{V}\, d\tau_2 = \oint_{\partial\Omega_2} \mathbf{V} \cdot d\mathbf{a}_2. \tag{6.48}$$

Adding these two equations together

$$\underbrace{\int_{\Omega_1} \nabla \cdot \mathbf{V}\, d\tau_1 + \int_{\Omega_2} \nabla \cdot \mathbf{V}\, d\tau_2}_{= \int_\Omega \nabla \cdot \mathbf{V}\, d\tau} = \oint_{\partial\Omega_1} \mathbf{V} \cdot d\mathbf{a}_1 + \oint_{\partial\Omega_2} \mathbf{V} \cdot d\mathbf{a}_2, \tag{6.49}$$

the left-hand side is just the integral over the entire domain $\Omega = \Omega_1 \cup \Omega_2$. On the right, we do not have the integral over the boundary $\partial\Omega$ because the sum of the boundary integrals includes the interior shared wall (see Figure 6.6 again). But that shared wall has $d\mathbf{a}_1 = -d\mathbf{a}_2$, the area vector points outward in both cases, so that they point in opposite directions at the shared face. Meanwhile, the value of $\mathbf{V}$ is the same at the shared boundary, so we get equal and opposite contributions, and

$$\oint_{\partial\Omega_1} \mathbf{V} \cdot d\mathbf{a} + \oint_{\partial\Omega_2} \mathbf{V} \cdot d\mathbf{a} = \oint_{\partial\Omega} \mathbf{V} \cdot d\mathbf{a}. \tag{6.50}$$

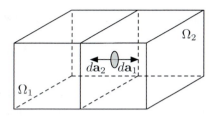

Fig. 6.6 A domain made out of two small cubical ones, Ω_1 and Ω_2 that meet at a shared internal face. The area vectors on the shared face point in opposite directions, so integrating **V** over each of the cube's surfaces will give an internal cancellation, leaving us with an integral over the external surfaces, i.e. $\partial\Omega$.

Putting this in (6.49), we have established the theorem for the pair of domains,

$$\int_\Omega \nabla \cdot \mathbf{V}\, d\tau = \oint_{\partial\Omega} \mathbf{V} \cdot d\mathbf{a}. \qquad (6.51)$$

Using additional small cubes, we can build any domain and add them up as we did here. The interior surface integrals will cancel in pairs, leaving us with the boundary of the composite domain Ω, and the theorem holds in general.

6.3.4 Conservation Laws in Three Dimensions

We developed the physics of conservation laws in one dimension in Section 4.8.1. Given a conserved quantity, like charge, which cannot be created or destroyed, we account for changes in the amount of charge in a region by looking at the flow of charge through the boundaries of the region. In three dimensions the same idea holds. Given a charge density $\rho(x, y, z, t)$, the charge per unit volume at a location x, y, z in space at time t, and a current density $\mathbf{J} = \rho\mathbf{v}$, also a function of position and time, conservation is expressed in differential form by

$$\frac{\partial\rho}{\partial t} = -\nabla \cdot \mathbf{J}, \qquad (6.52)$$

generalizing the one-dimensional form from (4.84).

If we consider a domain Ω with boundary $\partial\Omega$, then integrating both sides of (6.52) over the domain Ω gives

$$\int_\Omega \frac{\partial\rho}{\partial t}\, d\tau = -\int_\Omega \nabla \cdot \mathbf{J}\, d\tau. \qquad (6.53)$$

The quantity $\rho d\tau$ represents the amount of "stuff" (charge, mass, whatever the conserved quantity is) in an infinitesimal volume. For ρ a charge density, the total amount of charge in the domain Ω is

$$Q(t) = \int_\Omega \rho\, d\tau, \qquad (6.54)$$

and this is a function only of t since the spatial dependence has been integrated out.

Then the quantity on the left in (6.53) is really the time-derivative of[6] $Q(t)$,

$$\int_\Omega \frac{\partial \rho}{\partial t}\, d\tau = \frac{d}{dt} \int_\Omega \rho\, d\tau = \frac{dQ(t)}{dt}. \tag{6.55}$$

For the right-hand side of (6.53), we can use the divergence theorem (6.40) to rewrite the volume integral over the surface,

$$\int_\Omega \nabla \cdot \mathbf{J}\, d\tau = \oint_{\partial\Omega} \mathbf{J} \cdot d\mathbf{a}, \tag{6.56}$$

so the integrated form of the conservation law reads

$$\frac{dQ(t)}{dt} = \frac{d}{dt} \int_\Omega \rho\, d\tau = -\oint_{\partial\Omega} \mathbf{J} \cdot d\mathbf{a}. \tag{6.57}$$

The physical interpretation of this equation is that "the change in the amount of charge in the domain Ω can be accounted for by the current coming in (or leaving) through the boundary of the domain." The minus sign on the right makes physical sense: the area element points out of Ω, so if $\mathbf{J} \parallel d\mathbf{a}$ (with $\mathbf{J} \cdot d\mathbf{a} > 0$), current is carrying charge *out* of Ω, the amount of charge inside Ω should decrease.

6.3.5 Curl Theorem

Consider the surface integral of the curl of a vector function $\mathbf{V}$ over a domain S with boundary ∂S. For our version of the fundamental theorem of calculus here, we expect that the integral over the domain should be related to the evaluation of $\mathbf{V}$ on the boundary,

$$\int_S (\nabla \times \mathbf{V}) \cdot d\mathbf{a} = \oint_{\partial S} \mathbf{V} \cdot d\boldsymbol{\ell}. \tag{6.58}$$

We'll do a simple example of a curly function, $\mathbf{V} = \boldsymbol{\phi} \equiv -y\,\hat{\mathbf{x}} + x\,\hat{\mathbf{y}}$, with $\nabla \times \mathbf{V} = 2\,\hat{\mathbf{z}}$, as we have seen before. For the domain, S, take a square of side length 2ℓ centered at the origin, and lying in the xy plane as in Figure 6.7. This has area vector given by $d\mathbf{a} = dxdy\,\hat{\mathbf{z}}$. Then the area integral on the left of (6.58) is

$$\int_S (\nabla \times \mathbf{V}) \cdot d\mathbf{a} = \int_S 2\, dxdy = 2(2\ell)^2 = 8\ell^2. \tag{6.59}$$

For the line integral on the right of (6.58), we go around in the direction shown in Figure 6.7,

$$\oint_{\partial S} \mathbf{V} \cdot d\boldsymbol{\ell} = \int_{-\ell}^{\ell} V_y|_{x=\ell}\, dy - \int_{-\ell}^{\ell} V_x|_{y=\ell}\, dx - \int_{-\ell}^{\ell} V_y|_{x=-\ell}\, dy + \int_{-\ell}^{\ell} V_x|_{y=-\ell}\, dx \tag{6.60}$$

$$= \ell(2\ell) - (-\ell)(2\ell) - (-\ell)(2\ell) + (\ell)(2\ell) = 8\ell^2$$

and the equation (6.58) is true for this example.

We can again sketch the proof of this theorem using our approximate form (6.42) with S a small (so that the approximate expression for $\mathbf{V}$ holds) square of side length 2ℓ, centered

---

[6] Assuming, as usual, that the temporal derivative can be pulled out of the integral.

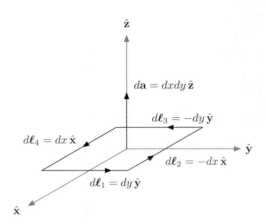

Fig. 6.7 The square domain S with boundary ∂S traversed using the direction given by the right-hand rule (point thumb in direction of **da**, then fingers curl in the direction of integration for the boundary).

at the origin, and lying in the xy plane. We will again assume that $\mathbf{r}_0 = 0$, and that $\mathbf{V}(0) = 0$ just to simplify the calculation. The curl of the approximate $\mathbf{V}$ is

$$\nabla \times \mathbf{V} = (c_y - b_z)\,\hat{\mathbf{x}} + (a_z - c_x)\,\hat{\mathbf{y}} + (b_x - a_y)\,\hat{\mathbf{z}}. \tag{6.61}$$

The surface integral is just the $\hat{\mathbf{z}}$ component of the curl multiplied by the surface area,

$$\int_S (\nabla \times \mathbf{V}) \cdot d\mathbf{a} = (b_x - a_y)(2\ell)^2. \tag{6.62}$$

We have chosen $\hat{\mathbf{z}}$ as the unit normal for the integration. That gives a direction for the line integral around ∂S obtained by the right-hand rule. Point the thumb of your right hand in the direction of the normal, then your fingers curl in the traversal direction. In this case, we want to go around the square counter-clockwise as shown in Figure 6.7.

For the boundary integral, we have $z = 0$, and we'll integrate over the four sides shown in Figure 6.7 (in the order shown)

$$\oint_{\partial S} \mathbf{V} \cdot d\boldsymbol{\ell} = \int_{-\ell}^{\ell} r \cdot \mathbf{b}|_{x=\ell}\, dy + \int_{-\ell}^{\ell} r \cdot \mathbf{a}|_{y=\ell}\,(-dx) + \int_{-\ell}^{\ell} r \cdot \mathbf{b}|_{x=-\ell}\,(-dy) + \int_{-\ell}^{\ell} r \cdot \mathbf{a}|_{y=-\ell}\, dx$$

$$= 4(b_x - a_y)\ell^2 = (b_x - a_y)(2\ell)^2 \tag{6.63}$$

so the theorem holds for this approximate, infinitesimal scenario.

We again combine integrals to finish the sketch. Take two platelets, S_1 and S_2 placed next to each other so that they share a line segment along their boundary. Call the union of the domains $S = S_1 \cup S_2$, with boundary ∂S. The theorem holds for each platelet,

$$\int_{S_1} (\nabla \times \mathbf{V}) \cdot d\mathbf{a}_1 = \oint_{\partial S_1} \mathbf{V} \cdot d\boldsymbol{\ell}_1 \qquad \int_{S_2} (\nabla \times \mathbf{V}) \cdot d\mathbf{a}_2 = \oint_{\partial S_2} \mathbf{V} \cdot d\boldsymbol{\ell}_2. \tag{6.64}$$

Adding the equations,

$$\int_{S_1} (\nabla \times \mathbf{V}) \cdot d\mathbf{a}_1 + \int_{S_2} (\nabla \times \mathbf{V}) \cdot d\mathbf{a}_2 = \oint_{\partial S_1} \mathbf{V} \cdot d\boldsymbol{\ell}_1 + \oint_{\partial S_2} \mathbf{V} \cdot d\boldsymbol{\ell}_2, \qquad (6.65)$$

$$\underbrace{\phantom{\int_{S_1} (\nabla \times \mathbf{V}) \cdot d\mathbf{a}_1 + \int_{S_2} (\nabla \times \mathbf{V}) \cdot d\mathbf{a}_2}}_{= \int_S (\nabla \times \mathbf{V}) \cdot d\mathbf{a}}$$

we have an integral over the union of the domains on the left, as desired. On the right, we have two line integrals around the boundaries of the platelets, including the shared interior line so that the integral is not over the boundary of S. Along the shared line, however, we have $d\boldsymbol{\ell}_1 = -d\boldsymbol{\ell}_2$ since we use the same counterclockwise direction for each, and the value of the function is the same, so the interior contributions cancel, and we have

$$\oint_{\partial S_1} \mathbf{V} \cdot d\boldsymbol{\ell}_1 + \oint_{\partial S_2} \mathbf{V} \cdot d\boldsymbol{\ell}_2 = \oint_{\partial S} \mathbf{V} \cdot d\boldsymbol{\ell}. \qquad (6.66)$$

Using this relation in (6.65),

$$\int_S (\nabla \times \mathbf{V}) \cdot d\mathbf{a} = \oint_{\partial S} \mathbf{V} \cdot d\boldsymbol{\ell}. \qquad (6.67)$$

So the theorem holds for S made up of the pair of platelet surfaces S_1 and S_2. You can continue to form any surface of interest by adding additional platelets with internal lines cancelling to leave the integral around ∂S, so the theorem holds for an arbitrary surface.

Problem 6.3.1 What path is described by the vector $\mathbf{w}(p) = \cos(p)\,\hat{\mathbf{x}} + \sin(p)\,\hat{\mathbf{y}} + p\,\hat{\mathbf{z}}$ for $p \in [0, 2\pi]$?

Problem 6.3.2 Generate the vector $\mathbf{w}(p)$ for a parabolic curve in two dimensions, $y = Ax^2$, for constant A. Construct $\mathbf{w}(p)$ so that: $\mathbf{w}(-1) = -A\,\hat{\mathbf{x}} + A\,\hat{\mathbf{y}}$. Compute the work done by the force $\mathbf{F} = k(x\,\hat{\mathbf{x}} - y\,\hat{\mathbf{y}})$ in going from $A(-\hat{\mathbf{x}} + \hat{\mathbf{y}})$ to $A(\hat{\mathbf{x}} + \hat{\mathbf{y}})$.

Problem 6.3.3 Is the force $\mathbf{F} = k(x\,\hat{\mathbf{x}} - y\,\hat{\mathbf{y}})$ (for constant k) conservative? How about $\mathbf{F} = k(y^2\,\hat{\mathbf{x}} - x\,\hat{\mathbf{y}})$ (again, with k some constant)?

Problem 6.3.4 Draw the picture, analogous to Figure 6.6, that shows the internal line cancellation between platelets allowing the boundaries of S_1 and S_2 to combine to form the boundary of $S \equiv S_1 \cup S_2$.

Problem 6.3.5 Check the divergence theorem for $\mathbf{v} = A(zy^2\,\hat{\mathbf{x}} - xz^2\,\hat{\mathbf{y}} + zx^2\,\hat{\mathbf{z}})$ (for constant A) with Ω a cube of side length 2ℓ centered at the origin (compute the two sides of (6.40) and check that they are the same).

Problem 6.3.6 One of Maxwell's equations for the electric field is $\nabla \cdot \mathbf{E} = \rho/\epsilon_0$ where ρ is a charge density (charge per unit volume) and ϵ_0 sets units for the electric field $\mathbf{E}$. Use the divergence theorem applied to an arbitrary closed domain Ω to show that

$$\frac{Q}{\epsilon_0} = \oint_{\partial\Omega} \mathbf{E} \cdot d\mathbf{a} \qquad (6.68)$$

where Q is the total charge contained in Ω. This is the integral form of Gauss's law.

Problem 6.3.7 Check the curl theorem using $\mathbf{v} = A(zy^2\,\hat{\mathbf{x}} - xz^2\,\hat{\mathbf{y}} + zx^2\,\hat{\mathbf{z}})$ and the domain S that is a square of side length 2ℓ lying centered at the origin at a height $z = 1$ (as with Problem 6.3.5, we are just checking here, this time using (6.58)).

6.4 Delta Functions in Three Dimensions

Recall the definition of the Dirac delta function in (2.116) from Section 2.6.2,

$$\delta(x) = \begin{cases} 0 & x \neq 0 \\ \infty & x = 0 \end{cases} \quad \text{with (for positive constants } a \text{ and } b) \int_{-a}^{b} \delta(x)\, dx = 1. \quad (6.69)$$

In three dimensions, we have

$$\delta^3(x, y, z) = \delta(x)\delta(y)\delta(z), \quad (6.70)$$

the product of delta functions in each of the three coordinate variables. Using the notation $f(\mathbf{r}) \equiv f(x, y, z)$ to denote a generic function of the three coordinates, we have

$$\int_{\Omega} \delta^3(\mathbf{r})\, d\tau = 1 \text{ if } \Omega \text{ contains the origin}, \quad (6.71)$$

and in general,

$$\int_{\Omega} \delta^3(\mathbf{r} - \mathbf{r}_0)f(\mathbf{r})\, d\tau = f(\mathbf{r}_0) \quad (6.72)$$

provided the domain Ω contains the point $\mathbf{r}_0$.

6.4.1 Divergence of $f = \hat{r}/r^2$

Remember one of our dilemmas from Section 6.2.2: For the function $\mathbf{f} = (1/r^2)\,\hat{\mathbf{r}}$, a highly divergent function, we found $\nabla \cdot \mathbf{f} = 0$ which seemed counterintuitive. The problem is with the value of the function at $r = 0$, where $\mathbf{f}$ blows up. Let's use the divergence theorem (6.40) for this $\mathbf{f}$ to make some progress.

Take the domain Ω to be a ball of radius R, where R can be as small as you like. The infinitesimal area element on the surface of the sphere, shown in Figure 6.8, is $d\mathbf{a} = R^2 \sin\theta d\theta d\phi\,\hat{\mathbf{r}}$ where θ is the polar angle of the patch with respect to the $\hat{\mathbf{z}}$ axis (see Appendix B for more on area elements in curvilinear coordinates). The area element is parallel to $\mathbf{f}$ so that $\mathbf{f} \cdot d\mathbf{a} = f da$, and the right-hand side of (6.40) is

$$\oint_{\partial\Omega} \mathbf{f} \cdot d\mathbf{a} = \int_0^{2\pi} \int_0^{\pi} \frac{1}{R^2} R^2 \sin\theta\, d\theta d\phi = 4\pi, \quad (6.73)$$

independent of the radius R.

For the left-hand side of (6.40), we need to integrate the divergence of $\mathbf{f}$ over the entire volume of the sphere, ending up with 4π to match the surface integral in the divergence theorem. The origin is included in the integration domain, and since the divergence of $\mathbf{f}$ is actually zero at all points except the origin, it must be the divergence at the origin that accounts for all of the 4π in the volume integral. That sounds like a job for the three-dimensional delta function. If we take

$$\nabla \cdot \mathbf{f} = 4\pi\delta^3(\mathbf{r}), \quad (6.74)$$

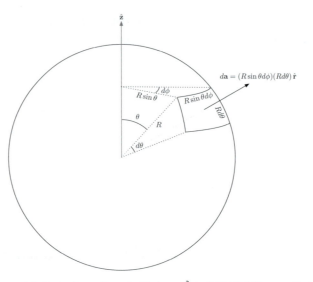

A sphere of radius R has an infinitesimal area element with $d\mathbf{a} = R^2 \sin\theta\, d\theta\, d\phi\, \hat{\mathbf{r}}$. The magnitude gives the area of the patch and the direction is normal to the sphere's surface, pointing outward.

then we have

$$\int_{\Omega} (\nabla \cdot \mathbf{f})\, d\tau = 4\pi \tag{6.75}$$

for our spherical domain. Away from the origin the divergence is zero, since the delta function is zero away from the origin. At the origin, the divergence is only as defined as the delta function, some sort of integrable infinity. The delta function is precisely the function we need to represent the divergence of this very special $\mathbf{f}$, while maintaining the divergence theorem.

6.4.2 Curl of $g \equiv \hat{\phi}/s$

We had a similar sort of problem with the curl of $\mathbf{g} = (1/s)\hat{\phi}$: $\nabla \times \mathbf{g} = 0$ for a function that clearly should have a curl. Once again, the problem is with the value of $\mathbf{g}$ at zero. We can pick a disk of radius R lying in the xy plane as our domain of integration S, and apply the curl theorem. The integral around the curve has $\mathbf{g} \cdot d\boldsymbol{\ell} = g\, d\ell$ since both $\mathbf{g}$ and $d\boldsymbol{\ell}$ point in the same, $\hat{\phi}$, direction. The magnitude of $d\ell$ is $R\, d\phi$, so that

$$\oint_{\partial S} \mathbf{g} \cdot d\boldsymbol{\ell} = \int_0^{2\pi} \frac{1}{R} R\, d\phi = 2\pi. \tag{6.76}$$

The integral of the curl over the disk appears to give zero, violating the curl theorem. But once again, the function blows up at the origin, and the origin is included in the S integration. So we guess

$$\nabla \times \mathbf{g} = 2\pi\delta(x)\delta(y)\,\hat{\mathbf{z}}, \tag{6.77}$$

where we only need a two-dimensional delta function on the right since it is an area integral that we will perform. Now the curl is zero away from zero, and integrable infinity at zero, so we recover

$$\int_S (\nabla \times \mathbf{g}) \cdot d\mathbf{a} = 2\pi \qquad (6.78)$$

noting that $d\mathbf{a}$ points in the $\hat{\mathbf{z}}$ direction (enforced by our choice to perform the boundary line integral in the counterclockwise direction). Taking (6.77) seriously saves the curl theorem.

Problem 6.4.1　Evaluate the divergence of $\hat{\mathbf{r}}/r$, which, like $\hat{\mathbf{r}}/r^2$ is infinite at the origin. Does the divergence of $\hat{\mathbf{r}}/r$ have a "hidden" delta function in it?

Problem 6.4.2　Evaluate the divergence of the gradient of $1/r$ (where $r \equiv \sqrt{x^2 + y^2 + z^2}$).

Problem 6.4.3　Evaluate the curl of the curl of $\log s\,\hat{\mathbf{z}}$ (for $s \equiv \sqrt{x^2 + y^2}$).

6.5　The Laplacian and Harmonic Functions

So far we have been working with first derivatives, the gradient, divergence and curl. But we can combine these to form second derivative operators. The most important of these is the "Laplacian," $\nabla^2 \equiv \nabla \cdot \nabla$. When acting on a function $f(x, y, z)$, it takes the form:

$$\nabla^2 f(x, y, z) = \frac{\partial^2 f}{\partial x^2} + \frac{\partial^2 f}{\partial y^2} + \frac{\partial^2 f}{\partial z^2}. \qquad (6.79)$$

The equation $\nabla^2 f(x, y, z) = 0$, known as Laplace's equation, is the static form of the wave equation in three dimensions, and itself has interesting solutions. With appropriate boundary conditions, solutions to Laplace's equation are called "harmonic functions." There are many ways to solve Laplace's equation, but more important than the solutions themselves are the general properties that we can prove directly from the defining equation. For example, we can show that solutions to Laplace's equation satisfy an "averaging property." For a function $f(x, y, z)$, the average value of f over a surface S is defined to be

$$\bar{f} \equiv \frac{1}{A_S} \oint_S f(x, y, z)\, da \qquad (6.80)$$

where A_S is the surface area of S,

$$A_S \equiv \oint_S da. \qquad (6.81)$$

Take S to be a sphere of radius R centered at the origin. We will show that for $\nabla^2 f = 0$, we have $\bar{f} = f(0, 0, 0)$: the average of the function f over the surface of the sphere is equal to the value of the function at the center of the sphere.

Let's Taylor expand $f(x, y, z)$ about the origin, this time including the quadratic terms:

$$f(x, y, z) \approx f(0, 0, 0) + \mathbf{r} \cdot \nabla f + \frac{1}{2}\left[\frac{\partial^2 f}{\partial x^2}x^2 + \frac{\partial^2 f}{\partial y^2}y^2 + \frac{\partial^2 f}{\partial z^2}z^2\right]$$
$$+ xy\frac{\partial^2 f}{\partial x \partial y} + xz\frac{\partial^2 f}{\partial x \partial z} + yz\frac{\partial^2 f}{\partial z^2} \qquad (6.82)$$

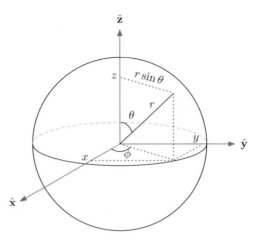

Fig. 6.9   Given r, θ, and ϕ as shown, we can identify the x, y, and z locations of a point: $z = r \cos \theta, x = r \sin \theta \cos \phi$, $y = r \sin \theta \sin \phi$.

where all derivatives are evaluated at the center. In the spherical coordinate setup with $\{r, \theta, \phi\}$, shown in Figure 6.9, we have

$$x = r \sin \theta \cos \phi \quad y = r \sin \theta \sin \phi \quad z = r \cos \theta. \tag{6.83}$$

The surface area element da, in spherical coordinates, is $da = R^2 \sin \theta d\theta d\phi$ from Figure 6.8 (see Appendix B for the development of the area element in spherical coordinates).

Looking back at the form of $f(x, y, z)$ in (6.82), the average over the sphere will involve integrals of the coordinates themselves, like

$$\oint_S x \, da = \int_0^{2\pi} \int_0^{\pi} R \sin \theta \cos \phi R^2 \sin \theta \, d\theta d\phi = 0 \tag{6.84}$$

and similarly for the y and z integrals. There are also integrals quadratic in the coordinates, like

$$\oint_S xy \, da = \int_0^{2\pi} \int_0^{\pi} R \sin \theta \cos \phi R \sin \theta \sin \phi R^2 \sin \theta \, d\theta d\phi dr = 0 \tag{6.85}$$

and the same happens for the xz and yz integrals; all of them collapse to zero. The terms quadratic in the same variable, like x^2 all contribute equally:

$$\oint_S x^2 \, da = \int_0^{2\pi} \int_0^{\pi} R^2 \sin^2 \theta \cos^2 \phi R^2 \sin \theta \, d\theta d\phi = \frac{4}{3} \pi R^4. \tag{6.86}$$

The average over the sphere is then

$$\bar{f} = \frac{1}{A_S} \oint_S f(x, y, z) \, da \approx \frac{1}{4\pi R^2} \left(f(0, 0, 0) 4\pi R^2 + 2\pi R^4 \nabla^2 f(0, 0, 0) \right) \tag{6.87}$$

with higher-order terms going like larger powers of R. As $R \to 0$, the only terms that remain, to leading order in R, are

$$\bar{f} = f(0,0,0) + \frac{1}{2}R^2 \nabla^2 f(0,0,0). \tag{6.88}$$

But the second term is zero, for $\nabla^2 f = 0$, so the average value of the function f over the sphere of radius R is just the value of f at the center of the sphere. We have used the fact that R is small to expand the function f, and then shown that for harmonic f, the averaging property holds through order R^2, and not just at order R, which is trivially true for all functions. In fact, the averaging property holds for spheres of any size.

That result can then be used to prove that for a domain Ω with $\nabla^2 f = 0$ in Ω, there can be no minimum or maximum values except on the boundary of the domain, $\partial \Omega$. Suppose you had a maximum value somewhere in Ω, and center a sphere of radius R at that maximum. Then by the averaging property, the average value of f over the sphere is equal to the value at the center, which is the maximum. But it is impossible for the average to equal the maximum, since the average must consist of values *less than* the maximum value, there is no way to take values less than the maximum and average them to get a value that is larger than the constituents. That impossibility is easiest to see in one dimension for a function $h(x)$ as in Figure 6.10. Taking the two points on either side of the maximum and averaging (the analogue of averaging over the surface of a sphere in higher dimension) gives $\bar{h} = h_0$ which is less than $h_{\max}$, you cannot have a maximum value together with the averaging property. The same argument holds for minima.

For harmonic functions, then, the maximum and minimum values must occur on the boundary of the domain, $\partial \Omega$. Finally, we can use this property to prove that the solution to $\nabla^2 f = 0$ in a domain Ω, with boundary values for f specified on $\partial \Omega$, is unique. We are given a function g on the boundary of the domain; the problem we want to solve is

$$\nabla^2 f = 0 \text{ in } \Omega \text{ with } f|_{\partial \Omega} = g. \tag{6.89}$$

Suppose there were two functions, f_1 and f_2 that solved this problem:

$$\nabla^2 f_1 = 0 \text{ in } \Omega \text{ with } f_1|_{\partial \Omega} = g, \qquad \nabla^2 f_2 = 0 \text{ in } \Omega \text{ with } f_2|_{\partial \Omega} = g. \tag{6.90}$$

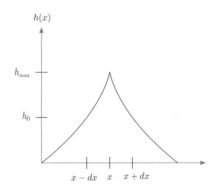

Fig. 6.10 A function h has a maximum value at x, and takes on value h_0 at nearby points $x \pm dx$.

Let $u \equiv f_1 - f_2$, then u satisfies

$$\nabla^2 u = 0 \text{ in } \Omega \text{ with } u|_{\partial\Omega} = 0. \tag{6.91}$$

Since u is harmonic, it must achieve its maximum value on the boundary $\partial\Omega$, but we know that $u = 0$ there. Since 0 is the maximum value, u must in fact be zero everywhere in Ω (including its boundary), and then $f_1 = f_2$ everywhere, so the solution to the Laplace problem (6.89) is unique.

Problem 6.5.1   Compute the Laplacian, $\nabla^2 f(x,y,z)$, for $f(x,y,z) = A(xy - x^2 + z)$ (A is a constant).

Problem 6.5.2   What is the Laplacian of $r \equiv \sqrt{x^2 + y^2 + z^2}$?

Problem 6.5.3   An example of a function that satisfies Laplace's equation is $f(x,y,z) = A + B(x^2 - y^2) + C(x - z)$ for constants A, B, and C. Compute the average value of this function over a sphere of radius R centered on the origin. It may help to use spherical coordinates from Figure 6.9 to carry out the integration.

Problem 6.5.4   Vectors like $\hat{\mathbf{r}}$ depend on position, and you can compute the Laplacian of them by writing the vector in terms of the Cartesian basis vectors $\hat{\mathbf{x}}$, $\hat{\mathbf{y}}$, and $\hat{\mathbf{z}}$ which do *not* depend on position, so that $\nabla^2 \hat{\mathbf{x}} = 0$, for example. Compute $\nabla^2 \hat{\mathbf{r}}$. Compute the Laplacian of $\hat{\mathbf{s}}$ where $\mathbf{s} \equiv x\hat{\mathbf{x}} + y\hat{\mathbf{y}}$.

Problem 6.5.5   Evaluate the second derivative $\nabla \times (\nabla \times \mathbf{A})$ for a vector function $\mathbf{A}(x,y,z)$ in terms of the Laplacian of $\mathbf{A}$ and the gradient of its divergence.

Problem 6.5.6   The "Poisson equation" is $\nabla^2 f(x,y,z) = \rho(x,y,z)$ for a source density ρ (charge per unit volume, for example, in E&M), a given function of position. This is just the Laplace equation with a nonzero right-hand side. Given a domain Ω in which the Poisson equation holds, and given a boundary value function g on the surface of Ω, we can solve the "Poisson problem." Show that the solution to the Poisson problem is unique.

6.6 Wave Equation

The wave equation we developed by taking a continuum limit of one-dimensional springs in Section 4.1 takes the three-dimensional form,

$$-\frac{\partial^2 \phi(\mathbf{r},t)}{\partial t^2} + v^2 \nabla^2 \phi(\mathbf{r},t) = 0, \tag{6.92}$$

where the second derivative with respect to x from (4.11) has turned into a sum of second derivatives, one in each direction. This form recovers the one-dimensional case. If we take $\phi(\mathbf{r},t) = \phi(x,t)$, so that ϕ has no y or z dependence, then we get (4.11) from (6.92), and similarly for the other two coordinates (omitting x and y dependence gives back the one-dimensional wave equation in z, for example).

6.6.1 Plane Waves

We can develop the three-dimensional form of the plane wave solutions from Section 4.5. Let $\mathbf{k}$ be a constant vector with dimension of inverse length. This vector is called the "wave vector" and its magnitude is the "wave number." For frequency, we'll use the angular $\omega = 2\pi f$, then a natural update for an expression like (4.67) is

$$\phi(\mathbf{r},t) = Ae^{i(-\omega t + \mathbf{k}\cdot\mathbf{r})} = Ae^{i(-\omega t + k_x x + k_y y + k_z z)} \tag{6.93}$$

for constant A that sets the scale and dimension of ϕ.

For the derivatives, we have

$$\nabla\phi = i\mathbf{k}Ae^{i(-\omega t + \mathbf{k}\cdot\mathbf{r})} = i\mathbf{k}\phi, \tag{6.94}$$

so that

$$\nabla^2\phi = -k^2\phi, \tag{6.95}$$

and similarly,

$$\frac{\partial^2\phi}{\partial t^2} = -\omega^2\phi. \tag{6.96}$$

Putting these expressions into the three-dimensional wave equation gives

$$\omega^2\phi - k^2 v^2\phi = 0 \longrightarrow \omega = kv. \tag{6.97}$$

Physically, $\phi(\mathbf{r},t)$ travels in the direction $\hat{\mathbf{k}}$, with wavelength $\lambda = 2\pi/k$, frequency $f = \omega/(2\pi)$ and speed v. The only difference between the solution ϕ here and the one in (4.67) is that the direction is more general.

The solution in (6.93) has only one direction of travel, along $\hat{\mathbf{k}}$, but by introducing a solution with the other sign, we can superimpose waves traveling in the $-\hat{\mathbf{k}}$ direction, a three-dimensional version of left and right traveling waves in one dimension as in (4.67). The full plane wave solution associated with a particular wave vector $\mathbf{k}$ is

$$\phi(\mathbf{r},t) = Ae^{i(-\omega t + \mathbf{k}\cdot\mathbf{r})} + Be^{i(-\omega t - \mathbf{k}\cdot\mathbf{r})}, \quad \omega = kv. \tag{6.98}$$

6.6.2 Longitudinal and Transverse Waves

In Chapter 4, we encountered two different types of wave motion. For one-dimensional masses connected by springs, the masses move parallel (or anti-parallel) to the wave's traveling direction, producing "longitudinal" waves. In the approximate wave equation that comes from a string under tension, the string moves up and down while the waves move left and right, so the string motion is perpendicular to the wave motion, and we call these "transverse" waves. It's possible to have a combination of transverse and longitudinal waves, too. The wave equation (6.93) refers to a single function $\phi(\mathbf{r},t)$ in three dimensions, but we can also apply the wave equation to a vector-valued function. Take $\mathbf{H}(\mathbf{r}, t) = H_x(\mathbf{r}, t)\,\hat{\mathbf{x}} + H_y(\mathbf{r}, t)\,\hat{\mathbf{y}} + H_z(\mathbf{r}, t)\,\hat{\mathbf{z}}$, then we could have a wave equation of the form

$$-\frac{\partial^2 \mathbf{H}(\mathbf{r}, t)}{\partial t^2} + v^2 \nabla^2 \mathbf{H}(\mathbf{r}, t) = 0, \tag{6.99}$$

which is really three equations, one for each component of the vector $\mathbf{H}$.

Referring to the plane wave solution (6.93), take $\mathbf{k} = k\hat{\mathbf{z}}$ so that the wave moves in the z direction. We can solve the wave equation (with appropriate boundary/initial conditions) with $H_y = 0$, then we just have the pair

$$-\frac{\partial^2 H_x(z, t)}{\partial t^2} + v^2 \frac{\partial^2 H_x(z, t)}{\partial z^2} = 0 \qquad -\frac{\partial^2 H_z(z, t)}{\partial t^2} + v^2 \frac{\partial^2 H_z(z, t)}{\partial z^2} = 0. \tag{6.100}$$

The solution for H_z is associated with the longitudinal component of $\mathbf{H}$, since H_z points in the $\pm\hat{\mathbf{z}}$ direction, and that is the direction of travel. The H_x solution provides a transverse component, pointing in the $\pm\hat{\mathbf{x}}$ direction, orthogonal to the direction of travel. We could write a plane wave solution as

$$\mathbf{H} = A e^{i(-\omega t + kz)}\,\hat{\mathbf{x}} + F e^{i(-\omega t + kz)}\,\hat{\mathbf{z}}, \tag{6.101}$$

for constants A and F. Because of the orthogonality of the Cartesian basis vectors, the two equations in (6.100) are independent, they are decoupled from one another. In Chapter 7 we shall see examples of wave equations in which the transverse and longitudinal components effect each other.

6.6.3 Spherical Waves

There are other wave geometries we can explore with our three-dimensional wave equation for vector $\mathbf{H}(\mathbf{r}, t)$. As an example, suppose we take $\mathbf{H}(\mathbf{r}, t) = H(r, t)\,\hat{\mathbf{r}}$ representing a vector that points radially away from the origin with magnitude that depends only on r, the distance from the origin. What happens if we put this assumed form into (6.99)? First let's think about the action of the ∇^2 operator on $\mathbf{H}$. The issue is that both the magnitude, $H(r, t)$, and the vector $\hat{\mathbf{r}}$ are position dependent[7] and the ∇^2 operator acts on both:

$$\nabla^2 \mathbf{H} = \left(\nabla^2 H(r, t)\right)\hat{\mathbf{r}} + H(r, t)\left(\nabla^2 \hat{\mathbf{r}}\right), \tag{6.102}$$

with

$$\nabla^2 \hat{\mathbf{r}} = \nabla^2 \left(\frac{x\hat{\mathbf{x}} + y\hat{\mathbf{y}} + z\hat{\mathbf{z}}}{\sqrt{x^2 + y^2 + z^2}}\right) = -2\frac{x\hat{\mathbf{x}} + y\hat{\mathbf{y}} + z\hat{\mathbf{z}}}{(x^2 + y^2 + z^2)^{3/2}} = -\frac{2}{r^2}\hat{\mathbf{r}} \tag{6.103}$$

from Problem 6.5.4.

The wave equation can be written in terms of $H(r, t)$ by itself, since all terms (including the Laplacian of $\hat{\mathbf{r}}$) point in the same direction,

$$-\frac{\partial^2 H(r, t)}{\partial t^2} + v^2 \left(\nabla^2 H(r, t) - \frac{2}{r^2} H(r, t)\right) = 0. \tag{6.104}$$

---

[7] Contrast this situation with a vector like $\hat{\mathbf{x}}$ which is position independent, so its derivatives vanish, $\nabla^2 \hat{\mathbf{x}} = 0$ automatically.

The Laplacian in these coordinates is given in (B.50), and the PDE of interest is

$$-\frac{\partial^2 H(r,t)}{\partial t^2} + \frac{v^2}{r^2}\left(\frac{\partial}{\partial r}\left(r^2\frac{\partial H(r,t)}{\partial r}\right) - 2H(r,t)\right) = 0. \tag{6.105}$$

We can proceed using multiplicative separation of variables: take $H(r,t) = R(r)T(t)$, then the PDE becomes (after dividing by $H(r,t)$)

$$-\frac{1}{T(t)}\frac{d^2 T(t)}{dt^2} + \frac{v^2}{R(r)r^2}\left(\frac{d}{dr}\left(r^2\frac{dR(r)}{dr}\right) - 2R(r)\right) = 0. \tag{6.106}$$

The first term depends only on t, the second only on r, so each must be equal to a constant. To get oscillatory behavior in time, it makes sense to set the first term equal to a^2 for a real constant a, then

$$T(t) = F\cos(at) + G\sin(at) \tag{6.107}$$

for constants F and G.

The spatial equation becomes, multiplying through by $r^2 R(r)/v^2$,

$$\frac{a^2 r^2}{v^2}R(r) + \left(\frac{d}{dr}\left(r^2\frac{dR(r)}{dr}\right) - 2R(r)\right) = 0. \tag{6.108}$$

We can introduce a new spatial coordinate $x \equiv ar/v$, to get (see Section 8.2.4 for a systematic look at this process)

$$\frac{d}{dx}\left(x^2\frac{dR(x)}{dx}\right) + (x^2 - 2)R(x) = 0. \tag{6.109}$$

This equation is the same one you solved using the method of Frobenius in Problem 1.4.5. The odd solution is the first spherical Bessel function, $j_1(x)$, and we can combine it with the temporal oscillation to get a spherically symmetric solution to the wave equation

$$H(r,t) = (F\cos(at) + G\sin(at))j_1(ar/v). \tag{6.110}$$

This is a standing wave solution where the profile is the first spherical Bessel function, with magnitude set by the temporal oscillation sitting out front. An example of $H(r,t)$ for a few different times is shown in Figure 6.11.

Problem 6.6.1   Show that for the Laplacian in spherical coordinates acting on a function of r only, $f(r)$, we have

$$\nabla^2 f(r) = \frac{1}{r}\frac{d^2}{dr^2}(rf(r)).$$

Problem 6.6.2   Find a solution to the wave equation for a function $f(r,t)$, a spherically symmetric function (not a vector this time, though).

Problem 6.6.3   Find a solution to the wave equation for a vector of the form $\mathbf{H}(r,t) = H(r,t)\,\hat{\mathbf{z}}$ with $r \equiv \sqrt{x^2 + y^2 + z^2}$ as usual.

Problem 6.6.4   Find a solution to the wave equation for a vector of the form $\mathbf{H}(s,t) = H(s,t)\,\hat{\mathbf{s}}$ where $s \equiv \sqrt{x^2 + y^2}$ and $\mathbf{s} = x\hat{\mathbf{x}} + y\hat{\mathbf{y}}$. This is a "cylindrically symmetric" solution. See Section 6.7.1 if you end up with an ODE you don't immediately recognize.

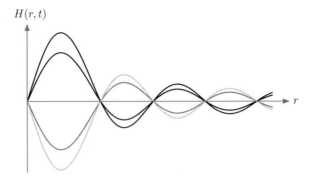

$H(r,t)$

Fig. 6.11   The magnitude of the spherically symmetric solution to the wave equation shown at a few different times (increasing from dark to light).

6.7 Laplace's Equation

In one (spatial) dimension, the wave equation reads

$$-\frac{\partial^2\phi(x,t)}{\partial t^2} + v^2\frac{\partial^2\phi(x,t)}{\partial x^2} = 0. \tag{6.111}$$

If we look for solutions that do not depend on t, $\phi(x,t) = \phi(x)$, we have the ODE

$$v^2\frac{d^2\phi(x)}{dx^2} = 0 \longrightarrow \phi(x) = Ax + B \tag{6.112}$$

an uninteresting linear function with little chance of satisfying realistic boundary conditions.

In higher dimensions, the static wave equation takes the form: $\nabla^2 f = 0$, Laplace's equation. Solutions to this equation are more interesting, as you have already seen in Problem 4.3.2, with general properties that we developed in Section 6.5. Let's look at the solutions in three dimensions for various boundary conditions. To enforce boundary conditions, it is often useful to employ coordinate systems other than Cartesian. If you are not used to the expressions for the Laplacian in cylindrical or spherical coordinates, take a look at Appendix B.

6.7.1 Cylindrical Coordinates

We want to solve $\nabla^2 f = 0$ with the value of f specified on an infinite cylinder of radius R. It is natural to use the cylindrical coordinates defined in Figure B.1, with

$$s = \sqrt{x^2 + y^2} \qquad \phi = \tan^{-1}\left(\frac{y}{x}\right) \qquad z \text{ the Cartesian } z, \tag{6.113}$$

so that the function f takes the form $f(s, \phi, z)$ and it is easy to impose the boundary condition: $f(R, \phi, z) = g(\phi, z)$ for boundary function $g(\phi, z)$. Laplace's equation in cylindrical coordinates is (see (B.19))

$$\nabla^2 f = \frac{1}{s}\frac{\partial}{\partial s}\left(s\frac{\partial f}{\partial s}\right) + \frac{1}{s^2}\frac{\partial^2 f}{\partial \phi^2} + \frac{\partial^2 f}{\partial z^2} = 0. \tag{6.114}$$

How should we solve this partial differential equation? The primary tool is separation of variables as described in Section 4.3.3. Assume that $f(s, \phi, z) = S(s)\Phi(\phi)Z(z)$ and run it through the Laplace equation, dividing by f to get

$$\frac{1}{s^2}\left(s\frac{\frac{d}{ds}\left(s\frac{dS(s)}{ds}\right)}{S(s)} + \frac{\frac{d^2\Phi(\phi)}{d\phi^2}}{\Phi(\phi)}\right) + \frac{\frac{d^2 Z(z)}{dz^2}}{Z(z)} = 0. \tag{6.115}$$

The first term depends on s and ϕ while the second depends only on z, so using the logic of separation of variables, we take $\frac{d^2 Z}{dz^2}/Z = \ell^2$, a constant. Inside the parentheses, there are functions that depend on s and on ϕ. Take $\frac{d^2\Phi}{d\phi^2}/\Phi = -m^2$ another constant (the minus sign is there to suggest an oscillatory solution, good for angular variables that require periodicity), to combine with the s-dependent piece. Then we have the triplet of equations

$$\frac{d^2 Z}{dz^2} = \ell^2 Z$$

$$\frac{d^2\Phi}{d\phi^2} = -m^2\Phi \tag{6.116}$$

$$\frac{d^2 S}{ds^2} + \frac{1}{s}\frac{dS}{ds} + \left(\ell^2 - \frac{m^2}{s^2}\right)S = 0.$$

The first two equations are familiar, and we can solve immediately for $Z(z)$ and $\Phi(\phi)$,

$$Z(z) = Ae^{\ell z} + Be^{-\ell z} \qquad \Phi(\phi) = Fe^{im\phi} + Ge^{-im\phi}, \tag{6.117}$$

where we require that m be an integer to get $\Phi(\phi) = \Phi(\phi + 2\pi)$, but there is no obvious restriction on ℓ. The third equation is not one we have encountered before. We can clean it up a bit by introducing a scaled s-coordinate, $x = \ell s$,

$$x^2\frac{d^2 S(x)}{dx^2} + x\frac{dS(x)}{dx} + \left(x^2 - m^2\right)S(x) = 0. \tag{6.118}$$

This is "Bessel's equation."

In order to solve it, we'll use the Frobenius approach from Section 1.4. Assume that

$$S(x) = x^p \sum_{j=0}^{\infty} a_j x^j. \tag{6.119}$$

Taking derivatives, putting them in (6.118), and collecting like powers of x^j, we get

$$a_0(p^2 - m^2) + a_1\left[(1+p)^2 - m^2\right]x + \sum_{j=2}^{\infty}\left[a_j\left((j+p)^2 - m^2\right) + a_{j-2}\right]x^j = 0. \tag{6.120}$$

To eliminate the first term, we take $p = \pm m$. Setting $a_1 = 0$ to get rid of the second term, we are left with the recursion relation

$$a_{j-2} = a_j\left(m^2 - (j \pm m)^2\right) = -a_j j(j \pm 2m) \longrightarrow a_j = -\frac{a_{j-2}}{j(j \pm 2m)}, \tag{6.121}$$

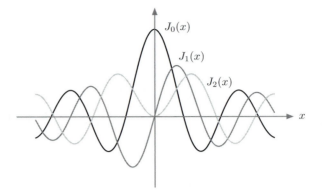

Fig. 6.12 The first three Bessel functions.

or letting $j \equiv 2k$,

$$a_{2k} = -\frac{a_{2k-2}}{2k(2k \pm 2m)} = -\frac{a_{2k-2}}{4k(k \pm m)}. \tag{6.122}$$

Writing out the first few terms, we can solve the recursion

$$a_2 = -\frac{a_0}{4(1 \pm m)}$$

$$a_4 = -\frac{a_2}{8(2 \pm m)} = \frac{a_0}{32(m \pm 1)(m \pm 2)}$$

$$a_6 = -\frac{a_0}{384(1 \pm m)(2 \pm m)(3 \pm m)}$$

$$a_8 = \frac{a_0}{2^8 4! \, (1 \pm m)(2 \pm m)(3 \pm m)(4 \pm m)} \tag{6.123}$$

$$a_{10} = -\frac{a_0}{2^{10} \times 5! \, (1 \pm m)(2 \pm m)(3 \pm m)(4 \pm m)(5 \pm m)}$$

$$\vdots$$

$$a_{2j} = (-1)^j \frac{a_0}{2^{2j} j! \, \prod_{k=1}^{j}(k \pm m)} = (-1)^j \frac{a_0 (-m)!}{2^{2j} j! \, (j \pm m)!}.$$

Using these coefficients in (6.119) leads to "Bessel's function" for both positive and negative values of [8]m:

$$J_{\pm m}(x) = a_0 \left(\frac{1}{2^{\pm m}(1 \pm m)!}\right) \sum_{j=0}^{\infty} (-1)^j \frac{(-m)!}{2^{2j} j! \, (j \pm m)!} x^{2j \pm m} \tag{6.124}$$

where the a_0 out front is just a constant, and the term in parentheses is a normalization convention. These functions are well-studied, and the first few are shown in Figure 6.12. They are oscillatory, and decay away from the origin. For even values of m, the functions are even, and for odd values of m, they are odd.

---

[8] This definition holds for integer values of m; there is a generalized expression that applies for noninteger values.

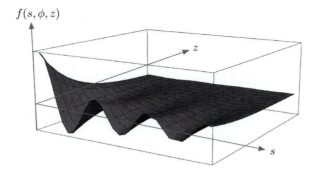

$f(s, \phi, z)$

 Fig. 6.13   An example of a function that has $\nabla^2 f = 0$.

Bessel's equation is linear, so sums of solutions are also solutions, and we can construct a general solution for $S(x)$ using the individual Bessel functions[9]:

$$S(x) = \sum_{k=0}^{\infty} c_k J_k(x). \tag{6.125}$$

The Bessel functions satisfy an orthogonality and completeness relation, much like sine and cosine, and we can build functions out of them by tuning the coefficients $\{c_k\}_{k=0}^{\infty}$.

Returning to our separation of variables solution, for a particular value of ℓ, we have

$$f(s, \phi, z) = \left(A e^{\ell z} + B e^{-\ell z}\right) \sum_{m=0}^{\infty} \left(f_m e^{im\phi} + g_m e^{-im\phi}\right) J_m(\ell s) \tag{6.126}$$

for constant A, B, and coefficients $\{f_m\}_{m=0}^{\infty}$, $\{g_m\}_{m=0}^{\infty}$. We can go further by adding together solutions with different values of ℓ. This general solution to $\nabla^2 f = 0$ can be used to enforce boundary conditions (in cylindrical coordinates) and obtain unique solutions.

The solutions to $\nabla^2 f = 0$ in three dimensions are much more interesting than the linear solution (6.112) in one dimension. As an example, take the solution $f(s, \phi, z) = e^{-z} J_0(s)$ (obtained from (6.126) with $m = 0$, $\ell = 1$) shown in Figure 6.13.

6.7.2 Spherical Boundary Conditions: Axial Symmetry

A function of spherical coordinates (see Figure 6.9), $f(r, \theta, \phi)$, is said to be "axially symmetric" if it is independent of the variable ϕ. The Laplacian in spherical coordinates, from (B.50), applied to an axially symmetric function becomes

$$\frac{1}{r^2} \frac{\partial}{\partial r} \left(r^2 \frac{\partial f}{\partial r}\right) + \frac{1}{r^2 \sin\theta} \frac{\partial}{\partial \theta} \left(\sin\theta \frac{\partial f}{\partial \theta}\right) = 0. \tag{6.127}$$

Taking the separation ansatz, $f(r, \theta) = R(r)\Theta(\theta)$, inserting in this equation, and performing some algebraic manipulation gives

$$\frac{1}{R(r)} \frac{d}{dr} \left(r^2 \frac{dR(r)}{dr}\right) + \frac{1}{\Theta(\theta) \sin\theta} \frac{d}{d\theta} \left(\sin\theta \frac{d\Theta(\theta)}{d\theta}\right) = 0. \tag{6.128}$$

---

[9] There is an independent second solution to Bessel's equation, called the Neumann function, but it blows up at the origin and hence is rejected in most physical applications.

The first term is a function only of r, while the second is a function of θ, so they must both be constant in order to satisfy the equation for all values of r and θ. Call that constant value $\ell(\ell+1)$,[10] then we have the pair:

$$\frac{d}{dr}\left(r^2\frac{dR(r)}{dr}\right) = \ell(\ell+1)R(r), \qquad \frac{d}{d\theta}\left(\sin\theta\frac{d\Theta(\theta)}{d\theta}\right) = -\ell(\ell+1)\sin\theta\,\Theta(\theta). \quad (6.129)$$

The first equation can be solved by taking $R(r) \propto r^q$ as a "guess" (see Appendix A), then we get

$$q(q+1)\,r^q = \ell(\ell+1)r^q \longrightarrow q = \ell \text{ or } q = -\ell - 1 \quad (6.130)$$

so the radial piece of the solution can be written as the combination of these two

$$R(r) = Ar^\ell + \frac{B}{r^{\ell+1}}. \quad (6.131)$$

For the angular piece, let $x \equiv \cos\theta$ (not to be confused with the Cartesian coordinate x which doesn't appear in this spherical setting); if we view $\Theta(\theta)$ as a function of x (i.e. depending on θ through $\cos\theta$), the chain rule gives

$$\frac{d\Theta(x)}{d\theta} = \frac{d\Theta(x)}{dx}\frac{dx}{d\theta} = -\sin\theta\frac{d\Theta(x)}{dx} = -\sqrt{1-x^2}\frac{d\Theta(x)}{dx} \quad (6.132)$$

and we can write the second ODE in (6.129) as

$$\left(1-x^2\right)\frac{d^2\Theta(x)}{dx^2} - 2x\frac{d\Theta(x)}{dx} + \ell(\ell+1)\Theta(x) = 0 \quad (6.133)$$

which is known as "Legendre's differential equation." We will again solve using the series solution method of Frobenius, taking

$$\Theta(x) = x^p \sum_{j=0}^{\infty} a_j x^j. \quad (6.134)$$

Putting this form into (6.133) and collecting as usual gives

$$\sum_{j=0}^{\infty} a_j(j+p)(j+p-1)x^{j-2} + \sum_{j=0}^{\infty}\left[-(j+p)(j+p-1) - 2(j+p) + \ell(\ell+1)\right]a_j x^j = 0,$$

$$(6.135)$$

or, re-indexing the first term,

$$0 = a_0(p)(p-1)\,x^{-2} + a_1(p+1)(p)\,x^{-1}$$

$$+ \sum_{j=0}^{\infty}\left[(-(j+p)(j+p-1) - 2(j+p) + \ell(\ell+1))\,a_j\right.$$

$$\left. +(j+2+p)(j+1+p)a_{j+2}\right]x^j = 0. \quad (6.136)$$

---

[10] We'll take ℓ to be an integer, although that is not strictly speaking necessary. This choice does simplify the solutions to Legendre's equation, and suffices in many cases. Noninteger values of ℓ lead to angular solutions that blow up.

Taking $p = 0$ so that a_0 and a_1 are free to take on any values, the recursion relation is

$$a_{j+2} = \frac{j(j+1) - \ell(\ell+1)}{(j+1)(j+2)} a_j. \tag{6.137}$$

This defines both the even (for $a_1 = 0$) and odd (for $a_0 = 0$) solutions. Notice that for a particular integer value of ℓ, the numerator of the recursion will be zero for $j = \ell$, we have a polynomial of degree ℓ, either even or odd depending on the value of ℓ. These polynomials are known as "Legendre polynomials," and indexed by the integer ℓ: $P_\ell(x)$.

For $\ell = 0$, we get the zeroth Legendre polynomial, $P_0(x) = a_0$. The Legendre polynomials are "normalized" as we shall see in a moment, and that normalization sets the value of a_0. Moving on to $\ell = 1$, working with the odd series, we have a_1 to start off, then $a_3 = 0$ as do all other coefficients, so the solution is $P_1(x) = a_1 x$ with a_1 again set by the normalization convention. For $\ell = 2$, we take a_0 to start off, and then $a_2 = -3a_0$ and a_4 and all higher coefficients are zero, giving $P_2(x) = (1 - 3x^2)a_0$.

The general solution to (6.133) is a superposition of the individual multiplicative terms

$$f(r, \theta) = \sum_{\ell=0}^{\infty} \left(A_\ell r^\ell + \frac{B_\ell}{r^{\ell+1}} \right) P_\ell(\cos \theta). \tag{6.138}$$

There is, again, an orthogonality relation for the Legendre polynomials, which can be obtained from the ODE itself. Take two solutions, $P_\ell(x)$ and $P_m(x)$ with $\ell \neq m$. These each satisfy the Legendre differential equation,

$$(1 - x^2)P_\ell''(x) - 2xP_\ell'(x) + \ell(\ell+1) P_\ell(x) = 0$$
$$(1 - x^2)P_m''(x) - 2xP_m'(x) + m(m+1) P_m(x) = 0. \tag{6.139}$$

The idea is to multiply the top equation by $P_m(x)$, the bottom by $P_\ell(x)$ and then integrate each with respect to x. Since the argument $x = \cos \theta$, we expect the domain here to be $x \in [-1, 1]$, and we'll integrate over all of it. For the top equation, multiplying and integrating gives

$$\int_{-1}^{1} (1 - x^2)P_\ell''(x)P_m(x)\, dx - \int_{-1}^{1} 2xP_\ell'(x)P_m(x)\, dx + \ell(\ell+1) \int_{-1}^{1} P_\ell(x)P_m(x)\, dx = 0. \tag{6.140}$$

Using integration by parts on the first term, we have

$$\int_{-1}^{1} (1 - x^2)P_\ell''(x)P_m(x)\, dx = \left. \left(1 - x^2 \right) P_\ell'(x)P_m(x) \right|_{x=-1}^{1}$$
$$- \int_{-1}^{1} \left(P_\ell'(x)P_m'(x) + P_\ell'(x)\left(x^2 P_m'(x) + 2xP_m(x) \right) \right)\, dx. \tag{6.141}$$

The boundary term vanishes,[11] and (6.140) becomes

$$-\int_{-1}^{1} P_\ell'(x)P_m'(x)(1+x^2)\,dx + \ell(\ell+1)\int_{-1}^{1} P_\ell(x)P_m(x)\,dx = 0. \qquad (6.142)$$

The equation for $P_m(x)$ in (6.139), when multiplied by $P_\ell(x)$ and integrated from $x = -1 \to 1$ is just (6.142) with $\ell \leftrightarrow m$,

$$-\int_{-1}^{1} P_m'(x)P_\ell'(x)(1+x^2)\,dx + m(m+1)\int_{-1}^{1} P_m(x)P_\ell(x)\,dx = 0. \qquad (6.143)$$

and subtracting this from (6.142) gives

$$(\ell(\ell+1) - m(m+1))\int_{-1}^{1} P_\ell(x)P_m(x)\,dx = 0 \qquad (6.144)$$

which means that

$$\int_{-1}^{1} P_\ell(x)P_m(x)\,dx = 0 \qquad (6.145)$$

since $\ell \neq m$ by assumption. The case $\ell = m$ will not give zero, and can be used to set the normalization of the Legendre polynomials.[12] A typical choice is

$$\int_{-1}^{1} P_\ell(x)^2\,dx = \frac{2}{2\ell+1} \qquad (6.146)$$

so that the full orthogonality relation reads

$$\int_{-1}^{1} P_\ell(x)P_m(x)\,dx = \frac{2}{2\ell+1}\delta_{\ell m}. \qquad (6.147)$$

If we return to the angular θ using $x = \cos\theta$, then the orthonormality condition is

$$\int_{0}^{\pi} P_\ell(\cos\theta)P_m(\cos\theta)\sin\theta\,d\theta = \frac{2}{2\ell+1}\delta_{\ell m}. \qquad (6.148)$$

Example

Suppose we solve $\nabla^2 f = 0$ in a domain Ω that is the interior of a sphere of radius R, and we are given a function $g(\theta)$ with $f(R,\theta) = g(\theta)$ on the boundary of the domain (i.e. at the spherical surface). The boundary condition and domain are axisymmetric, so it is safe to assume that f is as well.[13] If we want the solution $f(r,\theta)$ in Ω to be free of infinities, then

---

[11] The Legendre polynomials do not blow up at $x = \pm 1$.
[12] Many choices of normalization exist.
[13] An interesting assumption, why should the function f have the same symmetry as the domain and boundary condition function?

we must set $B_\ell = 0$ for all ℓ in the general solution (6.138), otherwise $r = 0$ will pose a problem. Starting from

$$f(r,\theta) = \sum_{\ell=0}^{\infty} A_\ell r^\ell P_\ell(\cos\theta), \qquad (6.149)$$

we want to find the set of coefficients $\{A_\ell\}_{\ell=0}^{\infty}$, and we can use the boundary condition at $r = R$, together with the orthogonality of the Legendre polynomials. At $r = R$,

$$f(R,\theta) = \sum_{\ell=0}^{\infty} A_\ell R^\ell P_\ell(\cos\theta) = g(\theta). \qquad (6.150)$$

Multiply both sides by $P_m(\cos\theta)\sin\theta$ and integrate as in (6.148),

$$A_m R^m \frac{2}{2m+1} = \int_0^\pi P_m(\cos\theta)g(\theta)\sin\theta\,d\theta \longrightarrow$$

$$A_m = \frac{2m+1}{2R^m} \int_0^\pi P_m(\cos\theta)g(\theta)\sin\theta\,d\theta. \qquad (6.151)$$

That does it, in principle, but the integral is difficult to carry out, especially since all we have is a recursive formula for the coefficients in the Legendre polynomials. One can sometimes (especially in textbook problems) get away with "inspection." For example, suppose that $g(\theta) = g_0\cos\theta$. This is just $g(\theta) = g_0 P_1(\cos\theta)$ and we know that no other Legendre polynomials can contribute by orthogonality. In that case, it is easy to read off the solution. We know $\ell = 1$ is the only contributing term, so

$$f(r,\theta) = A_1 r^1 P_1(\cos\theta) \qquad (6.152)$$

and imposing $f(R,\theta) = A_1 R\cos\theta = g_0\cos\theta$ gives us $A_1 = g_0/R$, with full solution

$$f(r,\theta) = \frac{g_0 r}{R}\cos\theta. \qquad (6.153)$$

Problem 6.7.1 Use the orthogonality relation (6.148) to set the coefficient for $P_3(x)$.

Problem 6.7.2 Show that the solutions to $f_n''(x) + (n\pi/L)^2 f_n(x) = 0$ for integer n, with $f_n(0) = f_n(L) = 0$ satisfy

$$\int_0^L f_n(x)f_m(x)\,dx = 0$$

for $n \neq m$.

Problem 6.7.3 The "Rodrigues formula" provides a way to find the nth Legendre polynomial. The formula reads

$$P_n(x) = \frac{1}{2^n n!}\left(\frac{d}{dx}\right)^n (x^2-1)^n, \qquad (6.154)$$

so that you take n derivatives of $(x^2-1)^n$ to get the nth polynomial, normalized according to the orthonormality convention in (6.147). Work out the first three Legendre polynomials using this formula and compare with the expressions from the last section.

Problem 6.7.4   There is an orthogonality relation for the Bessel functions that can be proved in a manner similar to the orthogonality of the Legendre polynomials we proved in the last section. Suppose α and β are zeroes of the nth Bessel function, $J_n(\alpha) = 0 = J_n(\beta)$. Show that

$$\int_0^1 xJ_n(\alpha x)J_n(\beta x)\, dx = 0$$

if $\alpha \neq \beta$, i.e. if they are not the same zero. Finding the zeroes of Bessel functions is not easy (unlike, for example, finding the regularly spaced zeroes of sine or cosine). The simplest approach is to isolate them numerically, and we'll see how to do that in Section 8.1.

Problem 6.7.5   In the example from the last section, we found the solution on the interior of a sphere of radius R. This time, find the solution to $\nabla^2 f = 0$ in the domain Ω defined by the *exterior* of the sphere of radius R given the boundary condition: $f(R, \theta) = g(\theta) = g_0 \cos \theta$. Just as we excluded one of the radial solutions because $r = 0$ was in the domain of the example, this time we must exclude one of the radial solutions because spatial infinity ($r \to \infty$) is in our domain.

7 Other Wave Equations

What is a wave? What is a wave equation? These are probably questions we should have started with. The "wave equation" we spent most of Chapter 4 developing and solving comes from the longitudinal motion of springs (exact), and the transverse motion of strings under tension (approximate). In the broader context of wave equations, we have studied conservation laws, and used them to predict some of the behavior that arises in the nonlinear setting (e.g., traffic flow). But, we have never really defined "waves" or the equations that govern them. There does not seem to be much in common between the applications, which become even more exotic in, for example, quantum mechanics. Physicists tend to have a fairly broad description of what constitutes a wave.

Turning to the professionals, in this case, physical oceanographers, it is interesting that the situation remains somewhat ambiguous. From [5], "Waves are not easy to define. Whitham (1974 [19]) defines a wave as 'a recognizable signal that is transferred from one part of a medium to another with recognizable velocity of propagation.'" This definition captures almost any physically relevant phenomenon, and is a parallel to Coleman's quote from the preface. What's worse, a "wave equation" is taken to be any equation that supports solutions that are waves. That seems tautological, but as we saw in the previous chapter, those wave equations have interesting solutions even when those solutions aren't themselves waves (as in the static cases).

In this chapter, we will look at some of the other places that wave and wave-like equations arise. In some cases, we can solve the wave equation, or find approximate solutions, as we have in previous chapters. But for a majority of the topics in this chapter, it is the journey that is the reward. Nonlinear partial differential equations are notoriously difficult to solve, and so we will simply work out the physics that develops the equations, letting further study address the challenges of solution. An exception comes at the end of the chapter, where Schrödinger's wave equation is introduced. There, many of the techniques we have studied so far will apply, and we will solve Schrödinger's equation for some familiar cases of interest.

7.1 Electromagnetic Waves

We have seen the wave equation in the form (4.11) (or its three-dimensional version (6.92)) emerge from coupled oscillators in Section 4.1, and as an approximation to the vertical motion of a string in Section 4.2, and there are many other places where this wave equation appears, either exactly or in approximation. Perhaps the most famous appearance is in electricity and magnetism, where the electromagnetic field satisfies the wave equation "in vacuum" (meaning away from the sources).

Given a charge density $\rho(\mathbf{r}, t)$, the charge-per-unit-volume in some domain, and a current density $\mathbf{J}(\mathbf{r}, t)$ which tells how that charge is moving around in time and space, Maxwell's equations relate the divergence and curl of an electric field $\mathbf{E}$ and a magnetic field $\mathbf{B}$ to ρ and $\mathbf{J}$:

$$\nabla \cdot \mathbf{E} = \frac{\rho}{\epsilon_0} \quad \nabla \times \mathbf{E} = -\frac{\partial \mathbf{B}}{\partial t} \quad \nabla \cdot \mathbf{B} = 0 \quad \nabla \times \mathbf{B} = \mu_0 \mathbf{J} + \frac{1}{c^2} \frac{\partial \mathbf{E}}{\partial t}, \tag{7.1}$$

where ϵ_0 is a constant associated with electric sources, μ_0 with magnetic ones, and $c = 1/\sqrt{\mu_0 \epsilon_0}$ is the speed of light. The electric field acts on a particle carrying charge q with a force $\mathbf{F} = q\mathbf{E}$ and the magnetic field acts on a particle of charge q moving with velocity vector $\mathbf{v}$ via the force $\mathbf{F} = q\mathbf{v} \times \mathbf{B}$. The equations can be understood geometrically using the intuition we developed in Section 6.2.2. For example, from $\nabla \cdot \mathbf{B} = 0$, we learn that magnetic fields cannot "diverge" from a point, they must be entirely "curly". Meanwhile, the divergence of $\mathbf{E}$ depends on ρ, so that where there is a large charge-per-unit-volume, the divergence of $\mathbf{E}$ at those points is itself large.

In vacuum, we have $\rho = 0$ and $\mathbf{J} = 0$. This does not mean that there are no sources for the electric and magnetic fields, just that they are distant from the domain of interest. Then Maxwell's equations read

$$\nabla \cdot \mathbf{E} = 0 \quad \nabla \times \mathbf{E} = -\frac{\partial \mathbf{B}}{\partial t} \quad \nabla \cdot \mathbf{B} = 0 \quad \nabla \times \mathbf{B} = \frac{1}{c^2} \frac{\partial \mathbf{E}}{\partial t}. \tag{7.2}$$

Taking the curl of the curl of the electric field and using the identity from Problem 6.5.5: $\nabla \times (\nabla \times \mathbf{E}) = \nabla(\nabla \cdot \mathbf{E}) - \nabla^2 \mathbf{E}$ gives

$$\nabla(\nabla \cdot \mathbf{E}) - \nabla^2 \mathbf{E} = -\frac{\partial}{\partial t}(\nabla \times \mathbf{B}) \tag{7.3}$$

and the divergence on the left is zero (from (7.2)), the curl of $\mathbf{B}$ on the right can be replaced by the time-derivative of $\mathbf{E}$, leaving

$$-\frac{\partial^2 \mathbf{E}}{\partial t^2} + c^2 \nabla^2 \mathbf{E} = 0, \tag{7.4}$$

the wave equation for the vector function $\mathbf{E}$. If we took the curl of the curl of $\mathbf{B}$, the same sort of simplification would occur, and we'd have the wave equation for $\mathbf{B}$ as well,

$$-\frac{\partial^2 \mathbf{B}}{\partial t^2} + c^2 \nabla^2 \mathbf{B} = 0. \tag{7.5}$$

Maxwell's equations in vacuum, then, lead to wave equations for $\mathbf{E}$ and $\mathbf{B}$, and the characteristic speed is c, the speed of light. This is surprising, since in our previous examples, the speed of the waves is set by physical properties of the medium that transmits the waves. But we have just developed the wave equation for electromagnetic fields *in vacuum*, with no obvious physical properties whatsoever. This paradox originally led to the idea of an "ether," a medium through which electromagnetic waves move, and which is responsible for setting the characteristic speed of the waves. That explanation is at odds with experiments that eventually supported the special relativistic interpretation, that the vacuum itself has a natural speed. The wave equation that appears here is *exact*, it does not come from any approximation (as with, for example, the wave equation governing waves on a string).

The wave equations (7.4) and (7.5) are not complete. They came from four first-derivative relations, Maxwell's equations, and we have lost some information in taking the derivatives. To see the problem, and its solution, take plane waves that solve the wave equations:

$$\mathbf{E} = E_0 e^{i(-\omega_E t + \mathbf{k}_E \cdot \mathbf{r})} \hat{\mathbf{e}} \qquad \mathbf{B} = B_0 e^{i(-\omega_B t + \mathbf{k}_B \cdot \mathbf{r})} \hat{\mathbf{b}} \tag{7.6}$$

where to satisfy the wave equation, we must have $\omega_E = ck_E$ and $\omega_B = ck_B$, relating frequencies and wave vector magnitude. The unit vector directions $\hat{\mathbf{e}}$ and $\hat{\mathbf{b}}$ are constant, and *a priori* unrelated. The constants E_0 and B_0, setting the size of the fields, are similarly unrelated (at this point).

Sending these wave equation solutions back in to Maxwell's equations in vacuum, we get

$$\nabla \cdot \mathbf{E} = 0 \longrightarrow i\mathbf{k}_E \cdot \hat{\mathbf{e}} E_0 e^{i(-\omega_E t + \mathbf{k}_E \cdot \mathbf{r})} = 0 \tag{7.7}$$

from which we learn that $\mathbf{k}_E \cdot \hat{\mathbf{e}} = 0$, the direction of wave travel, the wave vector $\hat{\mathbf{k}}_E$, is perpendicular to the direction of the electric field, $\hat{\mathbf{e}}$. The electric field plane wave is thus transverse. Running the divergence of $\mathbf{B}$ through $\nabla \cdot \mathbf{B} = 0$ gives the same orthogonality for the magnetic field's wave propagation direction and magnetic field direction, $\mathbf{k}_B \cdot \hat{\mathbf{b}} = 0$. The magnetic field also has a transverse plane wave solution. Remember, these constraints come from Maxwell's equations, they are not present in the wave equations themselves.

Moving on to the curls, we have

$$\nabla \times \mathbf{E} = -\frac{\partial \mathbf{B}}{\partial t} \longrightarrow i\mathbf{k}_E \times \hat{\mathbf{e}} E_0 e^{i(-\omega_E t + \mathbf{k}_E \cdot \mathbf{r})} = i\omega_B B_0 e^{i(-\omega_B t + \mathbf{k}_B \cdot \mathbf{r})} \hat{\mathbf{b}}. \tag{7.8}$$

This equation provides a wealth of information. In order for it to hold, for all time t and spatial locations $\mathbf{r}$, we must have $\omega_B = \omega_E$ and $\mathbf{k}_B = \mathbf{k}_E$ allowing us to clear out the exponentials. In addition, we see that the vector directions of $\mathbf{E}$ and $\mathbf{B}$ must be related: $\hat{\mathbf{b}} = \hat{\mathbf{k}}_E \times \hat{\mathbf{e}}$. Finally, the constants E_0 and B_0 are related

$$B_0 \omega_B = k_E E_0 \longrightarrow B_0 = \frac{k_E E_0}{\omega_E} = \frac{E_0}{c}. \tag{7.9}$$

With all these constraints in place, the plane waves $\mathbf{E}$ and $\mathbf{B}$ that solve Maxwell's equations are

$$\mathbf{E} = E_0 e^{i(-\omega_E t + \mathbf{k}_E \cdot \mathbf{r})} \hat{\mathbf{e}} \qquad \mathbf{B} = \frac{E_0}{c} e^{i(-\omega_E t + \mathbf{k}_E \cdot \mathbf{r})} \hat{\mathbf{k}}_E \times \hat{\mathbf{e}} \tag{7.10}$$

with $\hat{\mathbf{k}}_E \cdot \hat{\mathbf{e}} = 0$ and $\omega_E = ck_E$.

In this setting, the vector $\hat{\mathbf{e}}$ is called the "polarization" of the wave, and E_0 sets the magnitude of both $\mathbf{E}$ and $\mathbf{B}$. The electric and magnetic waves both move in the $\hat{\mathbf{k}}_E$

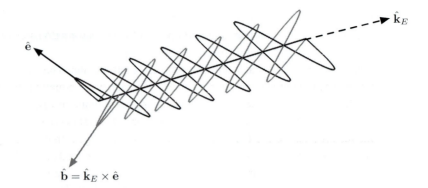

A snapshot of the real part of the electric and magnetic plane waves that solve Maxwell's equations. The electric field points in the $\hat{\mathbf{e}}$ direction (the "polarization"), the magnetic field points in $\hat{\mathbf{b}} = \hat{\mathbf{k}}_E \times \hat{\mathbf{e}}$, and the waves themselves travel in the $\hat{\mathbf{k}}_E$ direction.

direction, which is perpendicular to both the electric and magnetic field directions. Finally, the electric and magnetic fields are themselves perpendicular. An example of the real part of these fields is shown in Figure 7.1, where you can see the triumvirate of directions and their orthogonal relationships. These single-frequency electromagnetic waves are the mathematical representation of light, a phenomena produced by the mutual propagation of electric and magnetic fields, through the vacuum, at constant speed c.

Problem 7.1.1   From (7.2), show that $\mathbf{B}$ satisfies (7.5).

Problem 7.1.2   A plane wave has frequency $f = 5.1 \times 10^{14}$ Hz, what is its wavelength?

Problem 7.1.3   Give the form of both $\mathbf{E}$ and $\mathbf{B}$ for a plane wave of frequency f that is traveling in the $\hat{\mathbf{x}} - \hat{\mathbf{y}} + 2\,\hat{\mathbf{z}}$ direction and polarized in the $\hat{\mathbf{x}} + \hat{\mathbf{y}}$ direction with magnitude E_0.

Problem 7.1.4   The general solution to the wave equation (in one dimension)

$$-\frac{\partial^2 \phi(x,t)}{\partial t^2} + v^2 \frac{\partial^2 \phi(x,t)}{\partial x^2} = 0$$

is $\phi(x,t) = f(x - vt) + g(x + vt)$ as we saw in Section 4.3.1, with f and g chosen to satisfy initial conditions. Suppose you "complexify time" as in Problem 1.6.5 by taking $t = is$. What does the wave equation look like now? What happens to the general solution?

Problem 7.1.5   In two dimensions, the Laplace equation reads

$$\frac{\partial^2 u(x,y)}{\partial x^2} + \frac{\partial^2 u(x,y)}{\partial y^2} = 0.$$

A solution that is valid for all points except the origin is $u(x,y) = u_0 \log(\sqrt{x^2 + y^2})$. Check that this is a solution, and show that it separates into the "f" and "g" parts you established in the previous problem.

7.2 Fluids

One of the most familiar places to observe waves is water. The wave equation that governs water waves is more complicated than the linear wave equation that we get in the study of electricity and magnetism. Let's briefly develop a one-dimensional form of the "Euler" equations for fluids. There are two inputs here, the first is conservation of mass. As water moves around, the total amount of it remains unchanged. In a local region, the amount of water can fluctuate, but only because some mass has entered or left the domain, not because water is created or destroyed. Take ρ to be the density of water (mass per unit length here), a function of position and time. Then we have a current density $J = \rho v$, another function of position and time that describes how the water is moving. The conservation law, from Section 4.8.1, reads

$$\frac{\partial \rho}{\partial t} = -\frac{\partial J}{\partial x}. \tag{7.11}$$

The second piece of physics we need is Newton's second law,[1] $m\frac{dv}{dt} = -\frac{\partial U}{\partial x}$ for a conservative force with potential energy U. Here, we can replace mass with mass density, and then we'll substitute an energy density $\mathcal{U}$ for U, so we have

$$\rho \frac{dv}{dt} = -\frac{\partial \mathcal{U}}{\partial x}. \tag{7.12}$$

This equation applies to a small volume of fluid with density ρ and velocity v that is acted on by a conservative force density.

The velocity function v depends on position and time, and so the total time derivative can be written in terms of partial derivatives that take into account the change in v due to both changing position and changing time. In general, for a function $f(x, t)$, we have

$$\frac{df}{dt} = \frac{\partial f}{\partial x}\dot{x} + \frac{\partial f}{\partial t}. \tag{7.13}$$

The change in position, $\dot{x}$, is itself given by v, the function that tells us how the fluid is moving at different locations, so we can write

$$\frac{df}{dt} = \frac{\partial f}{\partial x}v + \frac{\partial f}{\partial t}. \tag{7.14}$$

This is called the "convective derivative" and accounts for the change in f that occurs because fluid is moving (first term) in addition to the change in f that comes from its explicit time dependence (second term). Using the convective derivative in (7.12) gives

$$\rho \frac{\partial v}{\partial t} + \rho v \frac{\partial v}{\partial x} = -\frac{\partial \mathcal{U}}{\partial x}. \tag{7.15}$$

Finally, we can put this equation into conservative form, i.e. into the form of a conservation law like (7.11), by writing the time-derivative of $J = \rho v$ in terms of a spatial derivative, employing (7.11) along the way:

---

[1] We're using a partial spatial derivative here as a reminder that in higher dimension, we could have a potential energy that depends on the other coordinates.

$$\frac{\partial(\rho v)}{\partial t} = \rho\frac{\partial v}{\partial t} + \frac{\partial \rho}{\partial t}v = -\rho v\frac{\partial v}{\partial x} - \frac{\partial \mathcal{U}}{\partial x} - v\frac{\partial(\rho v)}{\partial x} = -\frac{\partial}{\partial x}\left(\rho v^2 + \mathcal{U}\right). \tag{7.16}$$

In this context, what we have is conservation of momentum (density) ρv, with an effective force governed by both $\mathcal{U}$ and ρv^2. We can write the pair of equations now, both conservation laws, one for ρ, one for $J = \rho v$:

$$\frac{\partial \rho}{\partial t} = -\frac{\partial J}{\partial x}$$

$$\frac{\partial J}{\partial t} = -\frac{\partial}{\partial x}\left(\frac{J^2}{\rho} + \mathcal{U}\right). \tag{7.17}$$

This pair of equations is nonlinear in its variables (with J^2/ρ appearing in the second one), leading to immediate complications in its solution.

7.2.1 Three Dimensions

We can update the one-dimensional equations from (7.17) to three dimensions, where we have the mass conservation equation from (6.52). For ρ the mass per unit volume and $\mathbf{J} = \rho\mathbf{v}$ the current,

$$\frac{\partial \rho}{\partial t} = -\nabla \cdot \mathbf{J} \tag{7.18}$$

as in Section 6.3.4. The force equation becomes, for $\mathcal{U}$ now a potential energy density (see Problem 4.1.8) in three dimensions (energy per unit volume),

$$\rho\frac{d\mathbf{v}}{dt} = -\nabla\mathcal{U}. \tag{7.19}$$

In three dimensions, for a function $f(x, y, z, t)$, we again have the convective derivative,

$$\frac{df}{dt} = \frac{\partial f}{\partial x}\dot{x} + \frac{\partial f}{\partial y}\dot{y} + \frac{\partial f}{\partial z}\dot{z} + \frac{\partial f}{\partial t} = \mathbf{v}\cdot\nabla f + \frac{\partial f}{\partial t} \tag{7.20}$$

or, for $f \to \mathbf{v}$,

$$\frac{d\mathbf{v}}{dt} = (\mathbf{v}\cdot\nabla)\mathbf{v} + \frac{\partial\mathbf{v}}{\partial t}, \tag{7.21}$$

so that (7.19) becomes

$$\rho\frac{\partial\mathbf{v}}{\partial t} + \rho\mathbf{v}\cdot\nabla\mathbf{v} = -\nabla\mathcal{U}. \tag{7.22}$$

It is more difficult to put this equation into conservative form.

The potential energy density $\mathcal{U}$ has units of force per unit area, and in addition to any external potential energies governing the fluid (like gravity), this is where the fluid pressure comes into the picture. If two fluid elements exert a force on each other at their interface, that force per unit area acting on the interface is called the pressure p, and we can separate this from other external forces by writing $\mathcal{U} = p + \bar{\mathcal{U}}$ where $\bar{\mathcal{U}}$ represents other physical interactions in a particular problem. When we take $\bar{\mathcal{U}} = \rho g z$ for mass density ρ and gravity near the surface of the earth ($g \approx 9.8$ m/s^2), we get the "Euler" equations (for ρ and $\mathbf{v}$;

there is an additional equation enforcing conservation of energy which we omit here for simplicity),

$$\frac{\partial \rho}{\partial t} = -\nabla \cdot (\rho \mathbf{v})$$

$$\rho \frac{\partial \mathbf{v}}{\partial t} + \rho (\mathbf{v} \cdot \nabla) \mathbf{v} = -\nabla (p + \rho g z).$$

(7.23)

These equations govern, for example, gravity driven water waves. The pressure p must be specified, and it is typical to take $p \propto \rho^{\gamma}$, i.e., pressure is proportional to density raised to some power.

7.2.2 Shallow Water Wave Equation

In three dimensions, we want to find the density $\rho(x, y, z, t)$ and vector $\mathbf{v}(x, y, z, t)$ using (7.23) with appropriate initial and boundary conditions. This is a difficult task (to say the least), and there are many simplifying approximate forms for (7.23) that apply in particular situations. One special case is the "shallow water approximation." Suppose we have a volume of fluid with vertical depth (in the z direction, say) that is small compared to its horizontal extent, a puddle of water, if you like. In this case, we are interested in the horizontal motion of the water rather than its vertical motion. To impose the approximation, we assume that the vertical (z-) component of velocity is negligible compared to the horizontal ones, and that the details of the water below its surface are irrelevant (things like the geometry of the bottom of the puddle are ignorable, for example, we can consider a perfectly flat bottom). The approximation allows us to focus on the waves that form on the surface of the water, and ultimately the shallow water equations refer only to that surface.

We'll develop the shallow water equations in one horizontal direction (leaving two for Problem 7.2.3). Imagine a three-dimensional volume of water with a wave riding along the top surface as shown in Figure 7.2. The top surface is given by the function $h(y, t)$, and we'll assume, as shown in the figure, that the wave's shape is the same in the x direction so that we need only consider a slice in the yz plane to describe it. The relevant component of velocity is v_y: we have assumed that $v_z \approx 0$ is negligible compared to v_y, and because of the assumed symmetry along the x direction, there is no v_x. The function $v_y(y, t)$ describes the horizontal velocity of the wave, and we assume it is independent of z (again, because

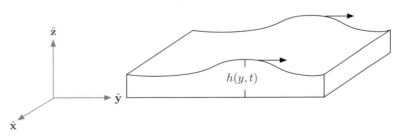

A wave traveling in the $\hat{\mathbf{y}}$ direction. The height of the top surface of the water is given by $h(y, t)$.

the water is shallow, the surface wave velocity holds at any of the necessarily negligible depths).

The density of the water is a function of y, z, and t, and can be written (recall the definition of the step function from (2.133))

$$\rho(y,z,t) = \rho_0 \theta(h(y,t) - z) \tag{7.24}$$

which says that the mass density of the water is only nonzero below its top surface $h(y,t)$ where it has uniform mass density (and no pressure variation). This is, again, a manifestation of the shallow water assumption – the depth isn't great enough to support changes in density beneath the water's surface.

From our assumptions about the velocity components and their dependencies, the conservation of density and $\hat{y}$ component of (7.22) read

$$\frac{\partial \rho}{\partial t} + \frac{\partial \rho}{\partial y} v_y + \rho \frac{\partial v_y}{\partial y} = 0$$

$$\rho \frac{\partial v_y}{\partial t} + \rho \left(v_y \frac{\partial v_y}{\partial y} \right) = -\frac{\partial \mathcal{U}}{\partial y}, \tag{7.25}$$

where the potential energy density $\mathcal{U} = \rho g z$ has implicit y dependence through ρ (and, again, we are ignoring the pressure contribution). Using the form for ρ from (7.24), together with its derivatives,

$$\frac{\partial \rho}{\partial t} = \rho_0 \delta(h(y,t) - z) \frac{\partial h(y,t)}{\partial t}$$

$$\frac{\partial \rho}{\partial y} = \rho_0 \delta(h(y,t) - z) \frac{\partial h(y,t)}{\partial y} \tag{7.26}$$

in (7.25) gives the pair

$$0 = \delta(h(y,t) - z) \left(\frac{\partial h(y,t)}{\partial t} + \frac{\partial h(y,t)}{\partial y} v_y(y,t) \right)$$

$$+ \theta(h(y,t) - z) \frac{\partial v_y(y,t)}{\partial y} \tag{7.27}$$

$$-\delta(h(y,t) - z) \frac{\partial h(y,t)}{\partial y} gz = \theta(h(y,t) - z) \left(\frac{\partial v_y(y,t)}{\partial t} + v_y(y,t) \frac{\partial v_y(y,t)}{\partial y} \right).$$

We are not interested in what happens below the surface of the fluid. In order to eliminate the "underwater" portion of these equations, we can integrate from $z = 0 \to \infty$, and use the delta and step functions to simplify the integrals,

$$\frac{\partial h(y,t)}{\partial t} + v_y(y,t) \frac{\partial h(y,t)}{\partial y} + \int_0^{h(y,t)} \frac{\partial v_y(y,t)}{\partial y} dz = 0$$

$$\int_0^{h(y,t)} \frac{\partial v_y(y,t)}{\partial t} dz + \int_0^{h(y,t)} v_y(y,t) \frac{\partial v_y(y,t)}{\partial y} dz = -gh(y,t) \frac{\partial h(y,t)}{\partial y}. \tag{7.28}$$

All of the remaining integrals are z-independent, so we get

$$\frac{\partial h(y,t)}{\partial t} + v_y(y,t)\frac{\partial h(y,t)}{\partial y} + h(y,t)\frac{\partial v_y(y,t)}{\partial y} = 0$$

$$\frac{\partial v_y(y,t)}{\partial t}h(y,t) + h(y,t)v_y(y,t)\frac{\partial v_y(y,t)}{\partial y} = -gh(y,t)\frac{\partial h(y,t)}{\partial y}.$$
(7.29)

Using the top equation, we can write this pair in "conservative" form,

$$\frac{\partial h(y,t)}{\partial t} = -\frac{\partial}{\partial y}(h(y,t)v_y(y,t))$$

$$\frac{\partial (h(y,t)v_y(y,t))}{\partial t} = -\frac{\partial}{\partial y}\left(h(y,t)(v_y(y,t))^2 + \frac{1}{2}gh(y,t)^2\right).$$
(7.30)

These are the "shallow water equations." The goal is to solve for the surface $h(y,t)$ and velocity at the surface, $v_y(y,t)$ given initial data and boundary conditions.

Problem 7.2.1   For a particle traveling along a one-dimensional trajectory, $x(t)$, the associated mass density is $\rho(x,t) = m\delta(x-x(t))$ and the particle travels with $v(x,t) = \dot{x}(t)$. Show that mass conservation, (7.11), is satisfied.

Problem 7.2.2   Write the one-dimensional shallow water equation height function as $h(y,t) = h_0 + \eta(y,t)$ where h_0 is some mean height and $\eta(y,t)$ rides on top. Take $\eta(y,t)$ and the velocity $v_y(y,t)$ to be small, and write the shallow water equations in terms of η and v_y "to first order" (meaning that you will eliminate η^2, v_y^2, and ηv_y terms once everything is expanded). This pair is called the "linearized" shallow water equations. Assuming a wave-like solution of the form $\eta(y,t) = F(y-vt)$, find the wave speed v that solves the linearized shallow water equations (hint: use one of the equations to express v_y in terms of η).

Problem 7.2.3   Work out the two-dimensional shallow water equations starting from (7.23) – keep $v_z = 0$ (and its derivatives), but assume both v_y and v_x are nonzero and can depend on the coordinates x, y, and the time t. Similarly, the height function can be written as $h(x,y,t)$. The derivation basically follows the one-dimensional case earlier, but with the additional x-velocity component. Don't worry about writing the resulting PDEs in conservative form.

Problem 7.2.4   The Korteweg–de Vries (KdV) equation is related to the shallow water equations and can be used to describe fluid motion in a long channel. In dimensionless form (see Section 8.2.4) it reads

$$\frac{\partial f(y,t)}{\partial t} + \frac{\partial^3 f(y,t)}{\partial y^3} - 6f(y,t)\frac{\partial f(y,t)}{\partial y} = 0.$$
(7.31)

There is a special class of solution called a "soliton" that behaves like a traveling wave. To find it, assume $f(y,t) = P(y-vt)$ a right-traveling waveform. This assumption turns the PDE into an ODE that can be solved for $P(u)$ – first integrate the ODE once assuming that $P(\pm\infty) \to 0$ as a boundary condition, then think about an ansatz of the form $P(u) = a/\cosh^2(bu)$ for constants a and b to be determined.

7.3 Nonlinear Wave Equation

When we developed the wave equation for strings in Section 4.2, we made several related and unreasonable assumptions, like constant tension throughout the string and the requirement that the string pieces move vertically with no longitudinal motion. The benefit was simplicity, resulting in a linear wave equation with known solutions. More realistic models lead to much more complicated nonlinear partial differential equations. These are interesting to develop even as they are difficult to solve. Here, we will work out the coupled longitudinal and transverse equations of motion for a string with spatially varying tension.

We'll take a "string" that has uniform mass density μ when in equilibrium, where the string lies straight along what we'll take to be the x axis. Focusing on a portion of the string between x and $x + dx$ as on the left in Figure 7.3, the total mass is μdx. Now suppose the string is not in equilibrium, as on the right in Figure 7.3. The portions of the string have moved both vertically and horizontally. The piece of string that was in equilibrium at x is now at vector location $\mathbf{r}(x, t)$ at time t, and the piece that was in equilibrium at $x + dx$ is at $\mathbf{r}(x + dx, t)$.

The force of tension acts along the string, pointing in the direction tangent to the string everywhere. That tangent direction is given by the x-derivative of the position vector, which we'll denote with primes. The tangent vector on the left, at time t, is $\mathbf{r}'(x, t)$, and the one on the right is $\mathbf{r}'(x + dx, t)$. We'll take the magnitude of the tension to be $T(x)$ at location x, so we need to make *unit* tangent vectors in order to express the forces acting on the segment of string. Define the unit vector

$$\hat{\mathbf{p}}(x, t) \equiv \frac{\mathbf{r}'(x, t)}{\sqrt{\mathbf{r}'(x, t) \cdot \mathbf{r}'(x, t)}} \tag{7.32}$$

which can be evaluated at $x + dx$ to describe the unit tangent vector on the right. The net force on the patch of string comes from adding the tension forces on the left and right. Since positive values of tension correspond to "pulling" (instead of "pushing"), the direction of the tension on the left is $-\hat{\mathbf{p}}(x, t)$, and on the right, $\hat{\mathbf{p}}(x + dx, t)$. The total force on the segment, at time t, is

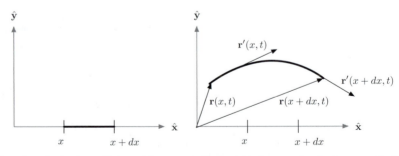

Fig. 7.3   On the left, a piece of string in equilibrium with mass μdx. On the right, the same piece has been stretched in both the horizontal and vertical directions. The vectors $\mathbf{r}(x, t)$ and $\mathbf{r}(x + dx, t)$ point to the new locations of the left and right end points of the piece of string at time t. The vectors tangent to those endpoints are given by the spatial derivatives $\mathbf{r}'(x, t)$ on the left, $\mathbf{r}'(x + dx, t)$ on the right.

$$\mathbf{F}(x) = -T(x)\,\hat{\mathbf{p}}(x,t) + T(x+dx)\,\hat{\mathbf{p}}(x+dx,t) \tag{7.33}$$

and then Newton's second law reads (using dots to denote partial time derivatives)

$$\mu\,dx\,\ddot{\mathbf{r}}(x,t) = T(x+dx)\,\hat{\mathbf{p}}(x+dx,t) - T(x)\,\hat{\mathbf{p}}(x,t). \tag{7.34}$$

For dx small, the term on the right can be Taylor expanded,

$$\mu\,dx\,\ddot{\mathbf{r}}(x,t) \approx \frac{\partial}{\partial x}\big(T(x)\,\hat{\mathbf{p}}(x,t)\big)\,dx. \tag{7.35}$$

Now we can cancel the dx from both sides to arrive at the general expression, written entirely in terms of the vector $\mathbf{r}(x,t)$ that, at time t, points from the origin to the piece of string that would be at x in equilibrium,

$$\mu\,\ddot{\mathbf{r}}(x,t) = \frac{\partial}{\partial x}\left(T(x)\frac{\mathbf{r}'(x,t)}{\sqrt{\mathbf{r}'(x,t)\cdot\mathbf{r}'(x,t)}} \right). \tag{7.36}$$

Next, let's think about how to express the vector $\mathbf{r}(x,t)$ in terms of its components. The vertical part is easy, we define the function $v(x,t)$ (not to be confused with velocity here) that gives the height above the x axis of the string segment at x. For the horizontal piece, we want a function $u(x,t)$ that gives the horizontal displacement of the piece of string that is in equilibrium at x, relative to x, just as we had with $\phi(x,t)$ in Section 4.1. Then the horizontal component of $\mathbf{r}(x,t)$ is $x + u(x,t)$ so that when $u(x,t) = 0$ (equilibrium) the horizontal location of the string is x. The position vector is

$$\mathbf{r}(x,t) = (x + u(x,t))\,\hat{\mathbf{x}} + v(x,t)\,\hat{\mathbf{y}}. \tag{7.37}$$

The equation of motion (7.36), written in terms of the functions $u(x,t)$ and $v(x,t)$ becomes the pair

$$\mu\,\ddot{u}(x,t) = \frac{\partial}{\partial x}\left(T(x)\frac{1 + u'(x,t)}{\sqrt{(1+u'(x,t))^2 + v'(x,t)^2}} \right)$$

$$\mu\,\ddot{v}(x,t) = \frac{\partial}{\partial x}\left(T(x)\frac{v'(x,t)}{\sqrt{(1+u'(x,t))^2 + v'(x,t)^2}} \right). \tag{7.38}$$

Here we have equations governing both longitudinal (top equation) and transverse (bottom equation) motion, and the motion is *coupled*, you can't have one without the other.

How do we recover the wave equation from Section 4.2? There, we only had transverse motion, so only the second equation in (7.38) is relevant (we took $u(x,t)$ and its derivatives to be zero). We assumed that the magnitude of tension was constant, $T(x) = T_0$, so that the equation governing $v(x,t)$ is

$$\ddot{v}(x,t) = \frac{T_0}{\mu}\frac{\partial}{\partial x}\left(\frac{v'(x,t)}{\sqrt{1 + v'(x,t)^2}} \right) \tag{7.39}$$

in this approximation. Finally, the small angle assumption (related to the tangent vector, $\sim v'(x, t)$) means that we can Taylor expand the term in parenthesis:

$$\left(\frac{v'(x, t)}{\sqrt{1 + v'(x, t)^2}} \right) \approx v'(x, t) + O(v'(x, t)^3) \tag{7.40}$$

giving

$$-\frac{\partial^2 v(x, t)}{\partial t^2} + \frac{T_0}{\mu} \frac{\partial^2 v(x, t)}{\partial x^2} = 0 \tag{7.41}$$

which is (4.23).

Going back to the full equations in (7.38), what form should we use for $T(x)$? There are a variety of options, but perhaps the simplest is to take a constant tension and introduce a "linear" correction, similar to Hooke's law itself. When the string is in its equilibrium configuration, lying along the x axis as in Figure 7.3, we have $u(x, t) = 0$, $u'(x, t) = 0$ and $v'(x, t) = 0$, and we could have constant tension T_0 that wouldn't cause a net force on the patch of string (equal and opposite tensions cancel). The length of the piece of string is dx. Once the string has been stretched, it has a new length approximated by

$$d\ell = \sqrt{\mathbf{r}'(x, t) \cdot \mathbf{r}'(x, t)} dx \tag{7.42}$$

just the length of the tangent vector $\mathbf{r}'(x, t)$ shown on the right in Figure 7.3. We will assume that the correction to the tension is linear in the difference $d\ell - dx$. Then the tension as a function of position takes the form

$$T(x) = T_0 + K \frac{d\ell - dx}{dx} = T_0 + K \left(\sqrt{\mathbf{r}'(x, t) \cdot \mathbf{r}'(x, t)} - 1 \right) \tag{7.43}$$

for K a constant with units of tension, since we made a dimensionless displacement measure. Putting this assumed form in to (7.36) gives

$$\mu \ddot{\mathbf{r}}(x, t) = (T_0 - K) \frac{\partial}{\partial x} \left(\frac{\mathbf{r}'(x, t)}{\sqrt{\mathbf{r}'(x, t) \cdot \mathbf{r}'(x, t)}} \right) + K \mathbf{r}''(x, t). \tag{7.44}$$

In this form, we can again see the wave equation emerge for $K = T_0$, a relation set by the material-dependent constants K and T_0. In that case, the transverse and longitudinal pieces of the wave equation decouple, effectively functioning independently.

Problem 7.3.1 The "curvature" of a curve specified by $\mathbf{w}(x)$ is defined to be the magnitude of the curvature vector,

$$\mathbf{k}(x) = \frac{1}{\sqrt{\mathbf{w}'(x) \cdot \mathbf{w}'(x)}} \frac{\partial}{\partial x} \left(\frac{\mathbf{w}'(x)}{\sqrt{\mathbf{w}'(x) \cdot \mathbf{w}'(x)}} \right), \tag{7.45}$$

proportional to the derivative of the unit tangent vector to the curve. Calculate the curvature of a circle with $\mathbf{w}(x) = R(\cos(x)\,\hat{\mathbf{x}} + \sin(x)\,\hat{\mathbf{y}})$. The inverse of the magnitude of the curvature defines the "radius of curvature" of a curve: Does that make sense for the curvature vector for a circle? Notice that the nonlinear wave equation in (7.44) can be written in terms of the curvature. Indeed, if you wrote the "wave" part of that equation on the left, you could even say that the nonlinear equation was a wave equation that had, as its "source," a term proportional to its own curvature.

Problem 7.3.2   Show, from its definition, that the curvature vector $\mathbf{k}(x)$ is orthogonal to the unit tangent vector $\hat{\mathbf{p}}(x)$ from (7.32). Check that this is true for the circular example from the previous problem.

Problem 7.3.3   Work out and keep the $v'(x,t)^3$ term from (7.40) in (7.39) to get the first corrective update to (7.41).

7.4 Schrödinger's Wave Equation

Schrödinger's equation governs the quantum mechanical "motion" of a particle of mass m that is acted on by a potential energy U. In one dimension, the equation reads

$$-\frac{\hbar^2}{2m}\frac{\partial^2\Psi(x,t)}{\partial x^2} + U(x)\Psi(x,t) = i\hbar\frac{\partial\Psi(x,t)}{\partial t} \tag{7.46}$$

where $\hbar$ is "Planck's constant." The goal is to solve for $\Psi(x,t)$ (which could be a complex function; i appears in the equation) given some boundary and initial conditions. This "wave equation" plays a central quantitative role in quantum mechanics, analogous to Newton's second law in classical mechanics. The interpretation of the target function $\Psi(x,t)$ is what makes quantum mechanics fundamentally different from all of classical physics. The "wave function" tells us about the probability of finding a particle in the dx vicinity of x at time t. That probability is given by[2] $dP = \Psi(x,t)^*\Psi(x,t)dx$. Quantum mechanics makes predictions about probabilities and averages obtained by performing experiments over and over from the same starting point. Given an ensemble of particles acted on by the same potential and starting from some initial probability distribution (in either space or momentum), the wave function tells us the likelihood of finding a particle near a particular location. Moving on from infinitesimals, if we want to know the probability of finding a particle between $x = a$ and $x = b$, we can integrate

$$P(a,b) = \int_a^b \Psi^*(x,t)\Psi(x,t)\,dx. \tag{7.47}$$

Because of the central role of probability here, and the statistical interpretation of the function $\Psi(x,t)$, we pause to review some statistics.

7.4.1 Densities and Averages

A probability density $\rho(x,t)$ in one dimension has dimension of inverse length (or, if you like, "probability" per unit length). The probability of an event occurring within a dx window of x is $dP = \rho(x,t)dx$, and the probability of an event occurring in a region $x \in [a,b]$ is

---

[2] Note that one peculiarity of densities is that the probability of finding a particle *at x* is zero. You can only have nonzero probability of finding the particle "in the vicinity," dx, of x.

$$P(a, b) = \int_a^b \rho(x, t)\, dx. \tag{7.48}$$

Probability densities are normalized so that the probability of an event occurring over the entire x axis must be one at all times t:

$$P(-\infty, \infty) = \int_{-\infty}^{\infty} \rho(x, t)\, dx = 1. \tag{7.49}$$

In quantum mechanics, we have $\rho(x, t) = \Psi^*(x, t)\Psi(x, t)$, and the event of interest is measuring a particle near location x.

We can use probability densities to compute "average" values of functions of position. Given some function of position, $f(x)$, the average or "expectation" value of the function is just the sum of all values of $f(x)$ weighted by the probability of the event occurring at x, so that

$$\langle f \rangle(t) \equiv \int_{-\infty}^{\infty} \rho(x, t) f(x)\, dx. \tag{7.50}$$

The probability density $\rho(x, t)$ may be a function of time, as is the case for densities that come from a wave function solving (7.46), or it may be static. As an example of an expectation value, we might be interested in the average location of a particle. The function $f(x) = x$, in that case, and we add up all the possible positions of the particle multiplied by the probability of finding the particle near each:

$$\langle x \rangle(t) = \int_{-\infty}^{\infty} \rho(x, t) x\, dx. \tag{7.51}$$

Again, the average value may or may not be time-dependent depending on the form of $\rho(x, t)$.

With the average value in hand, we can ask how much deviation from the average is associated with a particular probability density. We don't particularly care whether the deviation is to the left or right of the average value, so it makes sense to compute the average of the new function $f(x) = (x - \langle x \rangle)^2$, the "variance":

$$\sigma^2(t) \equiv \langle (x - \langle x \rangle)^2 \rangle(t) = \int_{-\infty}^{\infty} \rho(x, t)\left(x^2 - 2x\langle x \rangle + \langle x \rangle^2\right) dx. \tag{7.52}$$

The average $\langle x \rangle$ and its square $\langle x \rangle^2$ are just numbers (from the point of view of the position integral) and can be pulled outside the integral, leaving us with

$$\begin{aligned}
\sigma^2(t) &= \int_{-\infty}^{\infty} \rho(x, t) x^2\, dx - 2\langle x \rangle \int_{-\infty}^{\infty} \rho(x, t) x\, dx + \langle x \rangle^2 \int_{-\infty}^{\infty} \rho(x, t)\, dx \\
&= \langle x^2 \rangle - 2\langle x \rangle^2 + \langle x \rangle^2 \\
&= \langle x^2 \rangle - \langle x \rangle^2.
\end{aligned} \tag{7.53}$$

If you want a measure of the "spread" of the distribution, you can take the square root of the variance to obtain the "standard deviation," σ.[3]

---

[3] Note that there are other ways to measure the "spread." You could, for example, compute $\langle |x - \langle x \rangle| \rangle$, that is also a measure of the average distance to the mean and is insensitive to direction (above or below the mean). The variance is a nice, continuous, differentiable, function of the difference $x - \langle x \rangle$, and that makes it easier to work with than the absolute value (see Problem 7.4.3).

Example: Gaussian Density

Consider the well-known Gaussian probability density

$$\rho(x) = Ae^{-B(x-C)^2} \tag{7.54}$$

for constants A, B, and C – this density is time-independent, so we removed the t in $\rho(x,t)$ as a reminder. First, we will find out what the normalization requirement (7.49) tells us,

$$\int_{-\infty}^{\infty} \rho(x)\, dx = \int_{-\infty}^{\infty} Ae^{-B(x-C)^2}\, dx = A\sqrt{\frac{\pi}{B}} = 1 \tag{7.55}$$

so that $A = \sqrt{B/\pi}$ in order to normalize the density. Here, and for the integrals below, we have used the definite integral identities (for even and odd powers):

$$\int_0^{\infty} x^{2n} e^{-x^2}\, dx = \sqrt{\pi}\,\frac{(2n)!}{n!}\left(\frac{1}{2}\right)^{2n+1} \qquad \int_0^{\infty} x^{2n+1} e^{-x^2}\, dx = \frac{n!}{2} \tag{7.56}$$

which we will not prove, but you can test experimentally in Problem 8.3.4.

The average value of position is

$$\langle x \rangle = \int_{-\infty}^{\infty} x \sqrt{\frac{B}{\pi}}\, e^{-B(X-C)^2}\, dx = C \tag{7.57}$$

so that C is the average value here, the Gaussian is peaked at C, let $C \equiv \mu$ for "mean." For the variance, we have

$$\langle (x-\mu)^2 \rangle = \int_{-\infty}^{\infty} (x-\mu)^2 \sqrt{\frac{B}{\pi}}\, e^{-B(x-\mu)^2}\, dx = \frac{1}{2B} \equiv \sigma^2. \tag{7.58}$$

Since the variance is denoted σ^2, we sometimes write $B = 1/(2\sigma^2)$ and then the normalized Gaussian, with all constants tuned to their statistical meaning, becomes

$$\rho(x) = \sqrt{\frac{1}{2\pi\sigma^2}}\, e^{-\frac{(x-\mu)^2}{2\sigma^2}}. \tag{7.59}$$

The standard picture of the density, with mean, μ, and standard deviation, σ, marked on it, is shown in Figure 7.4.

Example: Constant Density

In quantum mechanics, we describe particle motion using probability densities. You can do this classically as well, although it is not as common. As a vehicle, think of a ball bouncing back and forth elastically between two walls (no gravity) separated by a distance a (one wall is at $x = 0$, the other at $x = a$), a classical "infinite square well." In one cycle of the ball's motion, what is the probability of finding the ball in the vicinity of any point $x \in [0, a]$? No point is preferred, the ball moves back and forth at constant speed, so it spends the same amount of time at every location. Then we know that the probability density is constant between the walls. Take $\rho(x) = 0$ (no time-dependence here) for $x < 0$

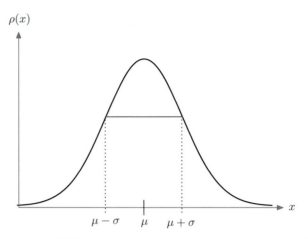

$\rho(x)$

$\mu - \sigma \quad \mu \quad \mu + \sigma$

Fig. 7.4 A plot of the Gaussian density from (7.59). The mean, μ, as well as one standard deviation to the left and right, $\mu \pm \sigma$, are shown.

and $x > a$, no chance of finding the particle outside of its walled box. Inside the walls, $\rho(x) = C$, and we can normalize,

$$\int_{-\infty}^{\infty} \rho(x)\, dx = \int_{0}^{a} C\, dx = aC = 1 \qquad (7.60)$$

from which we conclude that $C = 1/a$ and $\rho(x) = 1/a$. What is the average value of the ball's position? We expect it to be at $a/2$ on average, but let's calculate:

$$\langle x \rangle = \int_{-\infty}^{\infty} x\rho(x)\, dx = \int_{0}^{a} \frac{x}{a}\, dx = \frac{a}{2}. \qquad (7.61)$$

The variance is

$$\left\langle \left(x - \frac{a}{2}\right)^2 \right\rangle = \int_{-\infty}^{\infty} \left(x - \frac{a}{2}\right)^2 \rho(x)\, dx = \int_{0}^{a} \frac{1}{a}\left(x - \frac{a}{2}\right)^2 dx = \frac{a^2}{12}. \qquad (7.62)$$

The standard deviation, $\sigma = a/\sqrt{12}$ tells us, on average, how far from the mean the ball is. Since we are thinking about an average over one cycle to get $\rho(x)$, the actual speed doesn't matter to the deviation.

The statistical interpretation in the classical case represented by this and the next example is different from the quantum mechanical one. In classical mechanics, we are imagining a single particle moving under the influence of a potential, and the probability density has the interpretation of the probability of finding that particle at a particular location given the particle's time-averaged motion. For the ball bouncing back and forth, that time-averaging is carried out implicitly, we know the particle visits each point twice in each full cycle, and it spends the same amount of time in the vicinity of every point because the ball travels with constant speed. But we are making a prediction about the whereabouts of a single ball as it moves classically, not, as in quantum mechanics, about the whereabouts of a particle that is part of an ensemble of similarly prepared particles.

Example: Oscillator Density

Let's develop the classical probability density for our harmonic oscillator motion to contrast with the quantum mechanical result we will develop in Section 7.5. A particle of mass m starts from rest at location $-a/2$, and moves under the influence of the harmonic potential energy function $U(x) = m\omega^2 x^2/2$. For $x(t)$ the solution to Newton's second law with the associated force $F = -U'(x) = -m\omega^2 x$, the time-dependent spatial probability density is given by

$$\rho(x, t) = \delta(x - x(t)). \tag{7.63}$$

In classical mechanics, a particle is at a particular location at a particular time, and this form for $\rho(x, t)$ enforces that extreme localization and is normalized for all values of t, thanks to the delta function,

$$\int_{-\infty}^{\infty} \rho(x, t) \, dx = 1. \tag{7.64}$$

To define the time-independent density, we note that $x(t)$ is oscillatory as in our previous example, so we can average over a half period, during which the oscillating mass will visit all locations between $-a/2$ and $a/2$, so this suffices. Let $\rho(x)$ be the time average,

$$\rho(x) \equiv \frac{1}{\frac{T}{2}} \int_0^{T/2} \rho(x, t) \, dt = \frac{2}{T} \int_0^{T/2} \delta(x - x(t)) \, dt. \tag{7.65}$$

Using (2.132), we can change variables. Let $u \equiv x(t)$, then

$$\rho(x) = \frac{2}{T} \int_0^{T/2} \delta(x - x(t)) \, dt = \frac{2}{T} \int_{-a/2}^{a/2} \frac{\delta(x - u)}{|\dot{x}(t)|} \, du. \tag{7.66}$$

We must write the integrand in terms of $x(t)$, not its time-derivative, in order to use the delta function. From conservation of energy with $x(t) = u$ for use in the integral,

$$E = \frac{1}{2} m\dot{x}(t)^2 + \frac{1}{2} m\omega^2 u^2 \tag{7.67}$$

and from the initial condition, $E = m\omega^2 a^2/8$, so that

$$\dot{x}(t) = \pm\omega\sqrt{\left(\frac{a}{2}\right)^2 - u^2}. \tag{7.68}$$

Now we can replace the $\dot{x}(t)$ in the integral for $\rho(x)$ with its value in terms of $x(t) = u$ and evaluate the integral using the delta function,

$$\rho(x) = \frac{2}{T} \int_{-a/2}^{a/2} \frac{\delta(x - u)}{\omega\sqrt{\left(\frac{a}{2}\right)^2 - u^2}} \, du = \frac{2}{\omega T} \frac{1}{\sqrt{\left(\frac{a}{2}\right)^2 - x^2}}. \tag{7.69}$$

Finally, we use the expression for the period, $T = 2\pi/\omega$ to write

$$\rho(x) = \frac{1}{\pi\sqrt{\left(\frac{a}{2}\right)^2 - x^2}}. \tag{7.70}$$

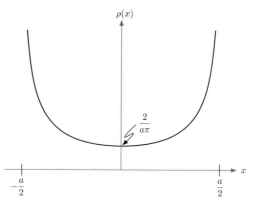

Fig. 7.5 The density from (7.70) as a function of position. The value at $x = 0$ is $2/(a\pi)$, and the function goes to infinity as $x \to \pm a/2$.

The classical harmonic oscillator is constrained to move between $-a/2$ and $a/2$, and cannot be found outside of that domain. If x is outside of $[-a/2, a/2]$, the density is complex, a warning sign here. We've defined the density to be zero for $|x| > a/2$, which is sensible. As you can check, the density is already normalized for $x \in [-a/2, a/2]$. A plot of $\rho(x)$ is shown in Figure 7.5. Notice that the density goes to infinity at the turning points. Because the mass stops at those points, it is more likely to be in their vicinity.

What do we expect from the average position and its variance? The average location should be $x = 0$ from the symmetry of the oscillator. The spread should be relatively wide since the value of the density is largest at the end points. Performing the integration, we get

$$\langle x \rangle = \int_{-a/2}^{a/2} \frac{x}{\pi \sqrt{\left(\frac{a}{2}\right)^2 - x^2}} \, dx = 0, \tag{7.71}$$

and since the mean is zero, the variance is easy to compute

$$\sigma^2 = \langle x^2 \rangle = \int_{-a/2}^{a/2} \frac{x^2}{\pi \sqrt{\left(\frac{a}{2}\right)^2 - x^2}} \, dx = \frac{1}{2} \left(\frac{a}{2}\right)^2 \tag{7.72}$$

so that $\sigma = (a/2)/\sqrt{2}$.

7.4.2 Separation and Schrödinger's Equation

Let's go back to Schrödinger's equation (7.46) and think about how to solve this partial differential equation. As written, the potential energy function $U(x)$ is time-independent, and that hints at a separation of variables approach. Let $\Psi(x, t) = \psi(x)T(t)$, the usual multiplication separation ansatz. Putting this into the Schrödinger equation and dividing by $\Psi(x, t)$ allows us to write

$$-\frac{\hbar^2}{2m\psi(x)} \frac{d^2\psi(x)}{dx^2} + U(x) = i\hbar \frac{\frac{dT(t)}{dt}}{T(t)}. \tag{7.73}$$

The left side depends only on x while the right depends only on t. For the equation to hold for all x and t, each side must be separately constant. That separation constant is traditionally called E, and we can use it to immediately solve for the time-dependent equation

$$i\hbar\frac{dT(t)}{dt} = ET(t) \longrightarrow T(t) = Ae^{-i\frac{Et}{\hbar}} \tag{7.74}$$

where A is a constant of integration, ultimately used to normalize the wave function.

Next we have to solve the "time-independent wave equation"

$$-\frac{\hbar^2}{2m}\frac{d^2\psi(x)}{dx^2} + U(x)\psi(x) = E\psi(x) \tag{7.75}$$

subject to some boundary conditions. This equation cannot be solved until the physical environment is specified by giving $U(x)$.

If we just take the solution we have so far, with

$$\Psi(x,t) = Ae^{-i\frac{Et}{\hbar}}\psi(x), \tag{7.76}$$

the associated position probability density is

$$\rho(x,t) = \Psi(x,t)^*\Psi(x,t) = |A|^2\psi(x)^*\psi(x), \tag{7.77}$$

which is time-independent. For this reason, these individual separable solutions are referred to as "stationary states." Schrödinger's equation is linear in $\Psi(x,t)$, and so superposition holds. It is only when we take solutions with multiple values for E, and add them together in a weighted sum to form $\Psi(x,t)$ that we get nontrivial dynamics for the probability density $\rho(x,t)$. We'll focus on the stationary solutions for a few different potential energy functions.

7.4.3 Time-Independent Solutions

The structure of the time-independent Schrödinger equation is that of a continuous eigenvalue problem (recall Problem 3.3.11) – we must find $\psi(x)$ and E with

$$\left[-\frac{\hbar^2}{2m}\frac{d^2}{dx^2} + U(x)\right]\psi(x) = E\psi(x), \tag{7.78}$$

where the term in brackets represents a linear differential operator, a continuous "matrix" of sorts. We typically require that the probability density, and hence ψ, vanish at spatial infinity. For any spatially localized potential energy, we don't expect to find the particle infinitely far away. Think of performing an experiment in a lab, the electron you're moving around with some electrostatic potential should remain in the lab, and not end up at the other side of the universe. This assumption gives us boundary conditions, $\psi(\pm\infty) = 0$, and we'll use those in what follows.

Infinite Square Well

Our first case will be the infinite square well, just a pair of impenetrable walls at $x = 0$ and a with potential energy

$$U(x) = \begin{cases} 0 & 0 < x < a \\ \infty & x < 0 \text{ or } x > a \end{cases} \tag{7.79}$$

The infinite potential outside the well allows us to set $\psi = 0$ for all points outside, automatically satisfying our boundary condition. Since (7.78) is second order, the wave function is continuous, and then we have $\psi(0) = \psi(a) = 0$. Our problem has been moved entirely inside the box, with localized boundary conditions. Inside, $U(x) = 0$, and we have to solve

$$-\frac{\hbar^2}{2m} \frac{d^2\psi(x)}{dx^2} = E\psi(x) \tag{7.80}$$

a familiar equation if there ever was one. Start with the general solution for constants F and G,

$$\psi(x) = F \cos\left(\sqrt{\frac{2mE}{\hbar^2}}x\right) + G \sin\left(\sqrt{\frac{2mE}{\hbar^2}}x\right). \tag{7.81}$$

Now for the boundaries, from the one at $x = 0$, $\psi(0) = F = 0$. Then the second is

$$\psi(a) = G \sin\left(\sqrt{\frac{2mE}{\hbar^2}}a\right) = 0 \tag{7.82}$$

and we can't take $G = 0$ to satisfy this equation, since then we get the trivial $\psi(x) = 0$, no particle anywhere. So instead, we must have

$$\sin\left(\sqrt{\frac{2mE}{\hbar^2}}a\right) = 0 \longrightarrow \sqrt{\frac{2mE}{\hbar^2}}a = n\pi \tag{7.83}$$

for integer n. This equation can be satisfied for an integer-indexed infinite family of values for E,

$$E_n = \frac{n^2\pi^2\hbar^2}{2ma^2}, \tag{7.84}$$

and then $\psi(x)$ is indexed by n as well,

$$\psi_n(x) = G \sin\left(\frac{n\pi x}{a}\right). \tag{7.85}$$

In terms of the standard interpretation of quantum mechanics, if you measured the energy of the particle in the infinite square well, you would get one of the $\{E_n\}_{n=1}^{\infty}$ as the result. These are the "allowed energies" that could be obtained upon energy measurement, and their integer index represents the quantization of energy. The solution to the full time-dependent wave for one of these stationary states is (absorbing the constant A from (7.74) into G in $\psi_n(x)$)

$$\Psi(x, t) = e^{-i\frac{E_n t}{\hbar}} \psi_n(x). \tag{7.86}$$

Again, there is no time dependence in $\rho(x,t)$ for this pure stationary state, where $\Psi(x,t)^*\Psi(x,t) = \psi_n(x)^*\psi_n(x)$. But we can use superposition to write a general solution for $\Psi(x,t)$,

$$\Psi(x,t) = \sum_{j=1}^{\infty} G_n e^{-i\frac{E_n t}{\hbar}} \sin\left(\frac{n\pi x}{a}\right). \tag{7.87}$$

To find the coefficients $\{G_n\}_{n=1}^{\infty}$, we need an initial function $f(x)$ where we require $\Psi(x,0) = f(x)$, then

$$\Psi(x,0) = \sum_{j=1}^{\infty} G_n \sin\left(\frac{n\pi x}{a}\right) = f(x), \tag{7.88}$$

and we would use the completeness and orthogonality of the sine functions to isolate the coefficients as in Section 4.3.3.

Focusing on the stationary states themselves, the building blocks of more general solutions, we can calculate the average position (expectation value) and variance to see how they compare with the classical case of a ball bouncing between two walls. First, we need to normalize the quantum mechanical probability density. Working from (7.85), the density is

$$\rho_n(x) = \psi_n(x)^*\psi_n = |G|^2 \sin^2\left(\frac{n\pi x}{a}\right) \tag{7.89}$$

and then normalization requires

$$\int_{-\infty}^{\infty} \rho_n(x)\,dx = 1 = \int_0^a |G|^2 \sin^2\left(\frac{n\pi x}{a}\right)\,dx = |G|^2 \frac{a}{2} \tag{7.90}$$

so that we could take $G = \sqrt{2/a}$. Then the position expectation value is

$$\langle x \rangle = \int_{-\infty}^{\infty} x\rho_n(x)\,dx = \int_0^a x\frac{2}{a}\sin^2\left(\frac{n\pi x}{a}\right)\,dx = \frac{a}{2} \tag{7.91}$$

as expected, and matching the classical result.

For the variance, we have

$$\sigma^2 \equiv \left\langle \left(x - \frac{a}{2}\right)^2 \right\rangle = \int_{-\infty}^{\infty} \left(x - \frac{a}{2}\right)^2 \rho_n(x)\,dx$$
$$= \int_0^a \left(x - \frac{a}{2}\right)^2 \frac{2}{a}\sin^2\left(\frac{n\pi x}{a}\right)\,dx = \frac{a^2}{12}\left(1 - \frac{6}{n^2\pi^2}\right). \tag{7.92}$$

Comparing this with (7.62), we recover the classical variance as $n \to \infty$.

Problem 7.4.1 Show that the Gaussian density in (7.59) is normalized using the provided identity in (7.56).

Problem 7.4.2 Using (7.56), compute the expectation values $\langle x \rangle$ and $\langle x^2 \rangle$ for the density (7.59) explicitly.

Problem 7.4.3 What is the derivative of the function $f(x) = |x|$ (absolute value for real x)? What is the second derivative? This is why it is preferable to compute the spread of a distribution using the standard deviation.

Problem 7.4.4 Suppose you had a quartic potential of the form: $U(x) = \alpha x^4$ for constant α (with what units?). Without worrying about normalization, what is the form of the classical, time-averaged probability density assuming a particle of mass m starts from rest at $-a$? Sketch $\rho(x)$ for $x = -a \to a$.

Problem 7.4.5 Given a potential energy, $U(x)$, that leads to oscillatory behavior for particles that move under its influence, use conservation of energy to solve for $\dot{x}(t)$ as a function of $x(t)$, and evaluate $\rho(x)$ from (7.66) to get the general expression.

Problem 7.4.6 The time-independent form of Schrödinger's equation is shown in (7.78), but the complex conjugate of that equation is also equally valid (the "energy" E is real here). Once the boundary conditions, $\psi(\pm\infty) = 0$, are in place, we have seen that the equations can be solved with quantized energy and wave functions. Suppose you have two solutions, indexed by integers n and m, and with $E_n \neq E_m$. Use (7.78), its complex conjugate, and the probabilistic interpretation of the wave function to prove the following orthonormality relation:

$$\int_{-\infty}^{\infty} \psi_n(x)^* \psi_m(x)\, dx = \delta_{nm} \qquad (7.93)$$

using, for example, the technique from Section 6.7.2.

Problem 7.4.7 Complexify time in the one-dimensional Schrödinger equation with generic potential $U(x)$ by taking $t = -is$ as in Problem 1.6.5 (although this time with a minus sign out front). Solve this modified form of Schrödinger's equation for the infinite square well and write the general solution, analogous to (7.87). What does the general solution become as $s \to \infty$ (can you see why we wanted the minus sign in $t = -is$?)?

7.5 Quantum Mechanical Harmonic Oscillator

We'll end by finding the stationary states of the harmonic oscillator in the quantum mechanical setting (this approach is carried out in many quantum mechanical texts, see [13], for example). Our goal is to solve Schrödinger's equation with the potential energy $U(x) = 1/2 m\omega^2 x^2$ for a spring with equilibrium at zero, and fundamental frequency of oscillation ω. Remember, we have to find both $\psi(x)$ and E in

$$-\frac{\hbar^2}{2m}\frac{d^2\psi(x)}{dx^2} + \frac{1}{2}m\omega^2 x^2 \psi(x) = E\psi(x) \qquad (7.94)$$

subject to the boundary conditions $\psi(\pm\infty) = 0$. Multiply both sides of Schrödinger's equation by $2/(\hbar\omega)$ to get

$$-\frac{1}{\frac{m\omega}{\hbar}}\frac{d^2\psi(x)}{dx^2} + \frac{m\omega}{\hbar}x^2\psi(x) = \frac{2E}{\hbar\omega}\psi(x), \qquad (7.95)$$

and we can define the dimensionless variable (see Section 8.2.4 for discussion of the systematic version of this process) $z \equiv \sqrt{m\omega/\hbar x}$,

$$-\frac{d^2\psi(z)}{dz^2} + z^2\psi(z) = \frac{2E}{\hbar\omega}\psi(z). \tag{7.96}$$

Define the dimensionless energy $W \equiv 2E/(\hbar\omega)$ to get the simplified equation

$$\frac{d^2\psi(z)}{dz^2} = z^2\psi(z) - W\psi(z). \tag{7.97}$$

We want to impose the boundary conditions at spatial infinity. For $x \to \infty$, the dimensionless z also goes to infinity, and for large enough value of z^2, the ODE we are trying to solve becomes W-independent. That simplifies our problem, at least in approximation, we have to solve

$$\frac{d^2\psi(z)}{dz^2} = z^2\psi(z). \tag{7.98}$$

This equation is similar in form to (1.107), consider the first-order equation for $p(z)$

$$\frac{dp(z)}{dz} = \pm zp(z), \tag{7.99}$$

then taking the derivative of both sides gives

$$\frac{d^2p(z)}{dz^2} = \pm z\frac{dp(z)}{dz} \pm p(z) = z^2p(z) \pm p(z) \tag{7.100}$$

where we used (7.99) to get the second equality. For large z, this second-order differential equation for $p(z)$ has right-hand side that is dominated by $z^2p(z)$, so that for $p(z)$ solving (7.99), we have an approximate solution to $p''(z) = z^2p(z)$. But we know the solution to (7.99) from (1.109), it's just

$$p(z) = p_0 e^{\pm\frac{z^2}{2}} \tag{7.101}$$

(for constant p_0) and so we have an approximate solution to (7.98), one that holds for large z ((7.98) is already a large z approximation to (7.97)),

$$\psi(z) \approx e^{-\frac{z^2}{2}} \tag{7.102}$$

where we have taken the solution that vanishes at spatial infinity in order to match our boundary conditions.

All we have done is make some approximate observations about the asymptotic form of the solution to (7.97) at large z in order to identify the correct boundary behavior. But we can use these observations to simplify the full problem by incorporating the asymptotic behavior from the start. Take (peeling off the desired behavior at spatial infinity)

$$\psi(z) = e^{-\frac{z^2}{2}}u(z) \tag{7.103}$$

and insert in (7.97) to get an ODE for $u(z)$,

$$\frac{d^2u(z)}{dz^2} - 2z\frac{du(z)}{dz} + (W - 1)u(z) = 0. \tag{7.104}$$

This is the equation we need to solve, and once we have $u(z)$, the solution to (7.97) will be given by (7.103).

We'll use the series solution approach to solve (7.104), take

$$u(z) = z^p \sum_{j=0}^{\infty} a_j z^j$$

$$\frac{du(z)}{dz} = z^p \sum_{j=0}^{\infty} a_j (j+p) z^{j-1} \tag{7.105}$$

$$\frac{d^2 u(z)}{dz^2} = z^p \sum_{j=0}^{\infty} a_j (j+p)(j+p-1) z^{j-2}$$

in (7.104). Performing the usual collection, we get

$$0 = \sum_{j=0}^{\infty} \left[a_{j+2}(j+p+2)(j+p+1) + a_j((W-1) - 2(j+p)) \right] z^j \tag{7.106}$$

$$+ a_0 p(p-1) z^{-2} + a_1 p(p+1) z^{-1}.$$

If we take $p = 0$, then the recursion relation is

$$a_{j+2} = \frac{2j - (W-1)}{(j+1)(j+2)} a_j \tag{7.107}$$

and there is an even series that starts at a_0 and an odd one that starts at a_1. As j gets large, this recursion relation becomes that of a function that goes to infinity for $z \to \infty$, as you will establish in Problem 7.5.2. Even when combined with the Gaussian in (7.103) this will give an overall $\psi(z)$ that does not go to zero at spatial infinity, violating our boundary condition. To preserve our solution, the recursion relation must truncate at some j. Then we will have a polynomial in z coming from $u(z)$ and that will get killed at spatial infinity by the Gaussian in $\psi(z)$. So there must be some $a_{j+2} = 0$, after which all other coefficients will be zero. To get $a_{j+2} = 0$, we must have

$$2j - (W-1) = 0 \longrightarrow W_j = 2j + 1. \tag{7.108}$$

Since $W = 2E/(\hbar\omega)$, our boundary condition requirement has once again quantized the energies in the problem,

$$E_j = \left(j + \frac{1}{2} \right) \hbar\omega. \tag{7.109}$$

To finish the job, let's look at the polynomials we get from the recursion. The equation for $u(z)$ in (7.104) is called "Hermite's equation" and its polynomial solution are the "Hermite polynomials," denoted $H_n(z)$ for integer n. If we start with $a_0 = 1$ and $a_1 = 0$, we would have $a_2 = 0$ corresponding to $E_0 = \hbar\omega/2$ the lowest possible energy here. All other coefficients vanish, and we have the zeroth-order polynomial, $H_0(z) = 1$. Taking[4] $a_0 = 0$, $a_1 = 2$, we would have $a_3 = 0$, giving us $E_1 = 3/2\hbar\omega$, and $H_1(z) = 2z$.

---

[4] The Hermite polynomials have a particular normalization convention, like the Legendre polynomials we encountered in Section 6.7.2.

For the higher-order polynomials, we have to use the recursion relation. Let's work out the second-order polynomial, for $W_2 = 5$, $E_2 = 5/2\hbar\omega$, we have $a_2 = -2a_0$, $a_4 = 0$ and all higher coefficients are zero, too. Taking $a_0 = -2$ (again, chosen for external normalization reasons), we get the second-order polynomial

$$H_2(z) = 4z^2 - 2. \tag{7.110}$$

For the third-order polynomial, associated with $W_3 = 7$, $E_3 = 7/2\hbar\omega$, we get $a_3 = -2a_1/3$, and taking $a_1 = -12$ (normalization convention), we get

$$H_3(z) = 8z^3 - 12z. \tag{7.111}$$

You can try getting a few of the higher-order forms in Problem 7.5.3. In these examples, we see that $W - 1$ is twice the order of the polynomial, and can rewrite (7.104) in terms of the polynomial order, n to get the traditional form of the "Hermite equation"

$$\frac{d^2u(z)}{dz^2} - 2z\frac{du(z)}{dz} + 2nu(z) = 0. \tag{7.112}$$

Putting the pieces back together, the stationary wave functions in dimensionless variables are

$$\psi_n(z) = e^{-\frac{z^2}{2}} H_n(z) \text{ with } W_n = 2n + 1, \tag{7.113}$$

and back in x coordinate,

$$\psi_n(x) = e^{-\frac{m\omega}{2\hbar}x^2} H_n\left(\sqrt{\frac{m\omega}{\hbar}}x\right) \text{ with } E_n = \left(n + \frac{1}{2}\right)\hbar\omega. \tag{7.114}$$

These wave functions are not normalized, but must be before any probabilistic interpretation is applied.

We can construct the stationary densities, $\rho_n(x) = \psi_n(x)^* \psi_n(x)$ (appropriately normalized), and the first three of these are plotted in Figure 7.6. Notice that there is a nonzero probability of finding the particle far from its classically allowed region (see Figure 7.5). In quantum mechanics, complete localization is never possible except in artificial cases like the infinite square well. There is almost always some probability of finding the particle far from where it is "supposed to be." You can see from the probability densities in Figure 7.6 that the effect is more pronounced for larger energies, the density is more and more spread out as n gets large.

Problem 7.5.1   For the ODE:

$$\frac{df(x)}{dx} = \left(2x + \frac{1}{x}\right)f(x),$$

identify the asymptotic, $x \to \infty$ ODE and solve it. Call that solution $g(x)$, and let $f(x) = g(x)h(x)$ for unknown $h(x)$. Running your assumed form through the ODE, write down and solve the ODE for $h(x)$. Finally, write the full solution for $f'(0) = 1$. This problem gives a first-derivative version of the procedure we carried out in the last section where we explicitly incorporated the asymptotic behavior of our solution to make life simpler.

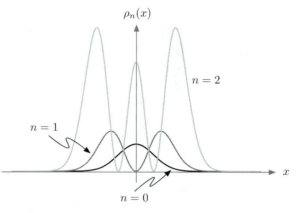

$\rho_n(x)$

$n = 2$

$n = 1$

$n = 0$

x

Fig. 7.6   Position probability densities for the first three stationary states of the quantum mechanical harmonic oscillator.

Problem 7.5.2   Write the Taylor expansion of the function $f(x) = e^{x^2}$. Identify the "coefficients" in the general form

$$f(x) = \sum_{j=0}^{\infty} a_j x^j$$

and find the recursion relation between successive terms for large j, i.e., what is a_{j+2} in terms of a_j (only even terms will contribute to the sum) when j is large? Compare with the large j limit of (7.107).

Problem 7.5.3   The Rodrigues formula for the Hermite polynomials is

$$H_n(x) = (-1)^n e^{x^2} \left(\frac{d}{dx}\right)^n e^{-x^2}. \tag{7.115}$$

Using this, verify that you get the first four polynomials as in the text and find $H_4(x)$.

Problem 7.5.4   Plot the first three stationary wave functions, $\psi_n(x)$ from (7.114). If you had a classical oscillator that started from rest at $x = \sqrt{\hbar/(m\omega)}$, what would the classically allowed region (in which motion could occur) be? Indicate this region on your plot.

Problem 7.5.5   Show that the Hermite polynomials have

$$\int_{-\infty}^{\infty} e^{-x^2} H_n(x) H_m(x)\, dx = 0 \text{ for } n \neq m, \tag{7.116}$$

an orthogonality relation. Use the Hermite differential equation (7.112) together with the technique from Section 6.7.2 that gave orthogonality for the Legendre polynomials.

Problem 7.5.6   For a classical particle of mass m attached to a spring with frequency ω released from rest at $x = \sqrt{\hbar/(m\omega)}$, we know oscillation occurs between $\pm\sqrt{\hbar/(m\omega)}$. In quantum mechanics, there is less restriction on where the particle can be. What is the probability of finding a particle outside the classically allowed region for the harmonic oscillator in its "ground state" (ψ_0 in (7.114))? How about in the "first excited state," ψ_1? (don't forget to normalize the probability densities before computing these probabilities).

Numerical Methods

In this final chapter, we look at a variety of numerical methods that can be applied to the problems we have encountered throughout the book. Each section is relatively self-contained, starting off with a familiar physical problem that cannot be solved analytically (at least, not completely). These problems are stated, or re-stated, as the case may be, and then a numerical method to solve them is introduced with some examples. The goal is to see how numerical techniques can augment the closed-form "analytical" approach taken in the rest of this book.[1]

At the end of each section are problems that are meant to be worked with paper and pencil, and some "lab" problems that require a computer. For some of the problems, a numerical package like `Mathematica` or Matlab should be used (for calculating eigenvalues/vectors, for example). I have described all algorithms in pseudo-code to avoid picking one package/language over another.

8.1 Root-Finding

The "root finding" problem is: Given a function $F(x)$, find some or all of the set $\{\bar{x}_i\}_{i=1}^n$ such that $F(\bar{x}_i) = 0$. This type of question shows up in a variety of settings.

8.1.1 E&M Example

For the time-dependent fields in electricity and magnetism, we know that "information" (the magnitude and direction of the electric and magnetic fields $\mathbf{E}$ and $\mathbf{B}$ at a particular point $\mathbf{r} = x\,\hat{\mathbf{x}} + y\,\hat{\mathbf{y}} + z\,\hat{\mathbf{z}}$ at time t) travels at a finite and specific speed: c (in vacuum). The immediate implication is that if we would like to know the electric field at our observation point (and time) due to a charge that is moving along some prescribed path, $\mathbf{w}(t)$, we need to evaluate the location of the particle not at time t, but at some earlier time t_r (the "retarded" time).

The setup is shown in Figure 8.1 – the particle of charge q is at $\mathbf{w}(t)$ at time t, but it is the earlier location $\mathbf{w}(t_r)$ that has information traveling at c to the observation point $\mathbf{r}$ at time t. The time it takes for the information from $\mathbf{w}(t_r)$ to reach $\mathbf{r}$ is $t - t_r$, and during that interval, the field information travels at c, so the distance travelled is $c(t - t_r)$. The distance

---

[1] Expanded discussion of these methods, with additional examples, can be found in [7].

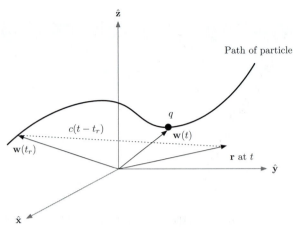

Fig. 8.1 A particle with charge q moves along the prescribed curve $\mathbf{w}(t)$. For the point $\mathbf{r}$ at time t, field information comes from the earlier time t_r, when the charge was at $\mathbf{w}(t_r)$.

between $\mathbf{r}$ and $\mathbf{w}(t_r)$ is, geometrically, $|\mathbf{r} - \mathbf{w}(t_r)|$. Putting these together, we have the defining equation for the retarded time:

$$c(t - t_r) = |\mathbf{r} - \mathbf{w}(t_r)| \equiv \sqrt{(\mathbf{r} - \mathbf{w}(t_r)) \cdot (\mathbf{r} - \mathbf{w}(t_r))}. \tag{8.1}$$

Notice that the retarded time t_r is a function of t and $\mathbf{r}$ (the observation time and location), and is defined implicitly in terms of $\mathbf{w}(t_r)$. We may or may not be able to solve (8.1) for t_r given $\mathbf{w}(t)$ analytically (this cannot be done for any but the simplest trajectories), but we can define a root-finding problem. The roots of the function

$$F(T) = c(t - T) - |\mathbf{r} - \mathbf{w}(T)| \tag{8.2}$$

give the retarded time.

8.1.2 Bisection

The procedure we will use is called "bisection." Start with a pair of values, x_ℓ^0 and x_r^0 with $x_\ell^0 < x_r^0$ (hence the ℓ and r designations), and $F(x_\ell^0)F(x_r^0) < 0$ (so that a root is in between the points). Then calculate the midpoint between this pair, $x_m^0 \equiv (x_\ell^0 + x_r^0)/2$, and evaluate the product $p \equiv F(x_\ell^0)F(x_m^0)$. If p is less than zero, then the root lies between the left and middle points, so we can move the left and right points over by setting $x_\ell^1 = x_\ell^0, x_r^1 = x_m^0$. If p is greater than zero, the root is between the middle and right points, and we update $x_\ell^1 = x_m^0, x_r^1 = x_r^0$. The iteration is continued until the absolute value of F at the current midpoint is smaller than some tolerance ϵ,

$$|F(x_m^n)| \le \epsilon, \tag{8.3}$$

and then x_m^n is an approximation to the root. The process is shown pictorially in Figure 8.2, and is described with pseudocode in Algorithm 8.1.

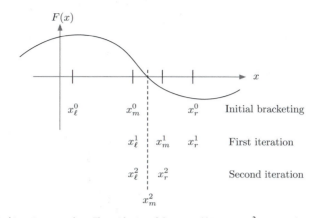

Fig. 8.2 Two iterations of the bisection procedure. The midpoint of the second bisection, x_m^2, is a good approximation to the root here.

Algorithm 8.1 Bisection$(F, x_\ell^0, x_r^0, \epsilon)$

$x_\ell \leftarrow x_\ell^0$
$x_r \leftarrow x_r^0$
$x_m \leftarrow 1/2(x_\ell + x_r)$
while $|F(x_m)| > \epsilon$ **do**
  if $F(x_m)F(x_r) < 0$ **then**
    $x_\ell \leftarrow x_m$
  else
    $x_r \leftarrow x_m$
  end if
  $x_m \leftarrow 1/2(x_\ell + x_r)$
end while
return x_m

8.1.3 Example: The Electric Potential of a Moving Charge

Given a vector describing the location of a charge q at time t, $\mathbf{w}(t)$, we can use root-finding to find the retarded time associated with a field point $\mathbf{r}$ at time t. From that, we can evaluate the full electric field, or we can focus on the (simpler) potential (see [12]):

$$V(\mathbf{r}, t) = \frac{q}{4\pi\epsilon_0} \frac{1}{|\mathbf{r} - \mathbf{w}(t_r)| - (\mathbf{r} - \mathbf{w}(t_r)) \cdot \dot{\mathbf{w}}(t_r)/c}, \tag{8.4}$$

which reduces, for slow source-charge motion ($\dot{w}(t_r) \ll c$), to

$$V(\mathbf{r}, t) = \frac{q}{4\pi\epsilon_0 |\mathbf{r} - \mathbf{w}(t_r)|}, \tag{8.5}$$

and we'll use this approximation.

Given a charge's trajectory through space, $\mathbf{w}(t)$, we can use the routine described in Algorithm 8.2 to generate the (approximate) electric potential at an individual field point, and by evaluating the potential at a variety of points (in a grid, for example), we can generate a contour plot of it (that process is described in Problem 8.1.5).

Algorithm 8.2 $\text{Vfield}(x, y, z, t, \mathbf{w}, \epsilon)$

$\mathbf{r} \leftarrow \{x, y, z\}$

$F(X) \leftarrow c(t - X) - |\mathbf{r} - \mathbf{w}(X)|$

$t_r \leftarrow \text{Bisection}(F, -\infty, t, \epsilon)$

$R \leftarrow \sqrt{(\mathbf{r} - \mathbf{w}(t_r)) \cdot (\mathbf{r} - \mathbf{w}(t_r))}$

$V \leftarrow \frac{q}{4\pi\epsilon_0 R}$

return V

Problem 8.1.1 The convergence of the bisection method to a root with tolerance ϵ is independent of the function being bisected. To see this, note that for an interval sufficiently close to the root, the interval width should be $\sim \epsilon$: think of the Taylor expansion of the function $F(x)$ near a root, $F(\bar{x} + \Delta x) \approx F(\bar{x}) + F'(\bar{x})\Delta x = F'(\bar{x})\Delta x$ so that the magnitude of the function in the Δx vicinity of $\bar{x}$ is itself $F'(\bar{x})\Delta x$, i.e., proportional to Δx. Then if we demand that $|F(\bar{x} + \Delta x)| \approx \epsilon$, we have $|F'(\bar{x})|\Delta x \approx \epsilon$ or $\Delta x \approx \epsilon$ for finite $F'(\bar{x})$. Starting with the initial interval, $\Delta^0 \equiv x_r^0 - x_\ell^0$, how many times, n, must you bisect the interval to get $\Delta^n \equiv x_r^n - x_r^\ell \approx \epsilon$?

Problem 8.1.2 Implement the bisection procedure, and test it by finding the roots of $f(x) = (x - \pi)(x + 1)$ and the first two roots of the 0th Bessel function, $J_0(x)$. Use $\epsilon = 10^{-10}$ as the tolerance in both cases.

Problem 8.1.3 In order to solve the delay differential equation from Problem 4.6.3, we needed to be able to solve the transcendental equation

$$x = e^{-\alpha x}.$$

This type of equation also shows up in quantum mechanics. Use bisection to find x given $\alpha = 0 \longrightarrow 10$ in steps of 1 using $\epsilon = 10^{-10}$ as the tolerance for bisection. Make a plot of x versus α and compare with a plot of the formal solution of this problem: $x = \text{ProductLog}(\alpha)/\alpha$ (which basically *defines* the product log function).

Problem 8.1.4 A charged particle moves along the trajectory described by

$$\mathbf{w}(t) = R\cos(\omega t)\,\hat{\mathbf{x}} + d\omega t\,\hat{\mathbf{z}}.$$

Use the bisection routine to find the retarded time for the point $\mathbf{r} = 0$ at $t = 0$ if $R = 10^5$ m, $\omega = 2000$ 1/s, $d = 1$ m. Use $\epsilon = 10^{-10}$ as your tolerance. What is $\mathbf{w}(t_r)$? How does this compare with $\mathbf{w}(0)$, the value we would use if we ignored the retarded time?

Problem 8.1.5 Given $\mathbf{w}(t) = d\cos(\omega t)\,\hat{\mathbf{y}}$ with $d = .05$ m, $\omega = 5$ s^{-1}, and taking $q/(4\pi\epsilon_0) \to 1$, $c \to 1$ for simplicity, implement Algorithm 8.2 to find V and make contour plots of its value for points between $-1 \to 1$ in x and $.1 \to 1$ in y, with $z = 0$ and t going from 0 to $8\pi/5$ in steps of $2\pi/(25)$. Use $\epsilon = 10^{-10}$ again, and take $-\infty \to -10$ for the bisection inside Algorithm 8.2.

8.2 Solving ODEs

Ordinary differential equations come up in physics all the time. The earliest example is Newton's second law, $m\ddot{x}(t) = F(x)$, or its relativistic version in (8.7). While the effect of some simple forces can be determined from these equations, most interesting cases cannot be solved analytically. We want a numerical method for solving Newton's second law. We'll start with some motivation, then introduce the method.

8.2.1 Pendulum

The nonlinearized pendulum has motion that can be described by an angle $\theta(t)$ which satisfies Newton's second law as we saw in Section 1.3.4:

$$\ddot{\theta}(t) = -\frac{g}{L}\sin(\theta(t)) \tag{8.6}$$

where g is the gravitational constant, L is the length of the pendulum, and $\theta(t)$ the angle the bob makes with respect to vertical at time t. We know how to find the period from Section 5.3.2, but the actual $\theta(t)$ can be found numerically by approximating the solution to (8.6).

8.2.2 Relativistic Spring

For a mass m attached to a wall by a spring with spring constant k, if we start the mass from rest with initial extension a, the mass moves according to $x(t) = a\cos(\sqrt{k/m}t)$ from Newton's second law with $F = -kx$. The maximum speed of the particle is then $\sqrt{k/m}a$ which can be greater than c. In its relativistic form, Newton's second law: $\frac{dp}{dt} = F$ is

$$\frac{d}{dt}\left[\frac{m\dot{x}}{\sqrt{1 - \frac{\dot{x}^2}{c^2}}}\right] = F \longrightarrow m\ddot{x} = \left(1 - \frac{\dot{x}^2}{c^2}\right)^{3/2} F, \tag{8.7}$$

and this comes from using the relativistic momentum for p (instead of $p = m\dot{x}$). The solution, in this setting, does not have speed greater than c. For the spring force, we have

$$m\ddot{x} = \left(1 - \frac{\dot{x}^2}{c^2}\right)^{3/2}(-kx). \tag{8.8}$$

As with the pendulum, we used integration to find the period in Section 5.3.2, and now we will use numerical methods to find $x(t)$.

8.2.3 Discretization and Approach

We want a numerical method that will allow us to generate an approximate solution for $\theta(t)$ in (8.6) or $x(t)$ in (8.8), and one way to go about doing that is to introduce a temporal "grid." Imagine chopping time up into small chunks of size Δt, then we could refer to the

jth chunk as $t_j \equiv j\Delta t$. We'll develop a numerical method for approximating the values of $x(t_j)$ at these special time points.

Let $x_j \equiv x(t_j)$, then we can relate x_j to x_{j+1} and x_{j-1} using you-know-what:

$$x_{j\pm 1} \approx x_j \pm \left.\frac{dx(t)}{dt}\right|_{t=t_j} \Delta t + \frac{1}{2} \left.\frac{d^2 x(t)}{dt^2}\right|_{t=t_j} \Delta t^2. \tag{8.9}$$

The second derivative of $x(t)$ at time t_j is just the acceleration at that time. If we are given a force $F(t)$ (or other form for the second derivative, as in (8.6)), we could write $x_{j\pm 1}$ as

$$x_{j\pm 1} \approx x_j \pm \left.\frac{dx(t)}{dt}\right|_{t=t_j} \Delta t + \frac{1}{2}\frac{F(t_j)}{m}\Delta t^2. \tag{8.10}$$

From here, we can eliminate velocity by adding x_{j+1} and x_{j-1}:

$$x_{j+1} + x_{j-1} = 2x_j + \Delta t^2 \frac{F_j}{m} \tag{8.11}$$

where F_j is short-hand for $F(t_j)$, or $F(x_j, t_j)$ for a force that depends on position and (explicitly on) time. This form allows us to define a method: Given x_j and x_{j-1}, set x_{j+1} according to:

$$x_{j+1} = 2x_j - x_{j-1} + \Delta t^2 \frac{F_j}{m}. \tag{8.12}$$

The update here defines the "Verlet method." If you know the previous and current positions of a particle, you can estimate the next position, and proceed.

In the Newtonian setting, we are typically given $x(0) = x_0$ and $v(0) = v_0$. How can we turn that initial data into $x(0) = x_0$ and $x(-\Delta t) = x_{-1}$? Using Taylor expansion, we can write:

$$x_{-1} \approx x_0 - \Delta t v_0 + \frac{1}{2}\Delta t^2 \frac{F_0}{m}. \tag{8.13}$$

Now, given both x_0 and v_0, and the force, you can estimate the trajectories. The Verlet method is shown in Algorithm 8.3: you provide the mass of the particle, the initial position x_0 and velocity v_0, the function that evaluates the force, F (that takes a position and time, $F(x,t)$), the time-step Δt and the number of steps to take, N. The Verlet method then sets the X_c, X_p (current and previous positions) and stores the initial position in "xout." Then the method goes from $j = 1$ to N, calculating X_n (the "next" position) according to (8.13), storing the new position, and updating the current and previous positions.

8.2.4 Rendering Equations Dimensionless

When working numerically, we have to be careful with the constants that appear in the relevant equations. The issue is that there are limits to the numbers we can represent on a computer – obviously infinity is out, but so, too, is zero. We'd like to stay away from zero and infinity numerically, keeping our problems safely in the range near 1.[2]

---

[2] The number one falls halfway between zero and infinity from this point of view.

Algorithm 8.3 Verlet$(m, x_0, v_0, F, \Delta t, N)$

$X_c \leftarrow x_0$

$X_p \leftarrow x_0 - \Delta t v_0 + \frac{1}{2}\Delta t^2 \frac{F(X_c, 0)}{m}$

xout $\leftarrow$ table of n zeroes

$\text{xout}_1 \leftarrow X_c$

for $j = 2 \rightarrow N$ **do**

    $X_n \leftarrow 2X_c - X_p + \Delta t^2 m^{-1} F(X_c, j\Delta t)$

    $\text{xout}_j \leftarrow X_n$

    $X_p = X_c$

    $X_c = X_n$

end for

return xout

In order to do this, we use a rescaling procedure to soak up unwieldy constants. That procedure is also interesting outside of numerical work, since the number and type of dimensionless quantities we can produce tells us something about the structure of the equations we are solving and can reveal the true number of degrees of freedom available to us. As an example, take the harmonic oscillator equation of motion, $m\ddot{x}(t) = -k(x(t) - a)$. We would say there are three constants here, the mass m, spring constant k and equilibrium spacing a. But we know from experience that only the ratio k/m shows up in the solution, so there are really only two independent constants, k/m and a.

The approach is to take every variable in a particular equation and make dimensionless versions of them. To demonstrate, we'll use the oscillator equation of motion

$$m\frac{d^2x(t)}{dt^2} = -k(x(t) - a) \tag{8.14}$$

as a model.

The "variables" here are t and $x(t)$, those are the things that can change. Let $t = t_0 T$ where t_0 has the dimension of time and the variable T is dimensionless (I like to use capitalized variable names to refer to the dimensionless form), and take $x = x_0 X$ where X is dimensionless and x_0 has dimension of length. We don't know what t_0 and x_0 are just yet, and we'll set them in order to simplify the final dimensionless equation of motion. Putting our expressions for x and t in gives

$$\frac{m}{t_0^2}\frac{d^2X}{dT^2} = -k\left(X - \frac{a}{x_0}\right). \tag{8.15}$$

If we take the dimensionless quantity $a/x_0 = 1$ and similarly $m/t_0^2 = k$, then we can write the differential equation governing $X(T)$ in simplified form:

$$\frac{d^2X(T)}{dT^2} = -(X(T) - 1) \tag{8.16}$$

where the equation is dimensionless, and governs dimensionless variables, with no constants in sight. We have fixed $x_0 = a$, $t_0 = \sqrt{m/k}$ which looks familiar, this is a characteristic time-scale for the problem.

The simplicity of (8.16) highlights another advantage of the dimensionless approach, we only need to solve the equation once.[3] The solution is, of course,

$$X(T) = F\cos(T) + G\sin(T), \tag{8.17}$$

for dimensionless constants of integration F and G. How do we restore the physical constants? Just invert the defining relations between t and T, x and X (i.e. $X = x/x_0$, $T = t/t_0$),

$$x(t) = a\left(F\cos\left(\sqrt{\frac{k}{m}}t\right) + G\sin\left(\sqrt{\frac{k}{m}}t\right)\right) \tag{8.18}$$

which we recognize.

The harmonic oscillator by itself doesn't have any tunable parameters left in its dimensionless form (8.16), there's nothing left to do but solve it. The damped harmonic oscillator has two time-scales in it, one set by the spring constant, one set by the damping parameter, and we can't pick a single value for t_0 that covers both of them. Let's see what happens in this case. Starting from

$$m\frac{d^2x(t)}{dt^2} = -m\omega^2(x(t) - a) - 2mb\frac{dx(t)}{dt}, \tag{8.19}$$

let $t = t_0T$, $x = x_0X$ as before. Putting these in gives

$$\frac{d^2X(T)}{dT^2} = -\omega^2 t_0^2\left(X(T) - \frac{a}{x_0}\right) - 2bt_0\frac{dX(T)}{dT}. \tag{8.20}$$

The length scale x_0 can again be set by demanding that $a/x_0 = 1$. But now we need to pick either $\omega^2 t_0^2 = 1$ or $bt_0 = 1$. Which should we choose? Either one is useful, and the particular choice depends on what you want to do. For example, if you would like to be able to probe the $b = 0$ (no damping) limit, then take $\omega^2 t_0^2 = 1$, allowing us to pick t_0 independent of the value of b. The equation of motion becomes

$$\frac{d^2X(T)}{dT^2} = -(X(T) - 1) - 2\frac{b}{\omega}\frac{dX(T)}{dT}. \tag{8.21}$$

The ratio b/ω is itself dimensionless, and clearly represents a parameter that can tune the influence of the damping from none, $b = 0$, to overdamped. It is useful to define the dimensionless ratios in the problem to highlight their role. Let $\alpha \equiv b/\omega$, then the final form of the dimensionless equation of motion governing the damped harmonic oscillator is

$$\frac{d^2X(T)}{dT^2} = -(X(T) - 1) - 2\alpha\frac{dX(T)}{dT}. \tag{8.22}$$

The initial values must also be dimensionless, for $x(0) = x_i$ and $\dot{x}(0) = v_i$ given, we have initial values for $X(T)$: $X(0) = x_i/x_0$ and $\frac{dX(T)}{dT}\big|_{T=0} = v_it_0/x_0$.

---

[3] That's no great savings if you solve an ODE symbolically, since all the constants are available in the single solution, but when working numerically, each new value of a constant like k in (8.14) requires a new numerical solution. With the dimensionless form, we solve once, and then change the "axis labels" in our plots to indicate the unit of measure.

Problem 8.2.1   Come up with a one-dimensional force that is familiar to you from your physics work so far, pose the Newton's second law problem (i.e. write Newton's second law and provide initial conditions). You will solve for the trajectory of a particle under the influence of this force (relativistically, if you like).

Problem 8.2.2   Pick the "other" value of t_0 for use in (8.20) and form the resulting dimensionless equation of motion. With this one, you could probe the "no oscillation" limit of the problem.

Problem 8.2.3   Make a dimensionless version of the equation of motion for a relativistic spring (8.8).

Problem 8.2.4   Using x_j and x_{j-1} (the "current" and "previous" positions), come up with an approximation for v_j (the velocity at t_j) using Taylor expansion.

Problem 8.2.5   Find the period for the nonrelativistic problem: $m\ddot{x} = -kx$ for $x(0) = a$ (the initial extension) and $\dot{x}(0) = 0$ (starting from rest). What is the maximum speed of the mass? For what initial extension will the maximum speed be c?

Problem 8.2.6   Implement and run Verlet for a spring with $\sqrt{k/m} = 2$ Hz (the spring constant is k and the mass is $m = 1$ kg). Go from $t = 0$ to $t = 4\pi$ s with $N = 1000$ for a mass that starts from rest at $x_0 = 2$ m. Plot the resulting position as a function of time. Make a table that takes each point in your numerical solution and subtracts the exact solution, plot this table of "residuals."

Problem 8.2.7   Run Verlet for a real pendulum (what will play the role of the "force" in Algorithm 8.3?) of length $L = 1$ m starting from rest at initial angles $\theta = 10°$, $45°$, and $80°$. Use $N = 1000$ and $t = 4\pi$ s as in the previous problem. For each of your solutions, plot the exact solution to the linearized problem, $\ddot{\theta}(t) \approx -g/L\theta(t)$ on top of your numerical solution for the full problem.

Problem 8.2.8   Modify Verlet to solve the relativistic problem (8.7) (you will need your result from Problem 8.2.4). Run your mass-on-a-spring from Problem 8.2.6 again, this time starting with a mass at rest at the critical initial extension that leads to a maximum speed of c from Problem 8.2.5 – use the same total time and N as in Problem 8.2.6. Try running at ten times the critical initial extension (you will need to increase the total time, and should also increase the number of steps). Is the position (as a function of time) what you expect? What is the period of the motion (just estimate from a plot)?

Problem 8.2.9   Solve your problem from Problem 8.2.1 numerically over some reasonable domain. Try solving the relativistic version using (8.7).

8.3 Integration

The one-dimensional integration problem is: Given a function $F(x)$, find the value of

$$I = \int_a^b F(x)\, dx \tag{8.23}$$

for provided integration limits a and b. There are any number of places this type of problem shows up. It can arise in a relatively direct setting: Given a rod with mass-per-unit-length $\lambda(x)$, how much mass is contained between x_0 and x_f? But evaluating (8.23) also occurs in the slightly disguised form of integral solutions to PDEs.

8.3.1 Biot–Savart Law

The Biot–Savart law (see [12]) governing the generation of a magnetic field from a closed loop of wire carrying a steady current I is

$$\mathbf{B}(\mathbf{r}) = \frac{\mu_0 I}{4\pi} \oint \frac{d\boldsymbol{\ell}' \times (\mathbf{r} - \mathbf{r}')}{|\mathbf{r} - \mathbf{r}'|^3} \tag{8.24}$$

where $\mathbf{r}$ is the vector pointing to the location at which we would like to know the field, $\mathbf{r}'$ points to the portion of the wire that is generating the field (we integrate over the closed wire loop, so we include all those source pieces), $d\boldsymbol{\ell}'$ is the vector line element that points locally along the wire, and there are constants out front. The setup is shown in Figure 8.3.

We can use the Biot–Savart law to find the magnetic field of a circular loop of radius R carrying steady current I. Set the loop in the xy plane, and use polar coordinates so that: $\mathbf{r}' = R\cos\phi'\,\hat{\mathbf{x}} + R\sin\phi'\,\hat{\mathbf{y}}$ points from the origin to points along the loop. Then the vector $d\boldsymbol{\ell}'$ is just:

$$d\boldsymbol{\ell}' = \frac{d\mathbf{r}'}{d\phi'}d\phi' = (-R\sin\phi'\,\hat{\mathbf{x}} + R\cos\phi'\,\hat{\mathbf{y}})\,d\phi'. \tag{8.25}$$

Take $\mathbf{r} = x\,\hat{\mathbf{x}} + y\,\hat{\mathbf{y}} + z\,\hat{\mathbf{z}}$, an arbitrary location, as the point at which we would like to know the magnetic field. The Biot–Savart law gives (once the dust has settled):

$$\mathbf{B}(\mathbf{r}) = \frac{\mu_0 I R}{4\pi} \int_0^{2\pi} \frac{z\cos\phi'\,\hat{\mathbf{x}} + z\sin\phi'\,\hat{\mathbf{y}} + (-x\cos\phi' - y\sin\phi' + R)\,\hat{\mathbf{z}}}{\left((x - R\cos\phi')^2 + (y - R\sin\phi')^2 + z^2\right)^{3/2}}\,d\phi'. \tag{8.26}$$

All that remains is the evaluation of the integral given a point of interest, $\mathbf{r}$.

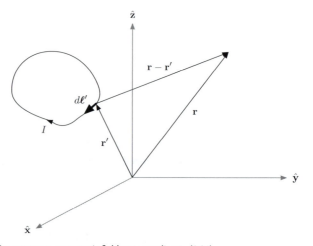

Fig. 8.3   A current carrying wire generates a magnetic field at $\mathbf{r}$ according to (8.24).

8.3.2 Quadrature

In order to numerically approximate I in (8.23), we'll introduce a grid in x as we did with time in Section 8.2. Let $x_j = a + j\Delta x$ with $x_n = b$ for given n (the number of points in the grid) so that j goes from zero to n. Let $F_j \equiv F(x_j)$. The first approximation to I comes from replacing the integral sign with a summation and dx with Δx:

$$I \approx \sum_{j=0}^{n-1} F_j \Delta x \equiv I_b. \tag{8.27}$$

You can think of this approximation as follows: Assuming the value of $F(x)$ is constant over the interval from x_j to x_{j+1}, the exact integral over the interval is

$$\int_{x_j}^{x_{j+1}} F_j \, dx = F_j(x_{j+1} - x_j) = F_j \Delta x, \tag{8.28}$$

so we are integrating a constant function exactly over each interval, then adding those up. The approximation comes from the fact that the function $F(x)$ does not take on the constant value F_j over the interval. The idea is shown geometrically in Figure 8.4.

We can refine the integral approximation (without changing Δx) by using better approximations to $F(x)$ on the interior of each interval. For example, suppose we make a linear approximation to $F(x)$ between x_j and x_{j+1} that matches $F(x)$ at x_j and x_{j+1}, then integrate that exactly. Take

$$F(x) \approx F_j + (x - x_j) \frac{F_{j+1} - F_j}{x_{j+1} - x_j} \tag{8.29}$$

for $x = x_j$ to x_{j+1}, then the exact integral of this linear function is

$$\int_{x_j}^{x_{j+1}} \left(F_j + (x - x_j) \frac{F_{j+1} - F_j}{x_{j+1} - x_j} \right) dx = \frac{1}{2}(F_j + F_{j+1})(x_{j+1} - x_j)$$
$$= \frac{1}{2}(F_j + F_{j+1}) \Delta x, \tag{8.30}$$

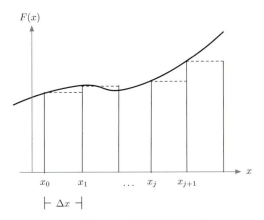

Fig. 8.4   A segment showing the approximation we make in using (8.27). The dashed lines represent the piecewise continuous function that (8.27) integrates exactly.

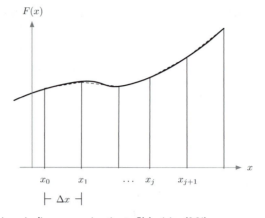

$F(x)$

x_0 x_1 $\cdots$ x_j x_{j+1}

x

$\vdash \Delta x \dashv$

Fig. 8.5 An exact integration of a piecewise linear approximation to $F(x)$, giving (8.31).

and if we add these up over the entire domain, we get a new approximation to I in (8.23)

$$I \approx \sum_{j=0}^{n-1} \frac{1}{2}(F_j + F_{j+1})\,\Delta x \equiv I_t. \tag{8.31}$$

This approximation for I is shown in Figure 8.5, and is known as the "trapezoidal approximation." Notice in Figure 8.5, the dashed lines indicating the piecewise linear function that we use to approximate $F(x)$ over each interval are a much better approximation to the function (and lie "on top of it" visually in places).

Continuing, we can replace our piecewise linear approximation with a quadratic one. This time, we need to include more points in our approximation of $F(x)$. To do linear interpolation, we only needed the values of $F(x)$ at x_j and x_{j+1} but now to set all the coefficients in a quadratic interpolation, we must use the values of $F(x)$ at x_j, x_{j+1}, and x_{j+2}. Over the interval x_j to x_{j+2}, then, we'll approximate $F(x)$ by

$$F(x) \approx \frac{1}{2\Delta x^2}\Big[(x - x_{j+1})(x - x_{j+2})F_j - 2(x - x_j)(x - x_{j+2})F_{j+1} \\ + (x - x_j)(x - x_{j+1})F_{j+2}\Big], \tag{8.32}$$

where the quadratic approximation matches $F(x)$ at x_j, x_{j+1}, and x_{j+2}. The exact integral of the quadratic approximation function, over the interval of interest, is

$$\int_{x_j}^{x_{j+2}} F(x)\,dx = \frac{1}{3}\Delta x(F_j + 4F_{j+1} + F_{j+2}), \tag{8.33}$$

giving us the "Simpson's rule" approximation to I,

$$I \approx \sum_{j=0,2,4,\ldots}^{n-2} \frac{1}{3}\Delta x(F_j + 4F_{j+1} + F_{j+2}) \equiv I_s. \tag{8.34}$$

The implementation of the Simpson's rule approximation is shown in Algorithm 8.4. There, you provide the function F, the limits of integration a to b and the number of grid points n (that *must* be a factor of two, as is clear from (8.34)).

Algorithm 8.4 Simpson(F, a, b, n)

Check that n is a factor of 2, exit if not.
$\Delta x \leftarrow (b - a)/n$
$I_s \leftarrow 0$
for $j = 0 \rightarrow n - 2$ in steps of 2 **do**
$\quad I_s \leftarrow I_s + \frac{1}{3}\Delta x(F(a + j\Delta x) + 4F(a + (j + 1)\Delta x) + F(a + (j + 2)\Delta x))$
end for
return I_s

You can use Simpson's method to find the magnetic field associated with (8.26) as shown:

Algorithm 8.5 Bfield(x, y, z, R, n)

$F = \frac{\mu_0 I R}{4\pi} \dfrac{z \cos\phi\, \hat{\mathbf{x}} + z \sin\phi\, \hat{\mathbf{y}} - (x\cos\phi + y\sin\phi - R)\, \hat{\mathbf{z}}}{\left((x - R\cos\phi)^2 + (y - R\sin\phi)^2 + z^2\right)^{3/2}}$
$Bout \leftarrow \text{Simpson}(F, 0, 2\pi, n)$
return $Bout$

Problem 8.3.1 Evaluate (8.26) for points along the z axis (i.e. for $x = y = 0$) analytically. We'll use this to compare with your numerical result.

Problem 8.3.2 Write (8.31) in terms of (8.27), i.e. relate I_t to I_b – you should need to evaluate two extra points, so that $I_t = I_b + X + Y$ where X and Y involve single evaluations of $F(x)$.

Problem 8.3.3 Implement Simpson's method, Algorithm 8.4, and use it to approximate the integral

$$I = \int_0^2 \frac{\sin(x)}{x}\, dx. \qquad (8.35)$$

The correct value, up to 10 digits, is 1.605412977. How many grid points do you need to match these ten digits?

Problem 8.3.4 Use Simpson's method to approximate the integral:

$$I(f) = \int_{-f}^{f} e^{-x^2}\, dx, \qquad (8.36)$$

and make a plot of $I(f)$ for f goes from $f = 1 \rightarrow 5$ in integer steps using $n = 100$ grid points for the integration. Does the integral converge to $\sqrt{\pi}$ as claimed in (7.56)? Try the same experiment for the integrand $x^2 e^{-x^2}$.

Problem 8.3.5 Use Simpson's method to approximate the sine integral

$$I(f) = \int_{-f}^{f} \frac{\sin(x)}{x}\, dx,$$

and make a plot of $I(f)$ for $f = 0 \to 100$ in integer steps, again using $n = 100$ grid points in Simpson's method. To what value does the integral converge?

Problem 8.3.6   A circular loop of wire carries steady current I, and sits in the xy plane. Take $\mu_0 I/(4\pi) \to 1$ and implement Algorithm 8.5. Calculate the magnetic field along the z axis and compare the magnetic field at the point $z = .5$ with your exact solution using $R = 1$. Use $n = 20$ grid points for the integration (here, the number of grid points won't change the result of the numerical integration, can you see why?).

Problem 8.3.7   Use your method from the previous problem, with $n = 20$, to generate the magnetic field at a set of points in three dimensions. Specifically, take $x = -2 \to 2$ in steps of .1, $y = -2 \to 2$ in steps of .1, and $z = -.5 \to .5$ in steps of $1/\pi$. If you have access to a program that will take this data and plot it as a vector field, make the plot.

8.4 Finite Difference

One can also "integrate" using a "finite difference" approximation, in which derivatives are replaced with differences (via Taylor expansion). This type of approach is well-suited to linear, boundary-value problems in mechanics. For example, solving the damped, driven harmonic oscillator ODE:

$$\frac{d^2 x(t)}{dt^2} + 2b\frac{dx(t)}{dt} + \omega^2 x(t) = \frac{1}{m}F(t), \tag{8.37}$$

in boundary value form, where the position of a mass is given at times $t = 0$ and $t = T$ is difficult using the Verlet method which is tailored to the initial value form of the problem (where $x(0)$ and $\dot{x}(0)$ are given). Of course, the damped, driven harmonic oscillator can also be solved using an explicit integral as in Section 5.1.1 (again, for the initial value problem), so the solution method here complements the integral solution to the same problem from the previous section.

In higher dimension, the same finite difference approach can be used to solve Poisson's problem, a partial differential equation familiar from E&M (and almost everywhere else). Because the method can be applied to problems beyond the familiar damped, driven oscillator, we'll replace (8.37) with a generic ODE governing a function $F(x)$ with "driving" source function $s(x)$. So we'll refer to the model problem (ODE plus boundary values):

$$\frac{d^2 F(x)}{dx^2} + \alpha\frac{dF(x)}{dx} + \beta F(x) = s(x)$$
$$F(0) = F_0 \tag{8.38}$$
$$F(X) = F_X,$$

where F_0 and F_X are the provided boundary values, and α, β are constants (they could be made into functions of x, an interesting extension that you should try).

8.4.1 One-Dimensional Form

As an example of the procedure, take the simplest case, with $\alpha = \beta = 0$. We want $F(x)$ solving

$$\frac{d^2F(x)}{dx^2} = s(x) \qquad F(0) = F_0 \qquad F(X) = F_X \qquad (8.39)$$

for given source $s(x)$ and boundary values provided at $x = 0$ and $x = X$. Introduce a grid in x: $x_j = j\Delta x$, for $j = 0 \to N+1$ and $x_{N+1} = X$, with $F_j \equiv F(x_j)$ as usual. The second derivative of $F(x)$ evaluated at x_j can be approximated by noting that

$$F_{j\pm1} = F_j \pm \Delta x \frac{dF}{dx}\bigg|_{x=x_j} + \frac{1}{2}\Delta x^2 \frac{d^2F}{dx^2}\bigg|_{x=x_j} \pm \frac{1}{6}\Delta x^3 \frac{d^3F}{dx^3}\bigg|_{x=x_j} + \cdots \qquad (8.40)$$

and then we can isolate the second derivative using a linear combination

$$\frac{d^2F(x)}{dx^2}\bigg|_{x=x_j} \approx \frac{F_{j+1} - 2F_j + F_{j-1}}{\Delta x^2} \qquad (8.41)$$

where the error we make in using this approximation is of order[4] Δx^2.

Putting the approximation (8.41) into (8.39), and writing $s_j \equiv s(x_j)$, we get a set of algebraic equations governing the approximations to $F(x)$ on the grid (F_j is now an unknown approximate numerical value for $F(x_j)$)

$$\frac{F_{j+1} - 2F_j + F_{j-1}}{\Delta x^2} = s_j \qquad (8.42)$$

for $j = 1 \to N$, and we need to be careful with the cases $j = 1$ and N, since those will involve the boundary values F_0 and F_X. The full set of equations, including those special cases, is

$$\frac{F_2 - 2F_1}{\Delta x^2} = s_1 - \frac{F_0}{\Delta x^2}$$

$$\frac{F_{j+1} - 2F_j + F_{j-1}}{\Delta x^2} = s_j \text{ for } j = 2 \to N-1 \qquad (8.43)$$

$$\frac{-2F_N + F_{N-1}}{\Delta x^2} = s_N - \frac{F_X}{\Delta x^2}$$

where we have moved the F_0 and F_X terms over to the right-hand side since those are known values. We want to solve this set of equations for $\{F_j\}_{j=1}^N$.

We can write (8.43) in matrix-vector form. Define the vector $\mathbf{F} \in \mathbb{R}^N$ with entries that are its unknown values:

$$\mathbf{F} \doteq \begin{pmatrix} F_1 \\ F_2 \\ \vdots \\ F_N \end{pmatrix} \qquad (8.44)$$

---

[4] Meaning, roughly, that the error is bounded by some constant times Δx^2. The constant that sits out front depends on the fourth derivative of $F(x)$ evaluated at x_j, in the present case.

and the tridiagonal matrix that acts on this to produce the left-hand side of (8.43) is[5]

$$
\mathbb{D} \doteq \frac{1}{\Delta x^2}
\begin{pmatrix}
-2 & 1 & 0 & \cdots & & 0 \\
1 & -2 & 1 & 0 & \cdots & \\
0 & 1 & -2 & 1 & 0 & \\
\vdots & 0 & \ddots & \ddots & \ddots & \\
0 & \cdots & 0 & 1 & -2
\end{pmatrix}.
\tag{8.45}
$$

Finally, define the slightly modified source "vector,"

$$
\mathbf{s} \doteq
\begin{pmatrix}
s_1 - \frac{F_0}{\Delta x^2} \\
s_2 \\
\vdots \\
s_{n-1} \\
s_N - \frac{F_X}{\Delta x^2}
\end{pmatrix},
\tag{8.46}
$$

and we can write (8.43) as

$$
\mathbb{D}\mathbf{F} = \mathbf{s}
\tag{8.47}
$$

which has solution obtained by inverting $\mathbb{D}$. Formally, $\mathbf{F} = \mathbb{D}^{-1}\mathbf{s}$ is what we want. How we obtain that matrix inverse depends on the particular problem (there are direct matrix inversion routines implemented in almost any programming language, and a suite of approximate inverses that can also be used).

The pseudocode that sets up the matrix $\mathbb{D}$ and vector $\mathbf{s}$ for this simplified one-dimensional problem and returns the solution is shown in Algorithm 8.6 – you send in the driving function $s(x)$, the end-point X, the number of grid points to use, N, and the initial and final values, F_0 and F_X, and this function constructs the matrix $\mathbb{D}$ and vector (including boundary values) $\mathbf{s}$, and returns $\mathbb{D}^{-1}\mathbf{s}$. In this segment of pseudocode, as in all of them, we use the convention that tables start at 1 (not zero).

For the more general model problem in (8.38), we also need an approximation to the derivative at x_j. From (8.40), we see that an appropriate combination is:

$$
\left. \frac{dF}{dx} \right|_{x=x_j} \approx \frac{F_{j+1} - F_{j-1}}{2\Delta x}.
\tag{8.48}
$$

The discretization of (8.38) (for points away from the boundaries), gives:

$$
\frac{F_{j+1} - 2F_j + F_{j-1}}{\Delta x^2} + \alpha \frac{F_{j+1} - F_{j-1}}{2\Delta x} + \beta F_j = s_j.
\tag{8.49}
$$

Problem 8.4.1   What is the exact solution to

$$
\frac{d^2 F(x)}{dx^2} = 4\sin(16x),
\tag{8.50}
$$

with $F(0) = 0$ and $F(1) = 1$?

---

[5] Note the similarity with the matrix for coupled oscillators, $\mathbb{Q}$ in (3.98) from Section 3.5.

Algorithm 8.6 FDint(s, X, N, F_0, F_X)

$\Delta x \leftarrow X/(N+1)$

Dmat $\leftarrow N \times N$ table of zeros.

svec $\leftarrow N$ table of zeros.

$\text{Dmat}_{11} \leftarrow -2/\Delta x^2$

$\text{Dmat}_{12} \leftarrow 1/\Delta x^2$

$\text{svec}_1 \leftarrow s(\Delta x) - F_0/\Delta x^2$

for $j = 2 \rightarrow N - 1$ **do**

   $\text{Dmat}_{jj-1} \leftarrow 1/\Delta x^2$

   $\text{Dmat}_{jj} \leftarrow -2/\Delta x^2$

   $\text{Dmat}_{jj+1} \leftarrow 1/\Delta x^2$

   $\text{svec}_j \leftarrow s(j\Delta x)$

end for

$\text{Dmat}_{NN-1} \leftarrow 1/\Delta x^2$

$\text{Dmat}_{NN} \leftarrow -2/\Delta x^2$

$\text{svec}_N \leftarrow s(N\Delta x) - F_X/\Delta x^2$

return Dmat^{-1} svec

Problem 8.4.2 What is the exact solution to the harmonic oscillator problem:

$$\frac{d^2 F(x)}{dx^2} + (2\pi)^2 F(x) = 0 \tag{8.51}$$

with $F(0) = 1$, $F(1) = 0$?

Problem 8.4.3 What is the exact solution to the damped harmonic oscillator problem:

$$\frac{d^2 F(x)}{dx^2} + 4\frac{dF(x)}{dx} + 5F(x) = 0, \tag{8.52}$$

with $F(0) = 0$ and $F(1) = 1$?

Problem 8.4.4 Write the entries of the matrix analogous to $\mathbb{D}$ in (8.45), that you will make for the full problem (8.49). What will you have to do to the first and last entries of the vector s from (8.46) (with $F(0) = F_0$, $F(X) = F_X$) in this expanded setting?

Problem 8.4.5 Implement Algorithm 8.6, use it to solve (8.50) with $N = 100$ grid points. Plot your numerical solution as points on top of the exact solution. Note that the output of Algorithm 8.6 is just a list of numbers, corresponding to our approximations on the grid. To plot the solution with the correct x axis underneath, you need to provide the grid points themselves.

Problem 8.4.6 Expand the content of Algorithm 8.6 (i.e. use the matrix you developed in Problem 8.4.4) so that it solves the full problem:

$$\frac{d^2 F(x)}{dx^2} + \alpha\frac{dF(x)}{dx} + \beta F(x) = s(x).$$

Find the solution for $\alpha = 4$, $\beta = 5$, and $s(x) = 0$ from (8.52), taking $F(0) = 0$, $F(1) = 1$, and using $N = 999$ gridpoints. Compare with your exact solution – what is the difference between the analytic solution and your numerical estimate at $x = .4$?

Problem 8.4.7   Use your expanded algorithm from the previous problem to solve (8.51) for $N = 100$ grid points. Compare with your solution to Problem 8.4.2.

Problem 8.4.8   Try solving

$$\frac{d^2F(x)}{dx^2} + 4\frac{dF(x)}{dx} + 5F(x) = 16\sin(16x),$$

with $F(0) = 0$, $F(1) = 1$ using your numerical method with $N = 999$. What is the value of your numerical solution at $x = .76$?

8.5 Eigenvalue Problems

The eigenvalue problem for matrices reads (referring to Section 3.3.2): Given a matrix $\mathbb{A} \in \mathbb{R}^{n \times n}$, find some/all of the set of vectors $\{\mathbf{v}^i\}_{i=1}^n$ and numbers $\{\lambda_i\}_{i=1}^n$ such that:

$$\mathbb{A}\mathbf{v}^i = \lambda_i \mathbf{v}^i. \tag{8.53}$$

In general, for a vector $\mathbf{y}$, the linear operation (matrix-vector multiplication) $\mathbb{A}\mathbf{y}$ can be thought of in terms of rotations and stretches of $\mathbf{y}$. The eigenvectors, $\mathbf{v}^i$, are the special vectors for which $\mathbb{A}\mathbf{v}^i$ is parallel to $\mathbf{v}^i$, only the length has changed.

There are continuous versions of the eigenvalue problem as we saw in Problem 3.3.11. The time-independent form of Schrödinger's equation, for a particle of mass m moving in the presence of a potential energy $U(x)$ (in one spatial dimension) is

$$\left[-\frac{\hbar^2}{2m}\frac{d^2}{dx^2} + U(x) \right]\psi(x) = E\psi(x), \tag{8.54}$$

where we have some sort of boundary condition, like $\psi(\pm\infty) \to 0$. We solve this equation for *both* $\psi(x)$ and E. The left-hand side of the equation represents a linear differential operator acting on $\psi(x)$, and if we can satisfy the equation, the effect of that linear operator is to scale $\psi(x)$ by E. This is a continuous eigenvalue problem, with $\psi(x)$ the "eigenfunction" and E the eigenvalue. What we will do is replace the differential operators with finite difference approximations, turning the continuous eigenvalue problem presented by Schrödinger's equation into a discretized matrix eigenvalue problem of the form (8.53).

Introduce a grid $x_j = j\Delta x$ with Δx fixed and $j = 0, 1, \ldots N + 1$ (the left-hand boundary is at $x = 0$ with $j = 0$, the right-hand boundary will be at x_f, with $j = N + 1$ so that $\Delta x = x_f/(N + 1)$). Then, letting $\psi_j \equiv \psi(x_j)$, and recalling the derivative approximation in (8.41), the projection of Schrödinger's equation onto the grid is:

$$-\frac{\hbar^2}{2m}\frac{\psi_{j+1} - 2\psi_j + \psi_{j-1}}{\Delta x^2} + U(x_j)\psi_j \approx E\psi_j, \tag{8.55}$$

and this holds for $j = 2 \to N - 1$. We'll assume that at the boundaries, the wave function vanishes (so that $-\infty \to \infty$ gets remapped to $0 \to x_f$), then for $j = 1$:

$$-\frac{\hbar^2}{2m}\frac{\psi_2 - 2\psi_1}{\Delta x^2} + U(x_1)\psi_1 = E\psi_1, \tag{8.56}$$

and for $j = N$,

$$-\frac{\hbar^2}{2m}\frac{-2\psi_N + \psi_{N-1}}{\Delta x^2} + U(x_N)\psi_N = E\psi_N, \tag{8.57}$$

and once again, we can box up the approximation in a matrix-vector multiplication. Let ψ_j be the approximation to $\psi(x_j)$ that is our target, and define the vector of unknown values:

$$\psi \doteq \begin{pmatrix} \psi_1 \\ \psi_2 \\ \vdots \\ \psi_N \end{pmatrix}. \tag{8.58}$$

The matrix that encapsulates (8.55) and the boundary points is, taking $p \equiv \hbar^2/(2m\Delta x^2)$,

$$\mathbb{H} \doteq - \begin{pmatrix} -2p + U(x_1) & p & 0 & \cdots & 0 \\ p & -2p + U(x_2) & p & 0 & \cdots \\ 0 & p & -2p + U(x_3) & p & 0 \\ \vdots & 0 & \ddots & \ddots & \ddots \\ 0 & \cdots & 0 & p & -2p + U(x_N) \end{pmatrix}. \tag{8.59}$$

This is a tridiagonal matrix, and can be made relatively easily using your function(s) from the previous section. Now we have turned the continuous eigenvalue problem into a discrete one, we want ψ and E that solve

$$\mathbb{H}\psi = E\psi, \tag{8.60}$$

a matrix eigenvalue problem. If you can construct the matrix $\mathbb{H}$, then you can use the built-in command "Eigensystem" in Mathematica, for example, to get the eigenvalues (the set of energies) and eigenvectors (the associated discrete approximations to the wave functions) of the matrix.

8.5.1 Infinite Square Well

For a particle in a box of length a, we have:

$$-\frac{\hbar^2}{2m}\frac{d^2\psi(x)}{dx^2} = E\psi(x), \tag{8.61}$$

with $\psi(0) = \psi(a) = 0$. Taking $x = x_0 X$, we can render the equation dimensionless using the approach from Section 8.2.4. Let $\tilde{E} \equiv 2mx_0^2 E/\hbar^2$ be the dimensionless energy, then (8.61) becomes

$$-\frac{d^2\psi(X)}{dX^2} = \tilde{E}\psi(X), \tag{8.62}$$

with $\psi(0) = 0$ and $\psi(a/x_0) = 0$, suggesting we take $x_0 = a$. The full solution to this problem is, of course:

$$\psi(X) = A\sin(\sqrt{\tilde{E}}X) \tag{8.63}$$

for constant A, which will be set by normalization, and with $\sqrt{\tilde{E}} = n\pi$ for integer n (to get $\psi(1) = 0$), so that in these units, $\tilde{E} = n^2\pi^2$, or, going back to the original units in the problem:

$$E = \frac{\hbar^2}{2ma^2}\tilde{E} = \frac{\hbar^2 n^2 \pi^2}{2ma^2}. \tag{8.64}$$

If we wanted to solve this problem numerically, we would approximate the derivative in (8.62) with a finite difference. The dimensionless spatial coordinate X runs from $0 \to 1$, so let $X_j = j\Delta X$ with $\Delta X = 1/(N+1)$ for N the number of grid points. Let $\psi_j = \psi(X_j)$ be the unknown values associated with ψ on the spatial grid. Then using our familiar finite difference approximation to the second derivative, (8.62) becomes:

$$-\frac{\psi_{j+1} - 2\psi_j + \psi_{j-1}}{\Delta X^2} = \tilde{E}\psi_j \tag{8.65}$$

for $j = 2 \to N - 1$. At $j = 1$ and $j = N$ we need to enforce our boundary conditions: $\psi_0 = 0$ and $\psi_{N+1} = 0$, so those two special cases satisfy:

$$-\frac{\psi_2 - 2\psi_1}{\Delta X^2} = \tilde{E}\psi_1$$
$$-\frac{-2\psi_N + \psi_{N-1}}{\Delta X^2} = \tilde{E}\psi_N. \tag{8.66}$$

Together (8.65) and (8.66) define a matrix $\mathbb{H}$ similar to (8.59), and we can again define $\boldsymbol{\psi}$ as in (8.58). Then the discretized problem can be written as $\mathbb{H}\boldsymbol{\psi} = \tilde{E}\boldsymbol{\psi}$. In Algorithm 8.7, we see the pseudocode that generates the matrix $\mathbb{H}$ given X_f (here, one) and the size of the grid, N.

Algorithm 8.7 Pbox(X_f, N)

$\Delta X \leftarrow X_f/(N+1)$
Hmat $\leftarrow N \times N$ table of zeros.
Hmat$_{11} \leftarrow 2/\Delta X^2$
Hmat$_{12} \leftarrow -1/\Delta X^2$
for $j = 2 \to N - 1$ **do**
$\quad$ Hmat$_{jj-1} \leftarrow -1/\Delta X^2$
$\quad$ Hmat$_{jj} \leftarrow 2/\Delta X^2$
$\quad$ Hmat$_{jj+1} \leftarrow -1/\Delta X^2$
end for
Hmat$_{NN-1} \leftarrow -1/\Delta X^2$
Hmat$_{NN} \leftarrow 2/\Delta X^2$
return Hmat

8.5.2 Hydrogen

To get the spectrum of hydrogen (see [13] for details), we just need to put the appropriate potential energy function in (8.59). For an electron and a proton separated by a distance x and interacting electrostatically, we have

$$U(x) = -\frac{e^2}{4\pi\epsilon_0 x} \tag{8.67}$$

where e is the charge of the electron. We can work on the "half"-line, letting $x = 0 \to \infty$, and we want to solve

$$-\frac{\hbar^2}{2m}\frac{d^2\psi(x)}{dx^2} - \frac{e^2}{4\pi\epsilon_0 x}\psi(x) = E\psi(x) \tag{8.68}$$

with $\psi(0) = 0$ and $\psi(\infty) = 0$.

Let $x = x_0 X$ for dimensionless X and where x_0 has dimension of length. Then (8.68) can be written:

$$-\frac{d^2\psi(X)}{dX^2} - \frac{mx_0 e^2}{4\pi\epsilon_0\hbar^2}\frac{2}{X}\psi(X) = \frac{2mx_0^2}{\hbar^2}E\psi(X), \tag{8.69}$$

and we'll clean up the potential term by requiring that

$$\frac{mx_0 e^2}{4\pi\epsilon_0\hbar^2} = 1, \tag{8.70}$$

which serves to define x_0. Letting $\tilde{E} = 2mx_0^2 E/\hbar^2$ (so that $\tilde{E}$ is dimensionless), the dimensionless form of (8.68) reads:

$$-\frac{d^2\psi(X)}{dX^2} - \frac{2}{X}\psi(X) = \tilde{E}\psi(X). \tag{8.71}$$

When we discretize in position, we cannot go all the way out to spatial infinity, so we pick a "large" value of X (since X is dimensionless, large means $X \gg 1$). Let X_f be the endpoint of our grid, then $\Delta X = X_f/(N+1)$. The discretized form of (8.71) is (for $\psi_j = \psi(X_j)$, an unknown value for ψ evaluated at $X_j = j\Delta X$)

$$-\frac{\psi_{j+1} - 2\psi_j + \psi_{j-1}}{\Delta X^2} - \frac{2}{X_j}\psi_j = \tilde{E}\psi_j \tag{8.72}$$

and this is for $j = 2 \to N-1$. For $j = 1$, we use the fact that $\psi_0 = 0$ (a boundary condition) to get

$$-\frac{\psi_2 - 2\psi_1}{\Delta X^2} - \frac{2}{X_1}\psi_1 = \tilde{E}\psi_1, \tag{8.73}$$

and similarly for $j = N$, with $\psi_{N+1} = 0$ (the boundary at "infinity"), we have

$$-\frac{-2\psi_N + \psi_{N-1}}{\Delta X^2} - \frac{2}{X_N}\psi_N = \tilde{E}\psi_N. \tag{8.74}$$

8.5.3 The Power Method

How does one obtain the eigenvalues and eigenvectors associated with a matrix? The approach we used in Section 3.3.2 is inefficient, and involves multiple root-finding expeditions just to start it off (and one must know something about the root structure of the relevant determinant in order to bracket the roots). The crux of almost any numerical eigenvalue-problem solver is the so-called "power method." Suppose you have a symmetric matrix $\mathbb{A} \in \mathbb{R}^{n \times n}$, so that we know that the eigenvectors span $\mathbb{R}^n$, and the eigenvalues are real. Assume, further, that the eigenvalues are all distinct and ordered, with $|\lambda_1| > |\lambda_2| > \ldots > |\lambda_n| > 1$ and we'll call the associated eigenvectors $\mathbf{v}^1$, $\mathbf{v}^2$, ..., $\mathbf{v}^n$. Pick a random, nonzero vector $\mathbf{x} \in \mathbb{R}^n$, then we know from Section 3.3.3 that there exist coefficients $\{\beta_j\}_{j=1}^n$ such that

$$\mathbf{x} = \sum_{j=1}^n \beta_j \mathbf{v}^j. \tag{8.75}$$

Now multiply the vector $\mathbf{x}$ by $\mathbb{A}$, the multiplication slips through the sum in (8.75) and because the vectors $\{\mathbf{v}^j\}_{j=1}^n$ are eigenvectors, we know that $\mathbb{A}\mathbf{v}^j = \lambda_j \mathbf{v}^j$, so that the multiplication yields

$$\mathbb{A}\mathbf{x} = \sum_{j=1}^n \beta_j \lambda_j \mathbf{v}^j. \tag{8.76}$$

Similarly, if we multiply by $\mathbb{A}^p$, i.e. multiply by $\mathbb{A}$ p-times, we get

$$\mathbb{A}^p \mathbf{x} = \sum_{j=1}^n \beta_j \lambda_j^p \mathbf{v}^j. \tag{8.77}$$

Because the largest eigenvalue is λ_1, we know the first term in the sum (8.77) will dominate, since $|\lambda_1|^p \gg |\lambda_2|^p \ldots$ for some p. Then for p "large enough," we get

$$\mathbb{A}^p \mathbf{x} \approx \beta_1 \lambda_1^p \mathbf{v}^1, \tag{8.78}$$

the product is parallel to $\mathbf{v}^1$. If you let $\mathbf{w} \equiv \mathbb{A}^p \mathbf{x}$, then $\hat{\mathbf{w}} \approx \mathbf{v}^1$, the unit vector $\mathbf{w}$ should approximate the normalized eigenvector $\mathbf{v}^1$. To get the associated eigenvalue, note that $\mathbb{A}\hat{\mathbf{w}} \approx \lambda_1 \hat{\mathbf{w}}$, and we can multiply both sides by $\hat{\mathbf{w}}^T$, isolating λ_1,

$$\lambda_1 \approx \hat{\mathbf{w}}^T \mathbb{A} \hat{\mathbf{w}}. \tag{8.79}$$

Now you have the first eigenvalue and associated eigenvector.

To continue the process, you basically start over with a new random vector, multiply by $\mathbb{A}$ over and over, but after each multiplication, you "project out" the component that lies along the now known $\mathbf{v}^1$ approximate. In this way, you can force the power method to converge to the eigenvector associated with the second-largest eigenvalue, and the process can be continued from there. While the method sketched here is specific to symmetric matrices, generalizations exist for nonsymmetric and complex matrices (see [11]).

Problem 8.5.1 Find the value of x_0 from (8.70), this is a characteristic length scale for hydrogen. For this value of x_0, find the energy scale $\hbar^2/(2mx_0^2)$ that you will use in converting the dimensionless $\tilde{E}$ back to E. What is the value of this energy scale in electron-volts?

Problem 8.5.2 Indicate the nonzero entries of the matrix associated with the dimensionless hydrogen problem – i.e. take (8.72), (8.73), and (8.74), and write the entries of the matrix $\mathbb{H}$ appearing in the discrete eigenvalue problem: $\mathbb{H}\psi = \tilde{E}\psi$.

Problem 8.5.3   The finite difference (8.65) can be solved analytically. Take the ansatz: $\psi_j = Ae^{i\pi nj\Delta X}$, insert in

$$-\frac{\psi_{j+1} - 2\psi_j + \psi_{j-1}}{\Delta X^2} = \tilde{E}_n \psi_j,$$

and find $\tilde{E}_n$. Take the limit as $\Delta X \to 0$ and show that you recover $n^2\pi^2$.

Problem 8.5.4   Generate the appropriate matrix $\mathbb{H}$ for the particle-in-a-box problem from Algorithm 8.7 using $X_f = 1$ and $N = 100$. Find the eigenvalues using a linear algebra package. Sort your list so that you get the values from smallest to largest. What is the difference between the first two smallest eigenvalues and their analytic values from your solution to Problem 8.5.3?

Problem 8.5.5   Modify your matrix from the previous problem so that its entries come from your solution to Problem 8.5.2, i.e. make a matrix that reflects the content of (8.72), (8.73), and (8.74). Using $X_f = 50$ and $N = 10000$, build the matrix and find its eigenvalues (using a linear algebra package). List the first four (sorted) eigenvalues (they should be negative). Guess the form of these eigenvalues if you associate the lowest one with the lowest energy, the second with next highest energy, and so on (i.e. find an expression for E_k with $k = 1, 2, 3$, and 4). These are dimensionless eigenvalues, restore dimensions of energy using your result from Problem 8.5.1, and express the spectrum of hydrogen, $E_k = ?$ in eV.

Problem 8.5.6   Use the power method to find the maximum eigenvalue and eigenvector for the matrix

$$A \doteq \begin{pmatrix} 1 & 2 & 3 & 4 \\ 2 & 5 & 6 & 7 \\ 3 & 6 & 8 & 9 \\ 4 & 7 & 9 & 10 \end{pmatrix}.$$

8.6 Discrete Fourier Transform

The Fourier series that we discussed in Section 2.3 decomposed a periodic function $p(t)$ into exponentials of the same period T:

$$p(t) = \sum_{j=-\infty}^{\infty} a_j e^{i2\pi jt/T}, \tag{8.80}$$

with coefficients given by the orthogonality, under integration, of the exponentials,

$$a_j = \frac{1}{T} \int_0^T p(t) e^{-i2\pi jt/T} \, dt. \tag{8.81}$$

If we discretize in time, letting $t_j \equiv j\Delta t$ for some time-step Δt,[6] and let $p_j \equiv p(t_j)$, then we can approximate the integral in (8.81), with $T = n\Delta t$,

---

[6] This discretization could be performed on known continuous functions $p(t)$ of course, but is meant to be reminiscent of the time-series obtained in the laboratory, where the cadence of data-taking is set by the particular instruments on which the data is taken, we poll the voltage in a circuit every $\Delta t = .1$ millisecond, for example.

$$a_j \approx \frac{1}{T} \sum_{k=0}^{n-1} p_k e^{-i2\pi j(k\Delta t)/T} \Delta t = \frac{1}{n} \sum_{k=0}^{n-1} p_k e^{-i2\pi jk/n}. \tag{8.82}$$

Assume that n is a multiple of 2, then looking at (8.80), it is clear we need both positive and negative values of j to recover (a truncated form of $p(t)$), so let $j = -n/2 \to n/2$ in evaluating a_j. That is, seemingly, $n + 1$ values for a_j, and this is one too many given the original set of n values for p_k. But note that the coefficients a_j are themselves periodic with period n, i.e. $a_{j+n} = a_j$,

$$a_{j+n} = \frac{1}{n} \sum_{k=0}^{n-1} p_k e^{-i2\pi(j+n)k/n} = \frac{1}{n} \sum_{k=0}^{n-1} p_k e^{-i2\pi jk/n} \underbrace{e^{-i2\pi k}}_{=1} = \frac{1}{n} \sum_{k=0}^{n-1} p_k e^{-i2\pi jk/n} = a_j. \tag{8.83}$$

Now we can see that $a_{-n/2} = a_{-n/2+n} = a_{n/2}$ so that there really are only n unique values for the coefficients a_j here.

To get the "inverse," recovering the set $\{p_k\}_{k=0}^{n-1}$, we need to evaluate (8.80) at the discrete temporal points

$$p_k = \sum_{j=-\infty}^{\infty} a_j e^{i2\pi jk/n}, \tag{8.84}$$

and again, since we only have n values of a_j, we cannot perform the infinite sum. Instead we truncate as before:

$$p_k = \sum_{j=-n/2}^{n/2} a_j e^{i2\pi jk/n}. \tag{8.85}$$

To summarize, given a set of data $\{p_j\}_{j=0}^{n-1}$ taken at a temporal spacing of Δt, the discrete Fourier transform (DFT) is given by the coefficients

$$a_j = \frac{1}{n} \sum_{k=0}^{n-1} p_k e^{-i2\pi jk/n} \text{ for } j = -n/2 \to n/2. \tag{8.86}$$

Given the coefficients $\{a_j\}_{j=-n/2}^{n/2}$, we can construct the inverse DFT,

$$p_k = \sum_{j=-n/2}^{n/2} a_j e^{i2\pi jk/n} \text{ for } k = 0 \to n-1. \tag{8.87}$$

If we imagine that the datum p_j came at time $t_j \equiv j\Delta t$ given the fixed temporal spacing Δt, what frequency should we associate with the coefficient a_k? The frequency spacing should, like the temporal one, be constant, call it Δf. Then we expect a_k to be the coefficient associated with the frequency $f_k = k\Delta f$. How do we determine Δf given Δt? Let's work out an example to see how to make the identification. Take a continuous signal, $p(t) = \sin(2\pi 5 t)$ so that the frequency f here is 5. We'll sample this function at $n = 10$ equally spaced steps of size $\Delta t = 1/25$ so that p_j has $j = 0 \to 9$. Computing the coefficients using (8.86), we find that the a_j are all zero except for $j = -2$ and $j = 2$ with $a_{-2} = .5i$ and $a_2 = -.5i$. Notice first that the positive and negative frequency components are related in the usual way for sine, which is both real and odd. Now we have $j = \pm 2$ as the integer index of the frequency, and we want $f_j = j\Delta f$, with $f_2 = 2\Delta f = 5$ the frequency of the original signal. Then $\Delta f = 5/2$ is the frequency spacing.

That's fine for calibration, but how do we generalize the result for an arbitrary Δt? We can't just send in a pure cosine or sine signal to find the spacing in each case, so we ask the general question: For a pure sinusoidal function, what is the minimum period signal we can represent with a set of data $\{p_j\}_{j=0}^{n-1}$ assumed to come from a temporal discretization Δt? Well, the simplest signal that is purely oscillatory has $p_j = 0$ for all values. Assuming we have not sent in "nothing," we must interpret this as a sine wave with zeroes at every grid point, meaning that for a signal of frequency $\bar{f}$, we must have

$$\sin(2\pi\bar{f}\Delta t) = 0 \longrightarrow 2\pi\bar{f}\Delta t = m\pi \tag{8.88}$$

for integer m. The smallest m could be is $m = 1$, and that represents the minimum period for a signal of frequency $\bar{f} = 1/(2\Delta t)$. Minimum period means maximum frequency and this special maximum is known as the "Nyquist frequency." Meanwhile, for a fixed Δf, the maximum frequency we can represent on our frequency grid is $(n/2)\Delta f$. Equating this with the Nyquist frequency allows us to solve for Δf:

$$\bar{f} = \frac{1}{2\Delta t} = \frac{n}{2}\Delta f \longrightarrow \Delta f = \frac{1}{n\Delta t} = \frac{1}{T}. \tag{8.89}$$

Let's check this spacing using our example, where $n = 10$, $\Delta t = 1/25$ so that $\Delta f = 5/2$ just as we expected.

There is a scale invariance in (8.86) and (8.87): Those equations do not depend on Δt or Δf, and hence refer to an infinite family of Fourier series, each with a different $\{\Delta t, \Delta f\}$ pair. Once one is given, the other is fixed, but *a priori* there is no preferred value. The algorithm for the DFT is shown in Algorithm 8.8, where the only input is the data itself, called "indata" there. You are in charge of keeping track of the input Δt and associated value of Δf which is necessary when plotting the power spectrum, for example. The input data must have length that is a multiple of 2, so that $n/2$ is an integer. The output of the DFT has an extra point in it as compared to the input data. That point is the redundant value of the DFT for the $n/2$ entry (identical to the $-n/2$ entry by periodicity). Note that, as with all pseudocode in this chapter, we assume that vector indices start at 1, hence the additional decorative indexing.

Algorithm 8.8 DFT(indata)

$n \leftarrow$ length(indata)
oput $\leftarrow$ zero table of length $n + 1$
for $j = -n/2 \rightarrow n/2$ **do**
    oput$_{n/2+j+1} = 1/n \sum_{k=1}^{n}$ indata$_k e^{-i2\pi j(k-1)/n}$
end for
return oput

The output is a vector of length $n + 1$ whose first entry corresponds to the frequency $-n/2\Delta f$. The inverse process is shown in Algorithm 8.9 where this time we send in the

Fourier transform data ("indata") and get back out the temporal signal, with first entry (at index location 1) associated with time $t = 0$.

Algorithm 8.9 iDFT(indata)

    $n \leftarrow$ length(indata) $- 1$
    oput $\leftarrow$ zero table of length n
    for $k = 0 \rightarrow n - 1$ **do**
        $\text{oput}_{k+1} = \sum_{j=-n/2}^{n/2} \text{indata}_{j+1+n/2} e^{i2\pi jk/n}$
    end for
    return oput

The DFT is a great start for signal processing applications. The only real problem with it is the amount of time it takes. Looking at Algorithm 8.8, each entry in the output vector, "oput," requires n additions (to form the sum), and there are $\sim n$ output entries so that the time it takes to compute the DFT (and the inverse) is proportional to n^2. That scaling with size can be avoided using a clever recursive structure that exploits the periodicity of the signal and the exponentials that make up the sum in the algorithm. The resulting recursive algorithm is called the "fast Fourier transform" abbreviated FFT. It produces the exact same results as the DFT, but scales with input vector size n like $n \log(n)$ which is much smaller than n^2 for large n.

Problem 8.6.1 Implement the DFT algorithm and use it to reproduce the power spectrum we used to find Δf as follows: Sample the signal function $\sin(2\pi 5t)$ for $n = 10$ and $\Delta t = 1/25$ (making sure to start with the value at $t = 0$), send that in to the DFT, find Δf and plot the magnitude-squared of the values from the DFT on the y axis, with the correct frequencies (negative and positive) on the x axis.

Problem 8.6.2 Implement the inverse DFT algorithm and use it to invert the Fourier transform you got in the previous problem. You should exactly recover the original discretized signal values.

Problem 8.6.3 Take the signal function

$$p(t) = e^{-t},$$

a decaying exponential, and discretize it using $\Delta t = 1/50$ and $n = 256$. Compute the DFT and plot the power spectrum (with correct x axis frequencies). Now take $p(t) = e^{-10t}$ and do the same thing. What happens to the power spectrum, does it become more or less sharply peaked (see Problem 2.6.4 and Problem 2.6.12)?

Problem 8.6.4 For the signal

$$p(t) = e^{-2t} \sin(2\pi 5t),$$

sample using $\Delta t = 1/50$, $n = 256$, compute and plot the power spectrum. Is the peak in the right place?

Problem 8.6.5 Make a signal with a variety of frequencies:

$$p(t) = \sin(2\pi(20t)^2),$$

and discretize using $\Delta t = 1/8000$, $n = 4096$. If possible, *play* the signal so you can hear what it sounds like. Compute the DFT, and attach the correct frequencies to the vector values. Now "zero out" all entries in the discrete transform that have absolute value frequency between 200 and 400 Hz, i.e. for any frequency values in this range, set the corresponding entry in the DFT vector to zero. Use the inverse DFT to recover a signal and try playing it. This problem gives an example of a frequency "filter," where you perform surgery on the Fourier transform to eliminate or augment frequencies in a certain range, then inverse transform to get a signal with those frequencies absent (or punched up).

Appendix A Solving ODEs
A Roadmap

We have covered both general and specialized methods for solving various ordinary differential equations, including: (1) definition: the solution is defined to be some named function, and properties of those functions are studied (think of cosine and sine); (2) series solution, we take the solution to be an infinite sum of powers (like a Taylor expansion) or exponentials (like Fourier series), and find the set of coefficients that solve the ODE; (3) separating out the homogeneous and sourced solutions (homogeneous solutions are used to set initial/boundary values). What should your plan of attack be when confronted with some new or unfamiliar ordinary differential equation? What I will suggest is a laundry list of things to try, but surprisingly many problems will yield to one of the approaches, and those that do not generally require numerical methods in order to make progress.

For an ODE of the form: $\mathcal{D}(f(x)) = g(x)$ where $\mathcal{D}(f(x))$ represents a differential operator acting on $f(x)$ and $g(x)$ is some provided function, and assuming the appropriate initial conditions are provided, the first thing to try is a guess of the form $f(x) \sim A e^{Bx}$ for constant A and B. This is a good guess because it represents a single term in a Fourier series expansion (where we would set $B = i2\pi b$ for some new constant b), and is particularly useful when the differential operator is linear. The goal is to find A and B that satisfy

$$\mathcal{D}\left(A e^{Bx}\right) = g(x) \tag{A.1}$$

for all x. Note that it may be impossible to find such an A and B – does that mean the problem is unsolvable? No, it just means that this initial guess does not yield a solution, and you should move on and try something else.

As an example, suppose we take $\mathcal{D}(f(x)) = f''(x) + pf'(x) - qf(x)$ (for constants p and q), and we have $g(x) = 0$, then

$$\mathcal{D}\left(A e^{Bx}\right) = 0 \longrightarrow B^2\left(A e^{Bx}\right) + pB\left(A e^{Bx}\right) - q\left(A e^{Bx}\right) = 0. \tag{A.2}$$

The equation can be divided by $A e^{Bx}$ leaving the algebraic equation $B^2 + pB - q = 0$. The values of B that solve this equation are $B_\pm = (-p \pm \sqrt{p^2 + 4q})/2$, but there is no constraint on A. Since the differential operator is linear in $f(x)$, we know that superposition holds, and we can take an arbitrary sum of the two solutions:

$$f(x) = A_1 e^{x\left(-p+\sqrt{p^2+4q}\right)/2} + A_2 e^{x\left(-p-\sqrt{p^2+4q}\right)/2}. \tag{A.3}$$

In this form, it is clear that the constants A_1 and A_2 will be set by initial (or boundary) values.

If we had the operator $\mathcal{D}(f(x)) = f'(x)^2 - pf(x)$ for constant p, and again $g(x) = 0$, then our guess would give

$$\mathcal{D}\left(Ae^{Bx}\right) = 0 \longrightarrow B^2\left(Ae^{Bx}\right)^2 - p\left(Ae^{Bx}\right) = 0. \tag{A.4}$$

This time, the exponentials cannot be cancelled, and we have an equation that cannot be solved for all x:

$$B^2\left(Ae^{Bx}\right) = p. \tag{A.5}$$

Again, the lesson here is that our initial guess is not robust enough. In order for the exponential ansatz to work, we need to be able to reliably cancel out the exponentials from the ODE, removing the x-dependence and leaving us with an equation for the constants A and B. Here, we cannot do that, so we move on.

Your next guess should be a polynomial in x, $f(x) = Ax^B$. This starting point represents a single term in an infinite series expansion of $f(x)$. We again run the assumed form through the ODE to see if we can solve for the constants A and B (possibly appealing to superposition)

$$\mathcal{D}\left(Ax^B\right) = g(x). \tag{A.6}$$

For the differential operator $\mathcal{D}(f(x)) = f'(x)^2 - pf(x)$ from (A.6), we have

$$\mathcal{D}\left(Ax^B\right) = 0 \longrightarrow A^2 B^2 x^{2(B-1)} - pAx^B = 0 \tag{A.7}$$

and we can see that for this equation to hold for all x, we must have $2(B-1) = B$ or $B = 2$. Then we are left with $4A^2 - pA = 0$ so that $A = p/4$, and our solution reads

$$f(x) = \frac{px^2}{4}, \tag{A.8}$$

but we are missing a constant of integration that would allow us to set the initial values. We have a particular solution here, and now we need to find "the rest" of the solution. In the present case, a clever observation allows us to make progress. When confronted with a nonlinear ODE, one should always look for a simple way to get a linear ODE out. Here, taking the derivative of the ODE itself gives

$$2f'(x)f''(x) - pf'(x) = 0 \longrightarrow 2f''(x) - p = 0 \longrightarrow f(x) = \frac{px^2}{4} + a + bx \tag{A.9}$$

where a and b are constants. The problem is that we have too many constants of integration, we expect to get just one. So take this $f(x)$ and run it through the original ODE:

$$\mathcal{D}\left(\frac{px^2}{4} + a + bx\right) = 0 \longrightarrow b^2 - ap = 0. \tag{A.10}$$

We can set $b = \pm\sqrt{ap}$, to get

$$f(x) = \frac{px^2}{4} + a \pm \sqrt{ap}x \tag{A.11}$$

with a waiting to be set by some provided initial value.

If we had the original $\mathcal{D}(f(x)) = f''(x) + pf'(x) - qf(x)$, we would not be able to solve using the polynomial guess:

$$\mathcal{D}(Ax^B) = 0 \longrightarrow A(B-1)Bx^{B-2} + ABpx^{B-1} - Aqx^B = 0, \tag{A.12}$$

and there is no way to get $B - 2 = B - 1 = B$ so as to cancel the x-dependence, and hence, no solution of this form with constant A and B.

Finally, when there is a nontrivial $g(x)$, you should try to find the homogeneous (setting $g(x) = 0$) and sourced solutions separately, although each solution benefits from the guesses in (A.12). If you had $\mathcal{D}(f(x)) = f''(x) + pf'(x) - qf(x)$ with $g(x) = g_0 x$, then you'd start by solving

$$\mathcal{D}(h(x)) = 0 \tag{A.13}$$

to get the homogeneous piece of the solution, and this would crack under $h(x) = Ae^{Bx}$ (the solution would be as in (A.3)). For the particular solution, we want

$$\mathcal{D}(\bar{f}(x)) = g_0 x \tag{A.14}$$

which we can get using the variation of parameters approach from Section 1.7.4 (and in particular, see Problem 1.7.3). It ends up being

$$\bar{f}(x) = -\frac{g_0 x}{q} - \frac{g_0 p}{q^2}, \tag{A.15}$$

and the full solution is the sum of the homogeneous and sourced solutions

$$f(x) = h(x) + \bar{f}(x) = A_1 e^{x\left(-p + \sqrt{p^2 + 4q}\right)/2} + A_2 e^{x\left(-p - \sqrt{p^2 + 4q}\right)/2} - \frac{g_0 x}{q} - \frac{g_0 p}{q^2}. \tag{A.16}$$

Problem A.0.1 Try putting $x(t) = At^p$ into the ODE

$$\ddot{x}(t) + \omega^2 x(t) + 2b\dot{x}(t) = 0,$$

and identify the problem with this attempted solution. (i.e. how can you tell it won't work?)

Problem A.0.2 Solve the second-order ODE (that comes up in both electricity and magnetism and gravity):

$$\frac{d^2 f(x)}{dx^2} + \frac{2}{x}\frac{df(x)}{dx} = \alpha$$

for constant α. In addition to writing the general solution, give the one that has $f(0) = 0$ and $f'(0) = 0$.

Problem A.0.3 Solve the second-order ODE (that comes up in both electricity and magnetism and gravity in two spatial dimensions):

$$\frac{d^2 f(x)}{dx^2} + \frac{1}{x}\frac{df(x)}{dx} = \alpha$$

for constant α. This time, find the general solution and the one with $f(1) = c_1$, $f'(1) = c_2$ for arbitrary constants c_1 and c_2.

Problem A.0.4   Solve the first-order ODE:

$$\frac{df(x)}{dx} = -\frac{1}{x}f(x)$$

for $f(x)$ – there should be one constant of integration.

Problem A.0.5   A driven harmonic oscillator has equation of motion:

$$m\ddot{x}(t) = -kx(t) + F_0 e^{-\mu t}$$

for $\mu > 0$. Find $x(t)$ given $x(0) = 0$ and $\dot{x}(0) = 0$. What is the solution as $t \to \infty$?

Appendix B **Vector Calculus**
Curvilinear Coordinates

When we developed the vector calculus operations in Chapter 6, we started with the gradient operator written in Cartesian coordinates,

$$\nabla = \hat{\mathbf{x}}\frac{\partial}{\partial x} + \hat{\mathbf{y}}\frac{\partial}{\partial y} + \hat{\mathbf{z}}\frac{\partial}{\partial z}, \tag{B.1}$$

which acted on functions $f(x, y, z)$. The divergence and curl were then naturally written in terms of Cartesian coordinates and basis vectors. In this appendix, we'll look at how to use the chain rule to express these vector derivative operations in other coordinate systems. Some of this material appears in the main text, but I'd like to develop it clearly and completely in one place to serve as a reference.

B.1 Cylindrical Coordinates

In cylindrical coordinates, we have the three coordinate variables s, ϕ, and z, defined as shown in Figure B.1.

Referring to the figure, we can write

$$x = s\cos\phi \quad y = s\sin\phi \quad z = z \tag{B.2}$$

(the cylindrical coordinate z is the same as the Cartesian z). These can be inverted, to give

$$s = \sqrt{x^2 + y^2} \quad \phi = \tan^{-1}\left(\frac{y}{x}\right) \quad z = z. \tag{B.3}$$

Along with the coordinate definitions, we need to develop the cylindrical basis vectors. A coordinate basis vector points in the direction of increasing coordinate value and is normalized to one. The basis vector associated with z is easy, that's $\hat{\mathbf{z}}$ as always. How about the basis vector that points in the direction of increasing s? Well, start with the vector that points from the origin to the point with coordinates x, y: $\mathbf{s} \equiv x\hat{\mathbf{x}} + y\hat{\mathbf{y}}$. At any point, the direction of $\mathbf{s}$ is parallel to the direction of increasing s.[1] The unit vector is

$$\hat{\mathbf{s}} = \frac{x\hat{\mathbf{x}} + y\hat{\mathbf{y}}}{\sqrt{x^2 + y^2}} = \cos\phi\,\hat{\mathbf{x}} + \sin\phi\,\hat{\mathbf{y}} \tag{B.4}$$

---

[1] Besides this geometrical approach, we could identify the direction of increasing s using the gradient as in Section 6.2.1,

$$\nabla s = \frac{x}{\sqrt{x^2 + y^2}}\,\hat{\mathbf{x}} + \frac{y}{\sqrt{x^2 + y^2}}\,\hat{\mathbf{y}} = \cos\phi\,\hat{\mathbf{x}} + \sin\phi\,\hat{\mathbf{y}}.$$

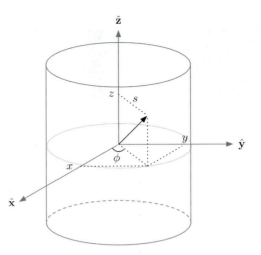

The cylindrical coordinates s, ϕ, and z are related to the Cartesian x, y, and z.

where we can express the components of this basis vector in Cartesian coordinates (middle equality) or in the new cylindrical coordinates (right-hand equality).

For the ϕ direction, we have increasing ϕ pointing tangent to a circle, and going around in the counter-clockwise direction. We know, from, for example Section 6.2.2, that $\phi = -y\hat{\mathbf{x}} + x\hat{\mathbf{y}}$ does precisely this,[2] and we can again normalize to get a unit vector

$$\hat{\boldsymbol{\phi}} = \frac{-y\hat{\mathbf{x}} + x\hat{\mathbf{y}}}{\sqrt{x^2 + y^2}} = -\sin\phi\,\hat{\mathbf{x}} + \cos\phi\,\hat{\mathbf{y}}, \tag{B.5}$$

writing the components in either Cartesian or cylindrical coordinates. We now have the relation between the cylindrical basis vectors and the Cartesian ones,

$$\hat{\mathbf{s}} = \cos\phi\,\hat{\mathbf{x}} + \sin\phi\,\hat{\mathbf{y}} \quad \hat{\boldsymbol{\phi}} = -\sin\phi\,\hat{\mathbf{x}} + \cos\phi\,\hat{\mathbf{y}} \quad \hat{\mathbf{z}} = \hat{\mathbf{z}}, \tag{B.6}$$

and we can algebraically invert these to get the relation between the Cartesian set and the cylindrical one

$$\hat{\mathbf{x}} = \cos\phi\,\hat{\mathbf{s}} - \sin\phi\,\hat{\boldsymbol{\phi}} \quad \hat{\mathbf{y}} = \sin\phi\,\hat{\mathbf{s}} + \cos\phi\,\hat{\boldsymbol{\phi}} \quad \hat{\mathbf{z}} = \hat{\mathbf{z}}. \tag{B.7}$$

For derivatives, let's start with the gradient. What happens if we take $\nabla f(s, \phi, z)$? We know the gradient in Cartesian coordinates and with respect to the Cartesian basis. We'll use the chain rule from Calculus to work out expressions like $\frac{\partial f(s,\phi,z)}{\partial x}$, and then we can use the basis relations in (B.7) to rewrite the gradient in terms of the cylindrical basis vectors. First, let's move the derivatives from Cartesian to cylindrical,

---

[2] The gradient of $\phi = \tan^{-1}(y/x)$ also gives a vector pointing in the direction of increasing ϕ,

$$\nabla\left[\tan^{-1}\left(\frac{y}{x}\right)\right] = -\frac{y}{x^2 + y^2}\,\hat{\mathbf{x}} + \frac{x}{x^2 + y^2}\,\hat{\mathbf{y}} = \frac{1}{s}(-\sin\phi\,\hat{\mathbf{x}} + \cos\phi\,\hat{\mathbf{y}}).$$

$$\frac{\partial f(s,\phi,z)}{\partial x} = \frac{\partial f}{\partial s}\frac{\partial s}{\partial x} + \frac{\partial f}{\partial \phi}\frac{\partial \phi}{\partial x} + \frac{\partial f}{\partial z}\frac{\partial z}{\partial x} = \frac{\partial f}{\partial s}\frac{x}{\sqrt{x^2+y^2}} + \frac{\partial f}{\partial \phi}\frac{-y}{x^2+y^2} + \frac{\partial f}{\partial z}0$$

$$= \frac{\partial f}{\partial s}\cos\phi - \frac{\partial f}{\partial \phi}\frac{\sin\phi}{s}$$

(B.8)

where again we can write everything in terms of Cartesian (top right equality) or cylindrical coordinates (bottom). Similarly, we have

$$\frac{\partial f(s,\phi,z)}{\partial y} = \frac{\partial f}{\partial s}\frac{\partial s}{\partial y} + \frac{\partial f}{\partial \phi}\frac{\partial \phi}{\partial y} + \frac{\partial f}{\partial z}\frac{\partial z}{\partial y} = \frac{\partial f}{\partial s}\frac{y}{\sqrt{x^2+y^2}} + \frac{\partial f}{\partial \phi}\frac{x}{x^2+y^2} + \frac{\partial f}{\partial z}0$$

$$= \frac{\partial f}{\partial s}\sin\phi + \frac{\partial f}{\partial \phi}\frac{\cos\phi}{s}.$$

(B.9)

Finally, the z-derivative is unchanged from its Cartesian form. Now putting it all together,

$$\nabla f(s,\phi,z) = \left(\frac{\partial f}{\partial s}\cos\phi - \frac{\partial f}{\partial \phi}\frac{\sin\phi}{s}\right)\hat{\mathbf{x}} + \left(\frac{\partial f}{\partial s}\sin\phi + \frac{\partial f}{\partial \phi}\frac{\cos\phi}{s}\right)\hat{\mathbf{y}} + \frac{\partial f}{\partial z}\hat{\mathbf{z}}, \quad \text{(B.10)}$$

or, using (B.7) to write the gradient in terms of $\hat{\mathbf{s}}$, $\hat{\boldsymbol{\phi}}$ and $\hat{\mathbf{z}}$,

$$\nabla f(s,\phi,z) = \frac{\partial f}{\partial s}\hat{\mathbf{s}} + \frac{1}{s}\frac{\partial f}{\partial \phi}\hat{\boldsymbol{\phi}} + \frac{\partial f}{\partial z}\hat{\mathbf{z}}. \quad \text{(B.11)}$$

We can write this in operator form similar to (B.1),

$$\nabla = \hat{\mathbf{s}}\frac{\partial}{\partial s} + \frac{1}{s}\hat{\boldsymbol{\phi}}\frac{\partial}{\partial \phi} + \hat{\mathbf{z}}\frac{\partial}{\partial z}. \quad \text{(B.12)}$$

The divergence of a vector function $\mathbf{V}(s,\phi,z) = V_s\hat{\mathbf{s}} + V_\phi\hat{\boldsymbol{\phi}} + V_z\hat{\mathbf{z}}$ comes directly from the application of the gradient operator dotted into $\mathbf{V}$. We have to be careful now that the basis vectors themselves are position dependent, and so have nontrivial derivatives. Start by applying ∇ from (B.12) to each term in $\mathbf{V}$ employing the product rule:

$$\nabla \cdot \mathbf{V} = \hat{\mathbf{s}} \cdot \left[\frac{\partial V_s}{\partial s}\hat{\mathbf{s}} + V_s\frac{\partial \hat{\mathbf{s}}}{\partial s} + \frac{\partial V_\phi}{\partial s}\hat{\boldsymbol{\phi}} + V_\phi\frac{\partial \hat{\boldsymbol{\phi}}}{\partial s} + \frac{\partial V_z}{\partial s}\hat{\mathbf{z}} + V_z\frac{\partial \hat{\mathbf{z}}}{\partial s}\right]$$

$$+ \frac{1}{s}\hat{\boldsymbol{\phi}} \cdot \left[\frac{\partial V_s}{\partial \phi}\hat{\mathbf{s}} + V_s\frac{\partial \hat{\mathbf{s}}}{\partial \phi} + \frac{\partial V_\phi}{\partial \phi}\hat{\boldsymbol{\phi}} + V_\phi\frac{\partial \hat{\boldsymbol{\phi}}}{\partial \phi} + \frac{\partial V_z}{\partial \phi}\hat{\mathbf{z}} + V_z\frac{\partial \hat{\mathbf{z}}}{\partial \phi}\right]$$

$$+ \hat{\mathbf{z}} \cdot \left[\frac{\partial V_s}{\partial z}\hat{\mathbf{s}} + V_s\frac{\partial \hat{\mathbf{s}}}{\partial z} + \frac{\partial V_\phi}{\partial z}\hat{\boldsymbol{\phi}} + V_\phi\frac{\partial \hat{\boldsymbol{\phi}}}{\partial z} + \frac{\partial V_z}{\partial z}\hat{\mathbf{z}} + V_z\frac{\partial \hat{\mathbf{z}}}{\partial z}\right]$$

$$= \left[\frac{\partial V_s}{\partial s} + V_s\hat{\mathbf{s}} \cdot \frac{\partial \hat{\mathbf{s}}}{\partial s} + V_\phi\hat{\mathbf{s}} \cdot \frac{\partial \hat{\boldsymbol{\phi}}}{\partial s} + V_z\hat{\mathbf{s}} \cdot \frac{\partial \hat{\mathbf{z}}}{\partial s}\right]$$

$$+ \frac{1}{s}\left[V_s\hat{\boldsymbol{\phi}} \cdot \frac{\partial \hat{\mathbf{s}}}{\partial \phi} + \frac{\partial V_\phi}{\partial \phi} + V_\phi\hat{\boldsymbol{\phi}} \cdot \frac{\partial \hat{\boldsymbol{\phi}}}{\partial \phi} + V_z\hat{\boldsymbol{\phi}} \cdot \frac{\partial \hat{\mathbf{z}}}{\partial \phi}\right]$$

$$+ \left[V_s\hat{\mathbf{z}} \cdot \frac{\partial \hat{\mathbf{s}}}{\partial z} + V_\phi\hat{\mathbf{z}} \cdot \frac{\partial \hat{\boldsymbol{\phi}}}{\partial z} + \frac{\partial V_z}{\partial z} + V_z\hat{\mathbf{z}} \cdot \frac{\partial \hat{\mathbf{z}}}{\partial z}\right].$$

(B.13)

We need to evaluate the derivatives of the basis vectors. To do this, we will express the cylindrical basis vectors in terms of the Cartesian ones, which do not depend on position, from (B.6):

$$\frac{\partial \hat{s}}{\partial s} = 0 \qquad \frac{\partial \hat{s}}{\partial \phi} = -\sin\phi\,\hat{x} + \cos\phi\,\hat{y} = \hat{\phi} \qquad \frac{\partial \hat{s}}{\partial z} = 0$$

$$\frac{\partial \hat{\phi}}{\partial s} = 0 \qquad \frac{\partial \hat{\phi}}{\partial \phi} = -\cos\phi\,\hat{x} - \sin\phi\,\hat{y} = -\hat{s} \qquad \frac{\partial \hat{\phi}}{\partial z} = 0 \qquad \text{(B.14)}$$

$$\frac{\partial \hat{z}}{\partial s} = 0 \qquad \frac{\partial \hat{z}}{\partial \phi} = 0 \qquad \frac{\partial \hat{z}}{\partial z} = 0.$$

Using these results in (B.13),

$$\nabla \cdot \mathbf{V} = \frac{\partial V_s}{\partial s} + \frac{1}{s}\left(V_s + \frac{\partial V_\phi}{\partial \phi}\right) + \frac{\partial V_z}{\partial z} \qquad \text{(B.15)}$$

which we can write, using the product rule, as

$$\nabla \cdot \mathbf{V} = \frac{1}{s}\frac{\partial}{\partial s}(sV_s) + \frac{1}{s}\frac{\partial V_\phi}{\partial \phi} + \frac{\partial V_z}{\partial z}. \qquad \text{(B.16)}$$

The final single derivative operator of interest is the curl. To calculate the curl, we use the same approach as for the divergence, just replacing the dot product with the cross product. We can use the derivates from (B.14) and the right-hand rule to perform the calculation:

$$\nabla \times \mathbf{V} = \hat{s} \times \left[\frac{\partial V_s}{\partial s}\hat{s} + V_s\frac{\partial \hat{s}}{\partial s} + \frac{\partial V_\phi}{\partial s}\hat{\phi} + V_\phi\frac{\partial \hat{\phi}}{\partial s} + \frac{\partial V_z}{\partial s}\hat{z} + V_z\frac{\partial \hat{z}}{\partial s}\right]$$

$$+ \frac{1}{s}\hat{\phi} \times \left[\frac{\partial V_s}{\partial \phi}\hat{s} + V_s\frac{\partial \hat{s}}{\partial \phi} + \frac{\partial V_\phi}{\partial \phi}\hat{\phi} + V_\phi\frac{\partial \hat{\phi}}{\partial \phi} + \frac{\partial V_z}{\partial \phi}\hat{z} + V_z\frac{\partial \hat{z}}{\partial \phi}\right]$$

$$+ \hat{z} \times \left[\frac{\partial V_s}{\partial z}\hat{s} + V_s\frac{\partial \hat{s}}{\partial z} + \frac{\partial V_\phi}{\partial z}\hat{\phi} + V_\phi\frac{\partial \hat{\phi}}{\partial z} + \frac{\partial V_z}{\partial z}\hat{z} + V_z\frac{\partial \hat{z}}{\partial z}\right]$$

$$= \left[\frac{\partial V_\phi}{\partial s}\hat{z} - \frac{\partial V_z}{\partial s}\hat{\phi}\right] + \frac{1}{s}\left[-\frac{\partial V_s}{\partial \phi}\hat{z} + V_\phi\hat{z} + \frac{\partial V_z}{\partial \phi}\hat{s}\right] + \left[\frac{\partial V_s}{\partial z}\hat{\phi} - \frac{\partial V_\phi}{\partial z}\hat{s}\right].$$
$$\text{(B.17)}$$

Collecting terms, and using the product rule, we have

$$\nabla \times \mathbf{V} = \left[\frac{1}{s}\frac{\partial V_z}{\partial \phi} - \frac{\partial V_\phi}{\partial z}\right]\hat{s} + \left[\frac{\partial V_s}{\partial z} - \frac{\partial V_z}{\partial s}\right]\hat{\phi} + \frac{1}{s}\left[\frac{\partial}{\partial s}(sV_\phi) - \frac{\partial V_s}{\partial \phi}\right]\hat{z} \qquad \text{(B.18)}$$

For the second derivatives in the form of the Laplacian, $\nabla^2 f = \nabla \cdot \nabla f$, we just apply the divergence formula in (B.16) to the gradient using the elements of the gradient vector $\mathbf{V} \equiv \nabla f$ from (B.11),

$$\nabla^2 f(s, \phi, z) = \frac{1}{s}\frac{\partial}{\partial s}\left(s\frac{\partial f}{\partial s}\right) + \frac{1}{s^2}\frac{\partial^2 f}{\partial \phi^2} + \frac{\partial^2 f}{\partial z^2}. \qquad \text{(B.19)}$$

B.2 A Better Way

All that chain ruling and basis switching is clear, but not always easy to carry out. Let's organize the calculation and broaden its applicability. Let the Cartesian coordinates be the numbered set $x_1 = x$, $x_2 = y$, $x_3 = z$. Suppose we introduce an arbitrary new set of coordinates, call them X_1, X_2, and X_3, each a function of x_1, x_2, and x_3 as in (B.3) (where we'd have $X_1 = s$, $X_2 = \phi$, $X_3 = z$). We'll assume that the coordinate transformation is invertible, so that you could write x_1, x_2, and x_3 in terms of the $X_1(x_1, x_2, x_3)$, $X_2(x_1, x_2, x_3)$, and $X_3(x_1, x_2, x_3)$ – we carried out that process explicitly in, for example (B.2).

With the coordinate transformation and its inverse given, our next job is to get the basis vectors. We know that the $\hat{\mathbf{X}}^1$ basis vector points in the direction of increasing X_1 coordinate, and similarly for the other basis vectors. If we take the usual $\mathbf{r} = x_1\,\hat{\mathbf{x}}^1 + x_2\,\hat{\mathbf{x}}^2 + x_3\,\hat{\mathbf{x}}^3$, and evaluate it at X_1 and $X_1 + \Delta$, then subtract, we'll get a vector parallel to $\hat{\mathbf{X}}^1$. That is, $(\mathbf{r}(X_1 + \Delta, X_2, X_3) - \mathbf{r}(X_1, X_2, X_3))/\Delta$ points from the original location to the new one, defining the direction of X_1 increase in the original Cartesian basis. If we take the parameter $\Delta \to 0$, we can write the difference in terms of the derivative of $\mathbf{r}$ with respect to X_1:

$$\mathbf{W}^1 \equiv \lim_{\Delta \to 0}\left[\frac{\mathbf{r}(X_1 + \Delta, X_2, X_3) - \mathbf{r}(X_1, X_2, X_3)}{\Delta}\right] = \frac{\partial \mathbf{r}}{\partial X_1}, \tag{B.20}$$

and then $\hat{\mathbf{X}}^1 = \mathbf{W}^1/W^1$ is the unit vector. We can similarly generate $\mathbf{W}^2 \equiv \frac{\partial \mathbf{r}}{\partial X_2}$ and $\mathbf{W}^3 \equiv \frac{\partial \mathbf{r}}{\partial X_3}$ with $\hat{\mathbf{X}}^2 = \mathbf{W}^2/W^2$ and $\hat{\mathbf{X}}^3 = \mathbf{W}^3/W^3$.

The process here can be written in matrix-vector form, define the matrix[3]

$$\mathbb{J} \doteq \begin{pmatrix} \frac{\partial x_1}{\partial X_1} & \frac{\partial x_2}{\partial X_1} & \frac{\partial x_3}{\partial X_1} \\ \frac{\partial x_1}{\partial X_2} & \frac{\partial x_2}{\partial X_2} & \frac{\partial x_3}{\partial X_2} \\ \frac{\partial x_1}{\partial X_3} & \frac{\partial x_2}{\partial X_3} & \frac{\partial x_3}{\partial X_3} \end{pmatrix}, \tag{B.21}$$

with entries $J_{ij} = \frac{\partial x_j}{\partial X_i}$. Now we can summarize the relation between the new basis vectors and the old ones

$$\begin{pmatrix} \hat{\mathbf{X}}^1 \\ \hat{\mathbf{X}}^2 \\ \hat{\mathbf{X}}^3 \end{pmatrix} = \underbrace{\begin{pmatrix} \frac{1}{\left|\frac{\partial \mathbf{r}}{\partial X_1}\right|} & 0 & 0 \\ 0 & \frac{1}{\left|\frac{\partial \mathbf{r}}{\partial X_2}\right|} & 0 \\ 0 & 0 & \frac{1}{\left|\frac{\partial \mathbf{r}}{\partial X_3}\right|} \end{pmatrix}}_{\equiv \mathbb{H}} \begin{pmatrix} \frac{\partial x_1}{\partial X_1} & \frac{\partial x_2}{\partial X_1} & \frac{\partial x_3}{\partial X_1} \\ \frac{\partial x_1}{\partial X_2} & \frac{\partial x_2}{\partial X_2} & \frac{\partial x_3}{\partial X_2} \\ \frac{\partial x_1}{\partial X_3} & \frac{\partial x_2}{\partial X_3} & \frac{\partial x_3}{\partial X_3} \end{pmatrix} \begin{pmatrix} \hat{\mathbf{x}}^1 \\ \hat{\mathbf{x}}^2 \\ \hat{\mathbf{x}}^3 \end{pmatrix} \tag{B.22}$$

where the matrix out front on the right serves to normalize the basis vectors, call it $\mathbb{H}$. This matrix-vector structure allows us to easily write the formal inverse to relate the original basis vectors to the new ones (i.e. we can invert to write $\hat{\mathbf{x}}^1$, $\hat{\mathbf{x}}^2$, and $\hat{\mathbf{x}}^3$ in terms of the new basis vectors).

---

[3] This matrix is called the "Jacobian" of the transformation. In a confusing abuse of notation, sometimes it is the determinant of this matrix that is called the Jacobian.

Let's make sure we recover the cylindrical basis vectors from the previous section using our new approach. We have $x_1 = x$, $x_2 = y$, and $x_3 = z$ as usual, with $X_1 = s$, $X_2 = \phi$, and $X_3 = z$ for the cylindrical coordinates. The matrices $\mathbb{H}$ and $\mathbb{J}$ become, in this specific setting,

$$\mathbb{H} \doteq \begin{pmatrix} 1 & 0 & 0 \\ 0 & 1/X_1 & 0 \\ 0 & 0 & 1 \end{pmatrix} \qquad \mathbb{J} \doteq \begin{pmatrix} \cos(X_2) & \sin(X_2) & 0 \\ -X_1 \sin(X_2) & X_1 \cos(X_2) & 0 \\ 0 & 0 & 1 \end{pmatrix} \tag{B.23}$$

and using these in (B.22) gives

$$\begin{pmatrix} \hat{\mathbf{X}}^1 \\ \hat{\mathbf{X}}^2 \\ \hat{\mathbf{X}}^3 \end{pmatrix} = \begin{pmatrix} \cos(X_2)\,\hat{\mathbf{x}}^1 + \sin(X_2)\,\hat{\mathbf{x}}^2 \\ -\sin(X_2)\,\hat{\mathbf{x}}^1 + \cos(X_2)\,\hat{\mathbf{x}}^2 \\ \hat{\mathbf{x}}^3 \end{pmatrix} \tag{B.24}$$

matching, in this notation, our expression from (B.6).

In order to write everything in a consistent matrix-vector form, I need an expression for the "vector of vectors" on the left and right in (B.22). Define

$$\mathbf{E} \doteq \begin{pmatrix} \hat{\mathbf{X}}^1 \\ \hat{\mathbf{X}}^2 \\ \hat{\mathbf{X}}^3 \end{pmatrix} \qquad \mathbf{e} \doteq \begin{pmatrix} \hat{\mathbf{x}}^1 \\ \hat{\mathbf{x}}^2 \\ \hat{\mathbf{x}}^3 \end{pmatrix} \tag{B.25}$$

These are just convenient collections of the symbols $\hat{\mathbf{X}}^1$, $\hat{\mathbf{X}}^2$, $\hat{\mathbf{X}}^3$, and the lower-case version, although it looks odd to have a bold face vector whose entries are themselves basis vectors. It will only happen in this section, and is only to express the fact that as objects, the basis vectors are related by linear combination. We can now write

$$\mathbf{E} = \mathbb{H}\mathbb{J}\mathbf{e} \longrightarrow \mathbf{e} = \mathbb{J}^{-1}\mathbb{H}^{-1}\mathbf{E}. \tag{B.26}$$

Moving on to the gradient. Remember there are two steps we have to carry out. The first is to use the chain rule to take derivatives. For $f(X_1, X_2, X_3)$, we have

$$\begin{aligned}
\frac{\partial f}{\partial x_1} &= \frac{\partial f}{\partial X_1}\frac{\partial X_1}{\partial x_1} + \frac{\partial f}{\partial X_2}\frac{\partial X_2}{\partial x_1} + \frac{\partial f}{\partial X_3}\frac{\partial X_3}{\partial x_1} \\
\frac{\partial f}{\partial x_2} &= \frac{\partial f}{\partial X_1}\frac{\partial X_1}{\partial x_2} + \frac{\partial f}{\partial X_2}\frac{\partial X_2}{\partial x_2} + \frac{\partial f}{\partial X_3}\frac{\partial X_3}{\partial x_2} \\
\frac{\partial f}{\partial x_3} &= \frac{\partial f}{\partial X_1}\frac{\partial X_1}{\partial x_3} + \frac{\partial f}{\partial X_2}\frac{\partial X_2}{\partial x_3} + \frac{\partial f}{\partial X_3}\frac{\partial X_3}{\partial x_3},
\end{aligned} \tag{B.27}$$

and we can write this in matrix-vector form

$$\begin{pmatrix} \frac{\partial f}{\partial x_1} \\ \frac{\partial f}{\partial x_2} \\ \frac{\partial f}{\partial x_3} \end{pmatrix} = \begin{pmatrix} \frac{\partial X_1}{\partial x_1} & \frac{\partial X_2}{\partial x_1} & \frac{\partial X_3}{\partial x_1} \\ \frac{\partial X_1}{\partial x_2} & \frac{\partial X_2}{\partial x_2} & \frac{\partial X_3}{\partial x_2} \\ \frac{\partial X_1}{\partial x_3} & \frac{\partial X_2}{\partial x_3} & \frac{\partial X_3}{\partial x_3} \end{pmatrix} \begin{pmatrix} \frac{\partial f}{\partial X_1} \\ \frac{\partial f}{\partial X_2} \\ \frac{\partial f}{\partial X_3} \end{pmatrix}. \tag{B.28}$$

The matrix appearing in this equation is structurally identical to $\mathbb{J}$ in (B.21), but with the roles of the two coordinates reversed (so that this matrix is also a Jacobian, but with the $\{X_1, X_2, X_3\}$ as the "original" coordinates, $\{x_1, x_2, x_3\}$ as the "new" set). Call the matrix in (B.28) $\mathbb{K}$ with entries $K_{ij} = \frac{\partial X_j}{\partial x_i}$. What is the relationship between $\mathbb{K}$ and $\mathbb{J}$ from (B.21)? Consider their product:

$$(\mathbb{K}\mathbb{J})_{mn} = \sum_{k=1}^{3} K_{mk} J_{kn} = \sum_{k=1}^{3} \frac{\partial X_k}{\partial x_m} \frac{\partial x_n}{\partial X_k} = \sum_{k=1}^{3} \frac{\partial x_n}{\partial X_k} \frac{\partial X_k}{\partial x_m}, \tag{B.29}$$

and think about the derivative $\frac{\partial x_n}{\partial x_m} = \delta_{mn}$ – the derivative is 1 if $n = m$ (an object like $\frac{\partial x_1}{\partial x_1}$) and 0 if $n \neq m$ (as with, for example, $\frac{\partial x_1}{\partial x_2}$). If we view x_n as a function of X_1, X_2, and X_3, then the chain rule says

$$\frac{\partial x_n}{\partial x_m} = \frac{\partial x_n}{\partial X_1} \frac{\partial X_1}{\partial x_m} + \frac{\partial x_n}{\partial X_2} \frac{\partial X_2}{\partial x_m} + \frac{\partial x_n}{\partial X_3} \frac{\partial X_3}{\partial x_m} = \sum_{k=1}^{3} \frac{\partial x_n}{\partial X_k} \frac{\partial X_k}{\partial x_m} \tag{B.30}$$

which is precisely what appears in (B.29). But this is just δ_{mn}, so we have

$$(\mathbb{K}\mathbb{J})_{mn} = \delta_{mn} \tag{B.31}$$

or, in terms of matrices (rather than their entries): $\mathbb{K}\mathbb{J} = \mathbb{I}$ the identity matrix. We have just learned that the matrix $\mathbb{K}$ is the matrix inverse of $\mathbb{J}$: $\mathbb{K} = \mathbb{J}^{-1}$.

Now for the second piece of the gradient, the basis vectors. In our current vector notation, the gradient can be expressed as a product of $\mathbf{e}^T$ with the vector appearing on the left in (B.28):

$$\nabla f = \hat{\mathbf{x}}^1 \frac{\partial f}{\partial x_1} + \hat{\mathbf{x}}^2 \frac{\partial f}{\partial x_2} + \hat{\mathbf{x}}^3 \frac{\partial f}{\partial x_3} = \underbrace{\left(\begin{array}{ccc} \hat{\mathbf{x}}^1 & \hat{\mathbf{x}}^2 & \hat{\mathbf{x}}^3 \end{array} \right)}_{= \mathbf{e}^T} \cdot \left(\begin{array}{c} \frac{\partial f}{\partial x_1} \\ \frac{\partial f}{\partial x_2} \\ \frac{\partial f}{\partial x_3} \end{array} \right). \tag{B.32}$$

If we multiply both sides of (B.28) by $\mathbf{e}^T$ from the left, we get the gradient:

$$\nabla f = \mathbf{e}^T \mathbb{J}^{-1} \left(\begin{array}{c} \frac{\partial f}{\partial X_1} \\ \frac{\partial f}{\partial X_2} \\ \frac{\partial f}{\partial X_3} \end{array} \right) \tag{B.33}$$

and now we can use $\mathbf{e} = \mathbb{J}^{-1}\mathbb{H}^{-1}\mathbf{E}$ to write everything in terms of the new basis set,

$$\nabla f = \mathbf{E}^T \mathbb{H}^{-1} \left(\mathbb{J}^{-1} \right)^T \mathbb{J}^{-1} \left(\begin{array}{c} \frac{\partial f}{\partial X_1} \\ \frac{\partial f}{\partial X_2} \\ \frac{\partial f}{\partial X_3} \end{array} \right). \tag{B.34}$$

The gradient operator that we will use in the divergence and curl is then

$$\nabla = \mathbf{E}^T \mathbb{H}^{-1} \left(\mathbb{J}^{-1} \right)^T \mathbb{J}^{-1} \left(\begin{array}{c} \frac{\partial}{\partial X_1} \\ \frac{\partial}{\partial X_2} \\ \frac{\partial}{\partial X_3} \end{array} \right). \tag{B.35}$$

Rather than keeping the matrix notation going throughout the divergence and curl calculations, we will assume that the gradient operator takes the general form, calculated using (B.35),

$$\nabla = F(X_1, X_2, X_3) \, \hat{\mathbf{X}}^1 \frac{\partial}{\partial X_1} + G(X_1, X_2, X_3) \, \hat{\mathbf{X}}^2 \frac{\partial}{\partial X_2} + H(X_1, X_2, X_3) \, \hat{\mathbf{X}}^3 \frac{\partial}{\partial X_3}. \tag{B.36}$$

For a vector function $\mathbf{V} = V_{X_1}\,\hat{\mathbf{X}}^1 + V_{X_2}\,\hat{\mathbf{X}}^2 + V_{X_3}\,\hat{\mathbf{X}}^3$, we can use the product rule to write the divergence

$$\nabla \cdot \mathbf{V} = F\left[\frac{\partial V_{X_1}}{\partial X_1} + V_{X_1}\,\hat{\mathbf{X}}^1 \cdot \frac{\partial \hat{\mathbf{X}}^1}{\partial X_1} + V_{X_2}\,\hat{\mathbf{X}}^1 \cdot \frac{\partial \hat{\mathbf{X}}^2}{\partial X_1} + V_{X_3}\,\hat{\mathbf{X}}^1 \cdot \frac{\partial \hat{\mathbf{X}}^3}{\partial X_1}\right]$$

$$+ G\left[\frac{\partial V_{X_2}}{\partial X_2} + V_{X_2}\,\hat{\mathbf{X}}^2 \cdot \frac{\partial \hat{\mathbf{X}}^2}{\partial X_2} + V_{X_1}\,\hat{\mathbf{X}}^2 \cdot \frac{\partial \hat{\mathbf{X}}^1}{\partial X_2} + V_{X_3}\,\hat{\mathbf{X}}^2 \cdot \frac{\partial \hat{\mathbf{X}}^3}{\partial X_2}\right] \quad \text{(B.37)}$$

$$+ H\left[\frac{\partial V_{X_3}}{\partial X^3} + V_{X_3}\,\hat{\mathbf{X}}^3 \cdot \frac{\partial \hat{\mathbf{X}}^3}{\partial X_3} + V_{X_2}\,\hat{\mathbf{X}}^3 \cdot \frac{\partial \hat{\mathbf{X}}^2}{\partial X_3} + V_{X_1}\,\hat{\mathbf{X}}^3 \cdot \frac{\partial \hat{\mathbf{X}}^1}{\partial X_3}\right].$$

These expressions require us to evaluate the derivatives of the basis functions, easy enough to do once you have their expressions from (B.26). The curl requires us to know all of the cross products of the basis vectors, which can again be calculated from (B.26). Then the curl reads,

$$\nabla \times \mathbf{V} =$$

$$F\left[V_{X_1}\,\hat{\mathbf{X}}^1 \times \frac{\partial \hat{\mathbf{X}}^1}{\partial X_1} + \frac{\partial V_{X_2}}{\partial X_1}\,\hat{\mathbf{X}}^1 \times \hat{\mathbf{X}}^2 + V_{X_2}\,\hat{\mathbf{X}}^1 \times \frac{\partial \hat{\mathbf{X}}^2}{\partial X_1} + \frac{\partial V_{X_3}}{\partial X_1}\,\hat{\mathbf{X}}^1 \times \hat{\mathbf{X}}^3 + V_{X_3}\,\hat{\mathbf{X}}^1 \times \frac{\partial \hat{\mathbf{X}}^3}{\partial X_1}\right]$$

$$+ G\left[V_{X_2}\,\hat{\mathbf{X}}^2 \times \frac{\partial \hat{\mathbf{X}}^2}{\partial X_2} + \frac{\partial V_{X_1}}{\partial X_2}\,\hat{\mathbf{X}}^2 \times \hat{\mathbf{X}}^1 + V_{X_1}\,\hat{\mathbf{X}}^2 \times \frac{\partial \hat{\mathbf{X}}^1}{\partial X_2} + \frac{\partial V_{X_3}}{\partial X_2}\,\hat{\mathbf{X}}^2 \times \hat{\mathbf{X}}^3 + V_{X_3}\,\hat{\mathbf{X}}^2 \times \frac{\partial \hat{\mathbf{X}}^3}{\partial X_2}\right]$$

$$+ H\left[V_{X_3}\,\hat{\mathbf{X}}^3 \times \frac{\partial \hat{\mathbf{X}}^3}{\partial X_3} + \frac{\partial V_{X_2}}{\partial X_3}\,\hat{\mathbf{X}}^3 \times \hat{\mathbf{X}}^2 + V_{X_2}\,\hat{\mathbf{X}}^3 \times \frac{\partial \hat{\mathbf{X}}^2}{\partial X_3} + \frac{\partial V_{X_1}}{\partial X_3}\,\hat{\mathbf{X}}^3 \times \hat{\mathbf{X}}^1 + V_{X_1}\,\hat{\mathbf{X}}^3 \times \frac{\partial \hat{\mathbf{X}}^1}{\partial X_3}\right].$$

$$\text{(B.38)}$$

Finally, the general Laplacian acting on $f(X^1, X^2, X^3)$ comes from applying the divergence to the gradient, as usual:

$$\nabla^2 f = F\left[\frac{\partial}{\partial X_1}\left(F\frac{\partial f}{\partial X_1}\right) + F\frac{\partial f}{\partial X_1}\,\hat{\mathbf{X}}^1 \cdot \frac{\partial \hat{\mathbf{X}}^1}{\partial X_1} + G\frac{\partial f}{\partial X_2}\,\hat{\mathbf{X}}^1 \cdot \frac{\partial \hat{\mathbf{X}}^2}{\partial X_1} + H\frac{\partial f}{\partial X_3}\,\hat{\mathbf{X}}^1 \cdot \frac{\partial \hat{\mathbf{X}}^3}{\partial X_1}\right]$$

$$+ G\left[\frac{\partial}{\partial X_2}\left(G\frac{\partial f}{\partial X_2}\right) + G\frac{\partial f}{\partial X_2}\,\hat{\mathbf{X}}^2 \cdot \frac{\partial \hat{\mathbf{X}}^2}{\partial X_2} + F\frac{\partial f}{\partial X_1}\,\hat{\mathbf{X}}^2 \cdot \frac{\partial \hat{\mathbf{X}}^1}{\partial X_2} + H\frac{\partial f}{\partial X_3}\,\hat{\mathbf{X}}^2 \cdot \frac{\partial \hat{\mathbf{X}}^3}{\partial X_2}\right]$$

$$+ H\left[\frac{\partial}{\partial X_3}\left(H\frac{\partial f}{\partial X_3}\right) + H\frac{\partial f}{\partial X_3}\,\hat{\mathbf{X}}^3 \cdot \frac{\partial \hat{\mathbf{X}}^3}{\partial X_3} + G\frac{\partial f}{\partial X_2}\,\hat{\mathbf{X}}^3 \cdot \frac{\partial \hat{\mathbf{X}}^2}{\partial X_3} + F\frac{\partial f}{\partial X_1}\,\hat{\mathbf{X}}^3 \cdot \frac{\partial \hat{\mathbf{X}}^1}{\partial X_3}\right].$$

$$\text{(B.39)}$$

B.3 Spherical Coordinates

Spherical coordinates are defined as shown in Figure 6.9, reproduced in this section as Figure B.2. From the geometry of the definition, we have $x = r\sin\theta\cos\phi$, $y = r\sin\theta\sin\phi$, $z = r\cos\theta$.

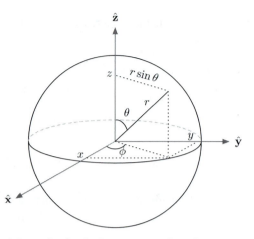

Fig. B.2   Spherical coordinates, r, θ, and ϕ are related to the Cartesian x, y, and z as shown.

To use the formalism from the previous section, take $X_1 \equiv r$, $X_2 \equiv \theta$, $X_3 \equiv \phi$, with $x_1 \equiv x$, $x_2 \equiv y$, $x_3 \equiv z$ as before. Our first job is to compute $\mathbb{J}$ and $\mathbb{H}$ from (B.21) and (B.22)

$$\mathbb{J} \doteq \begin{pmatrix} \sin(X_2)\cos(X_3) & \sin(X_2)\sin(X_3) & \cos(X_2) \\ X_1 \cos(X_2)\cos(X_3) & X_1 \cos(X_2)\sin(X_3) & -X_1 \sin(X_2) \\ -X_1 \sin(X_2)\sin(X_3) & X_1 \cos(X_3)\sin(X_2) & 0 \end{pmatrix}$$

$$\mathbb{H} \doteq \begin{pmatrix} 1 & 0 & 0 \\ 0 & \frac{1}{X_1} & 0 \\ 0 & 0 & \frac{1}{X_1 \sin(X_2)} \end{pmatrix}. \tag{B.40}$$

Using $\mathbf{E} = \mathbb{H}\mathbb{J}\mathbf{e}$, we can write the spherical basis vectors, in terms of the Cartesian ones:

$$\hat{\mathbf{X}}^1 = \sin(X_2)\cos(X_3)\,\hat{\mathbf{x}}^1 + \sin(X_2)\sin(X_3)\,\hat{\mathbf{x}}^2 + \cos(X_2)\,\hat{\mathbf{x}}^3$$

$$\hat{\mathbf{X}}^2 = \cos(X_2)\cos(X_3)\,\hat{\mathbf{x}}^1 + \cos(X_2)\sin(X_3)\,\hat{\mathbf{x}}^2 - \sin(X_2)\,\hat{\mathbf{x}}^3 \tag{B.41}$$

$$\hat{\mathbf{X}}^3 = -\sin(X_3)\,\hat{\mathbf{x}}^1 + \cos(X_3)\,\hat{\mathbf{x}}^2.$$

The gradient, according to (B.35) is

$$\nabla = \hat{\mathbf{X}}^1 \frac{\partial}{\partial X_1} + \frac{1}{X_1}\hat{\mathbf{X}}^2 \frac{\partial}{\partial X_2} + \frac{1}{X_1 \sin(X_2)}\hat{\mathbf{X}}^3 \frac{\partial}{\partial X_3}, \tag{B.42}$$

and we identify $F \equiv 1$, $G \equiv 1/X_1$, and $H \equiv 1/(X_1 \sin(X_2))$ in (B.36).

From here, we need to know the derivatives of the basis vectors. The nonzero derivatives are:

$$\frac{\partial \hat{\mathbf{X}}^1}{\partial X_2} = \hat{\mathbf{X}}^2 \qquad \frac{\partial \hat{\mathbf{X}}^2}{\partial X_2} = -\hat{\mathbf{X}}^1 \qquad \frac{\partial \hat{\mathbf{X}}^1}{\partial X_3} = \sin(X_2)\,\hat{\mathbf{X}}^3 \qquad \frac{\partial \hat{\mathbf{X}}^2}{\partial X_3} = \cos(X_2)\,\hat{\mathbf{X}}^3$$

$$\frac{\partial \hat{\mathbf{X}}^3}{\partial X_3} = -\sin(X_2)\,\hat{\mathbf{X}}^1 - \cos(X_2)\,\hat{\mathbf{X}}^2. \tag{B.43}$$

With these, we can write the divergence

$$\nabla \cdot \mathbf{V} = \left[\frac{\partial V_{X_1}}{\partial X_1}\right] + \frac{1}{X_1}\left[\frac{\partial V_{X_2}}{\partial X_2} + V_{X_1}\right] + \frac{1}{X_1 \sin(X_2)}\left[\frac{\partial V_{X_3}}{\partial X_3} + V_{X_2}\cos(X_2) + V_{X_1}\sin(X_2)\right]$$

$$= \frac{1}{(X_1)^2}\frac{\partial}{\partial X_1}\left((X_1)^2 \, V_{X_1}\right) + \frac{1}{X_1 \sin(X_2)}\frac{\partial}{\partial X_2}\left(\sin(X_2)V_{X_2}\right).$$

$$(B.44)$$

For the curl, we must compute the cross products of the new basis vectors

$$\hat{\mathbf{X}}^1 \times \hat{\mathbf{X}}^2 = \hat{\mathbf{X}}^3 \qquad \hat{\mathbf{X}}^1 \times \hat{\mathbf{X}}^3 = -\hat{\mathbf{X}}^2 \qquad \hat{\mathbf{X}}^2 \times \hat{\mathbf{X}}^3 = \hat{\mathbf{X}}^1 \qquad (B.45)$$

and then

$$\nabla \times \mathbf{V} = \left[\frac{\partial V_{X_2}}{\partial X_1}\hat{\mathbf{X}}^3 - \frac{\partial V_{X_3}}{\partial X_1}\hat{\mathbf{X}}^2\right] + \frac{1}{X_1}\left[V_{X_2}\hat{\mathbf{X}}^3 - \frac{\partial V_{X_1}}{\partial X_2}\hat{\mathbf{X}}^3 + \frac{\partial V_{X_3}}{\partial X_2}\hat{\mathbf{X}}^1\right]$$

$$+ \frac{1}{X_1 \sin(X_2)}\left[V_{X_3}\left(-\sin(X_2)\hat{\mathbf{X}}^2 + \cos(X_2)\hat{\mathbf{X}}^1\right) - \frac{\partial V_{X_2}}{\partial X_3}\hat{\mathbf{X}}^1 + \frac{\partial V_{X_1}}{\partial X_3}\hat{\mathbf{X}}^2\right]$$

$$= \frac{1}{X_1 \sin(X_2)}\left[\frac{\partial}{\partial X_2}\left(\sin(X_2)V_{X_3}\right) - \frac{\partial V_{X_2}}{\partial X_3}\right]\hat{\mathbf{X}}^1 + \frac{1}{X_1}\left[\frac{1}{\sin(X_2)}\frac{\partial V_{X_1}}{\partial X_3} - \frac{\partial}{\partial X_1}(X_1 V_{X_3})\right]\hat{\mathbf{X}}^2$$

$$+ \frac{1}{X_1}\left[\frac{\partial}{\partial X_1}(X_1 V_{X_2}) - \frac{\partial V_{X_1}}{\partial X_2}\right]\hat{\mathbf{X}}^3.$$

$$(B.46)$$

Finally, the Laplacian here is

$$\nabla^2 f = \left[\frac{\partial^2 f}{\partial (X_1)^2}\right] + \frac{1}{X_1}\left[\frac{\partial}{\partial X_2}\left(\frac{1}{X_1}\frac{\partial f}{\partial X_2}\right) + \frac{\partial f}{\partial X_1}\right]$$

$$+ \frac{1}{X_1 \sin(X_2)}\left[\frac{\partial}{\partial X_3}\left(\frac{1}{X_1 \sin(X_2)}\frac{\partial f}{\partial X_3}\right) + \frac{1}{X_1}\frac{\partial f}{\partial X_2}\cos(X_2) + \frac{\partial f}{\partial X_1}\sin(X_2)\right]$$

$$= \frac{1}{(X_1)^2}\frac{\partial}{\partial X_1}\left((X_1)^2\frac{\partial f}{\partial X_1}\right) + \frac{1}{(X_1)^2 \sin(X_2)}\frac{\partial}{\partial X_2}\left(\sin(X_2)\frac{\partial f}{\partial X_2}\right)$$

$$+ \frac{1}{(X_1)^2 \sin^2(X_2)}\frac{\partial^2 f}{\partial (X_3)^2}.$$

$$(B.47)$$

It is useful to record these results in terms of the more naturally named r, θ, and ϕ coordinate labels from Figure B.2. In this notation, the basis vectors are $\hat{\mathbf{r}} = \hat{\mathbf{X}}^1$, $\hat{\boldsymbol{\theta}} = \hat{\mathbf{X}}^2$, and $\hat{\boldsymbol{\phi}} = \hat{\mathbf{X}}^3$. The gradient is

$$\nabla f(r, \theta, \phi) = \frac{\partial f}{\partial r}\hat{\mathbf{r}} + \frac{1}{r}\frac{\partial f}{\partial \theta}\hat{\boldsymbol{\theta}} + \frac{1}{r\sin\theta}\frac{\partial f}{\partial \phi}\hat{\boldsymbol{\phi}}.$$

$$(B.48)$$

The divergence and curl are

$$\nabla \cdot \mathbf{V} = \frac{1}{r^2} \frac{\partial}{\partial r} \left(r^2 V_r \right) + \frac{1}{r \sin \theta} \frac{\partial}{\partial \theta} (\sin \theta V_\theta) + \frac{1}{r \sin \theta} \frac{\partial V_\phi}{\partial \phi}$$

$$\nabla \times \mathbf{V} = \frac{1}{r \sin \theta} \left[\frac{\partial}{\partial \theta} (\sin \theta V_\phi) - \frac{\partial V_\theta}{\partial \phi} \right] \hat{\mathbf{r}} + \frac{1}{r} \left[\frac{1}{\sin \theta} \frac{\partial V_r}{\partial \phi} - \frac{\partial}{\partial r} (r V_\phi) \right] \hat{\boldsymbol{\theta}} \quad \text{(B.49)}$$

$$+ \frac{1}{r} \left[\frac{\partial}{\partial r} (r V_\theta) - \frac{\partial V_r}{\partial \theta} \right] \hat{\boldsymbol{\phi}},$$

with Laplacian

$$\nabla^2 f = \frac{1}{r^2} \frac{\partial}{\partial r} \left(r^2 \frac{\partial f}{\partial r} \right) + \frac{1}{r^2 \sin \theta} \frac{\partial}{\partial \theta} \left(\sin \theta \frac{\partial f}{\partial \theta} \right) + \frac{1}{r^2 \sin^2 \theta} \frac{\partial^2 f}{\partial \phi^2}. \quad \text{(B.50)}$$

B.4 Integral Elements

At the heart of integrals in vector calculus is the vector "line element." We'll start in Cartesian coordinates, where it is easy to to define the line element, and then focus on its geometric meaning in order to build the other cases. Imagine sitting at a point in three dimensions, and you are told to go out a distance dx in the $\hat{\mathbf{x}}$ direction. Then the infinitesimal vector representing your direction of motion would be $dx\,\hat{\mathbf{x}}$. If you also went dy in the $\hat{\mathbf{y}}$ direction and dz in the $\hat{\mathbf{z}}$ direction, then the vector that points from your starting location to your new one is the line element:

$$d\boldsymbol{\ell} = dx\,\hat{\mathbf{x}} + dy\,\hat{\mathbf{y}} + dz\,\hat{\mathbf{z}}. \quad \text{(B.51)}$$

Since the directions are orthogonal, we just add together the "moves" in each direction. The magnitude of the line element gives the Pythagorean length:

$$d\ell = \sqrt{d\boldsymbol{\ell} \cdot d\boldsymbol{\ell}} = \sqrt{dx^2 + dy^2 + dz^2}. \quad \text{(B.52)}$$

In cylindrical coordinates, we'd like to use the cylindrical basis vectors $\hat{\mathbf{s}}$, $\hat{\boldsymbol{\phi}}$, and $\hat{\mathbf{z}}$ together with the infinitesimal changes in coordinates, ds, $d\phi$, and dz to express the line element $d\boldsymbol{\ell}$. Imagine making a move ds in the $\hat{\mathbf{s}}$ direction. If you take $d\phi = 0$ and $dz = 0$, then $d\boldsymbol{\ell} = ds\,\hat{\mathbf{s}}$. If you moved only in the $\hat{\mathbf{z}}$ direction a distance dz, then $d\boldsymbol{\ell} = dz\,\hat{\mathbf{z}}$. Since the $\hat{\mathbf{s}}$ and $\hat{\mathbf{z}}$ direction are orthogonal (at *all* points), we could combine these to get $d\boldsymbol{\ell} = ds\,\hat{\mathbf{s}} + dz\,\hat{\mathbf{z}}$. How about a $d\phi$ move in the $\hat{\boldsymbol{\phi}}$ direction? If we set $ds = 0 = dz$, and consider a move only in the $\hat{\boldsymbol{\phi}}$ direction, what is $d\boldsymbol{\ell} =$? We need to know the length associated with an angular arc $d\phi$, and that length depends on s. Referring to Figure B.3, if the starting point is at a height z, and a distance s from the $\hat{\mathbf{z}}$ axis, then the length of the arc spanned by $d\phi$ is $sd\phi$ so that $d\boldsymbol{\ell} = sd\phi\,\hat{\boldsymbol{\phi}}$ for a pure angular move. Putting all of these together,

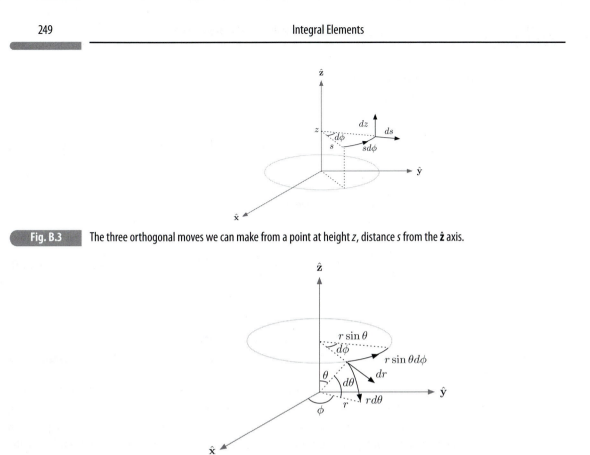

Fig. B.3   The three orthogonal moves we can make from a point at height z, distance s from the $\hat{\mathbf{z}}$ axis.

Fig. B.4   Infinitesimal moves in the $\hat{\mathbf{r}}$, $\hat{\boldsymbol{\theta}}$, and $\hat{\boldsymbol{\phi}}$ directions from the point with spherical coordinates r, θ, ϕ.

we take a step in each of the three directions as shown in Figure B.3, the cylindrical line element is

$$d\boldsymbol{\ell} = ds\,\hat{\mathbf{s}} + sd\phi\,\hat{\boldsymbol{\phi}} + dz\,\hat{\mathbf{z}}. \tag{B.53}$$

In spherical coordinates, infinitesimal displacements in the three independent directions for a point at r, θ, ϕ are shown in Figure B.4. From the picture, we can read off the line element

$$d\boldsymbol{\ell} = dr\,\hat{\mathbf{r}} + rd\theta\,\hat{\boldsymbol{\theta}} + r\sin\theta d\phi\,\hat{\boldsymbol{\phi}}. \tag{B.54}$$

These line elements are easy to pick off when we have a clear geometrical picture of the infinitesimals and their orthogonal directions at a point. But many coordinate transformations are much harder to draw, and we don't want to rely on always being able to make a picture and identify the line elements "by eye." The analytical approach is the obvious one, just take the Cartesian definition (B.51) and rewrite dx, dy and dz in terms of the new coordinate infinitesimals and the derivatives of the new coordinates with respect to the Cartesian ones, and express the Cartesian basis vectors in terms of the new ones.

As an example, let's work out the cylindrical line element directly from the Cartesian form. From the chain rule, we have

$$dx = \frac{\partial x}{\partial s}ds + \frac{\partial x}{\partial \phi}d\phi + \frac{\partial x}{\partial z}dz$$

$$= \cos\phi ds - s\sin\phi d\phi$$

$$dy = \frac{\partial y}{\partial s}ds + \frac{\partial y}{\partial \phi}d\phi + \frac{\partial y}{\partial z}dz$$

$$= \sin\phi ds + s\cos\phi d\phi$$

$$dz = \frac{\partial z}{\partial s}ds + \frac{\partial z}{\partial \phi}d\phi + \frac{\partial z}{\partial z}dz$$

$$= dz.$$

(B.55)

Using these, together with the basis vector relations from (B.7), we have

$$d\boldsymbol{\ell} = dx\,\hat{\mathbf{x}} + dy\,\hat{\mathbf{y}} + dz\,\hat{\mathbf{z}} = ds\,\hat{\mathbf{s}} + sd\phi\,\hat{\boldsymbol{\phi}} + dz\,\hat{\mathbf{z}}$$

(B.56)

as expected.

B.4.1 Volume Element

The line element defines an infinitesimal box with side lengths given by the infinitesimal lengths in each of the independent directions. If you want to build a volume element in preparation for performing a volume integral, you just form the volume of the box by multiplying the side lengths. For Cartesian coordinates with line element from (B.51), the infinitesimal cube has volume

$$d\tau = dxdydz.$$

(B.57)

For the cylindrical line element in (B.53), the product of the side lengths gives

$$d\tau = sdsd\phi dz,$$

(B.58)

and for the spherical line element, (B.54), the product is

$$d\tau = r^2\sin\theta dr d\theta d\phi.$$

(B.59)

B.4.2 Area Elements

An infinitesimal area element has direction that is parallel to the surface normal, and magnitude that is set by the area of the infinitesimal platelet on the surface. In Cartesian coordinates, for a surface lying in the xy plane, the area element points in the $\pm\hat{\mathbf{z}}$ direction with magnitude $dxdy$, the rectangular area spanned by the infinitesimal line element in the plane. If the surface was in the xz plane, the area element would be $d\mathbf{a} = dxdz\,\hat{\mathbf{y}}$ (picking

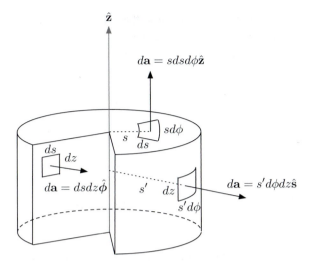

Three natural surface area elements for the cylinder. Each one has direction that is normal to the surface, and magnitude that is given by the infinitesimal surface patch.

the positive direction arbitrarily for this open surface). For a flat surface in the yz plane, the area element is $d\mathbf{a} = dydz\,\hat{\mathbf{x}}$.

When we move to curvilinear coordinates like cylindrical ones, there are three "natural" area elements we can make, pointing in each of the three directions. The algorithm is: Identify the direction of the normal, that gives the direction of $d\mathbf{a}$, then multiply the other two line element magnitudes together to get the infinitesimal area magnitude. For a cylinder, if you take the top surface, with $d\mathbf{a} \parallel \hat{\mathbf{z}}$, then the magnitude is $d\mathbf{a} = sdsd\phi$. The curved surface of the cylinder has $d\mathbf{a} = sd\phi dz\,\hat{\mathbf{s}}$. If you cut the cylinder in half, the interior face has $d\mathbf{a} = dsdz\,\hat{\boldsymbol{\phi}}$.

In spherical coordinates, the most natural surface of interest is the surface of a sphere, and this has $d\mathbf{a} \parallel \hat{\mathbf{r}}$. Multiplying the other two line element magnitudes together gives $d\mathbf{a} = r^2 \sin\theta d\theta d\phi\,\hat{\mathbf{r}}$.

Problem B.4.1 Evaluate $\nabla^2\hat{\boldsymbol{\theta}}$ (for the spherical unit vector pointing in the direction of increasing θ) and $\nabla \cdot \hat{\boldsymbol{\theta}}$.

Problem B.4.2 Work out the basis vectors, gradient, divergence, curl, and Laplacian for "elliptical cylindrical" coordinates $\{p, q, z\}$, related to the Cartesian ones via,

$$x = a\cosh p\cos q \quad y = a\sinh p\sin q \quad z = z \tag{B.60}$$

for constant parameter a.

Problem B.4.3 Work out the basis vectors, gradient, divergence, curl and Laplacian for "prolate spheroidal" coordinates, $\{u, v, \phi\}$, related to the Cartesian ones via,

$$x = a\sinh u\sin v\cos\phi \quad y = a\sinh u\sin v\sin\phi \quad z = a\cosh u\cos v \tag{B.61}$$

for constant parameter a.

Problem B.4.4 Four-dimensional Cartesian coordinates consist of the three usual coordinates, $\{x, y, z\}$ augmented with a fourth coordinate w. Work out the basis vectors, gradient, divergence, curl, and Laplacian for four-dimensional spherical coordinates where we introduce a new angle ψ to go along with θ and ϕ, and the four Cartesian coordinates are related to the spherical ones $\{r, \theta, \phi, \psi\}$ by

$$x = r \sin \psi \sin \theta \cos \phi \quad y = r \sin \psi \sin \theta \sin \phi \quad z = r \sin \psi \cos \theta \quad w = r \cos \psi.$$
$$\text{(B.62)}$$

References

[1] G. B. Arfken, H. J. Weber, and F. E. Harris, "Mathematical Methods for Physicists: A Comprehensive Guide," Academic Press, 7th ed., 2012.

[2] C. M. Bender and S. A. Orszag, "Advanced Mathematical Methods for Scientists and Engineers: Asymptotic Methods and Perturbation Theory," Springer, 1999.

[3] M. L. Boas, "Mathematical Methods in the Physical Sciences," Wiley, 3rd ed., 2005.

[4] F. W. Byron, Jr. and R. W. Fuller, "Mathematics of Classical and Quantum Physics," Dover Publications, revised ed., 1992.

[5] D. C. Chapman and P. M. Rizzoli, "Wave Motions in the Ocean: Myrl's View." Technical Report, MIT/WHOI Joint Program, Woods Hole, MA, 1989.

[6] D. Clark, J. Franklin, and N. Mann, "Relativistic Linear Restoring Force," *European Journal of Physics*, 33, 1041–1051, 2012.

[7] J. Franklin, " Computational Methods for Physics," Cambridge University Press, 2013.

[8] J. Franklin, "Classical Field Theory," Cambridge University Press, 2017.

[9] A. P. French, "Vibrations and Waves," CBS Publishers & Distributors, 2003.

[10] H. Georgi, "The Physics of Waves," Prentice Hall, Inc., 1993.

[11] G. H. Golub and C. F. Van Loan, "Matrix Computations," Johns Hopkins University Press, 4th ed., 2013.

[12] D. J. Griffiths, "Introduction to Electrodynamics," Cambridge University Press, 4th ed., 2017.

[13] D. J. Griffiths and D. F. Schroeter, "Introduction to Quantum Mechanics," Cambridge University Press, 3rd ed., 2018.

[14] R. J. LeVeque, "Numerical Methods for Conservation Laws," Lectures in Mathematics, Birkhäuser, 1992.

[15] K. F. Riley, M. P. Hobson, and S. J. Bence, "Mathematical Methods for Physics and Engineering," Cambridge University Press, 3rd ed., 2006.

[16] I. G. Main, "Vibrations and Waves in Physics," Cambridge University Press, 3rd ed., 1993.

[17] P. McCord Morse and H. Feschbach, "Methods of Theoretical Physics," McGraw-Hill, 1953.

[18] J. J. Sakurai and J. Napolitano, "Modern Quantum Mechanics," Cambridge University Press, 2nd ed., 2017.

[19] G. B. Whitham, "Linear and Nonlinear Waves," John Wiley & Sons, Inc., 1999.

Index